ALIPHATIC ORGANIC CHEMISTRY

By

Amit Arora

DISCOVERY PUBLISHING HOUSE
NEW DELHI-110002

First Published-2006

ISBN 81-8356-188-8

Published by

DISCOVERY PUBLISHING HOUSE
4831/24, Ansari Road, Prahlad Street,
Darya Ganj, New Delhi-110002 (India)
Phone: 23279245 • Fax: 91-11-23253475
E-mail:dphtemp@indiatimes.com

Printed at:
Amit Enterprises

Preface

This book is written for B.Sc., B.Sc. (Hons.) and M.Sc. students of various universities of India. In this book my aim has been describe the fundamental principles of organic chemistry. Since I do not consider the chemistry of natural products to be fundamental chemistry but rather the application of fundamental principles. The subject matter described in this book covers much of the basic organic chemistry that is needed by a student who wish to study chemistry as a main subject at degree level. The arrangement of the subject-matter is based on homologous series and in general, descriptions of reactions are followed by discussion of their mechanisms and these includes an elementary account of the sort of evidence that led workers to suggest mechanisms that are acceptable at the present time.

We are sure you will appreciate the thought and care that has gone into making of this text. This book has everything you want to have in your advanced organic chemistry text.

The author wish to thank all those who helped in the preparation of the text and illustration, suggestion for improvement and criticism will be gratefully received and acknowledge.

Author

CONTENTS

1

MONOHYDRIC AND POLYHYDRIC ALCOHOLS

INTRODUCTION TO MONOHYDRIC ALCOHOLS

An alcohol is a compound that contains one or more *hydroxyl* groups, *i.e.*, alcohols are hydroxy-derviatives of the alkanes. They are classified according to the number of hydroxyl groups present. Monohydric alcohols contain one hydroxyl group; dihydric, two; trihydric, three; etc. When the alcohols contain four or more hydroxyl groups, they are usually called polyhydric alcohols.

The monohydric alcohols from an homologous series with the general formula $C_nH_{2n+2}O$, but, since their functional group is the hydroxyl group, their general formula is more satisfactorily written as $C_nH_{2n+1}OH$ or ROH.

Nomenclature

The simpler alcohols are commonly known by their trivial names, which are obtained by naming the alcohol as a derivative of the alkyl group attached to the hydroxyl group, *e.g.*, CH_3OH, methyl alcohol; $CH_3CH_2CH_2OH$. n-propyl alcohol; $CH_3CH(OH)CH_3$, isopropyl alcohol; $(CH_3)_3COH$, t-butyl alcohol.

Another system of nomenclature considers the alcohols as derivatives of methyl alcohol, which is named *carbinol* or *methanol*, *e.g.*, $CH_3CH_2CHOHCH_3$, ethylmethylcarbinol or ethylmethylmethanol.

In the I.U.P.A.C. system of nomenclature, the longest carbon chain containing the hydroxyl group is chosen as the parent hydrocarbon. The class suffix is-*ol*, and the positions of side-chains and the hydroxyl group are indicated by numbers, the lowest possible number being given to the hydroxyl group, *e.g.*, CH_3OH, methanol; C_2H_5OH, ethanol; CH_3CH_2-CH_2OH, Propan-1-ol; $(CH_3)_2CHCHOHCH_3$, 3-methylbuttan-2-ol. According to this system the alcohols are referred to as the *alkanols*.

Monohydric alcohols are subdivided into primary, secondary and tertiary alcohols according as the alkyl group attached to the hydroxyl group is a primary, secondary or tertiary group, respectively. Primary alcohols contain the *primary alcoholic group* ---CH_2OH, *e.g.*, ethanol, CH_3CH_2OH; secondary alcohols, the *secondary alcoholic group* CH(OH), *e.g.*, isoprepanol, $(CH_3)_2CHOH$; and tertiary alcohols the *tertiary alcoholic group* $\equiv$*C(OH)*, *e.g.*, t-butanol, $(CH_3)_3COH$.

GENERAL METHODS OF PREPARATION

1. By the hydrolysis of an alkyl halide with aqueous alkali or silver oxide suspended in water:

$$RX + \text{‘AgOH’} \longrightarrow ROH + AgX$$

2. By the hydrolysis of esters with alkali:

$$R^1CO_2R^2 + KOH \longrightarrow R^1CO_2K + R^2OH$$

This method is important industrially for preparing certain alcohols that occur naturally as esters. Since tertiary alkyl halides give large amounts of alkene on hydrolysis, they are best converted into alcohols as follows:

$$R_3CX + CH_3CO_2Ag \xrightarrow[\text{heat}]{\text{EtOH}} AgX + CH_3CO_2CR_3$$

$$\xrightarrow{\text{NaOH}} CH_3CO_2Na + R_3COH$$

3. By heating ethers with dilute sulphuric acid under pressure, *e.g.*, diethyl ether forms ethanol:

$$(C_2H_5)_2O + H_2O \xrightarrow{H_2SO_4} 2C_2H_5OH$$

This method is important industrially, since ethers are formed as by-products in the preparation of certain alcohols (see ethanol and propanols).

4. By the reduction of aldehydes, ketones or esters by means of excess of sodium and ethanol or n-butanol as the reducing agent (*Bouveault-Blanc reduction*, 1903), *e.g.*,

(*i*) *Aldehydes:* $RCHO \xrightarrow{e;\,H^+} RCH_2OH$ (g.-v.g.)

(*ii*) *Esters:* $R^1CO_2R^2 \xrightarrow{e;\,H^+} R^1CH_2OH + R^2OH$ (g.)

(*iii*) *Ketones:* $R_2CO \xrightarrow{e;\ H^+} R_2CHOH$ (g.)

The Bouvealult-Blanc reduction is believed to occur in steps involving the transfer of one electron at a time; *e.g.*,

$$\underset{\underset{O\cdot}{\|}}{RC}{-}OEt \xrightarrow{Na} \underset{\underset{O\cdot}{|}}{R\ddot{C}}{-}OEt \xrightarrow{EtOH} \underset{\underset{O\cdot}{|}}{RCH}{-}OEt \xrightarrow{Na} \underset{\underset{O^-}{|}}{RCH}{-}OEt \longrightarrow$$

$$OEt^- + RCH{=}O \xrightarrow{Na} R\ddot{C}H{-}O\cdot \xrightarrow{EtOH} RCH_2{-}O\cdot \xrightarrow{Na}$$

$$RCH_2{-}O^- \xrightarrow{EtOH} RCH_2OH$$

Reduction with Metallic Hydrides

Many complex metallic hydrides reduce various functional groups, and their umber is increasing rapidly. The most versatile reagent is **lithium aluminium hydride** (LAH). This reduces most functional groups, Table 6.1, but does not normally reduce the olefinic bond. An unusual feature of this regent is its reduction of the *carboxyl group* to primary alcohol.

Reductions with lithium aluminium hydride are usually carried out in ethereal solutions, the compound in ether being added to the lithium aluminium hydride solution. In certain cases the reverse addition is necessary, *i.e.*, the hydride solution is added to the solution of the compound to be reduced.

Sodium borohydride, which is insoluble in ether, is used in *ethanolic* solution to reduce *carbonyl* compounds, the only important exception being the carboxyl group. It also does not normally reduce esters, but reduction to primary alcohol can often be effected by use of a large excess of reagent in methanol.

The reduction of the carbonyl group by LAH or sodium borohydride occurs in a stepwise manner, each step involving hydride ion transfer, *e.g.*,

$$\underset{R_2C}{\overset{O}{\|}}\ H{-}AlH_3 \longrightarrow \left[\underset{R_2CH}{\overset{O}{|}} + AlH_3\right] \longrightarrow R_2CH{-}O{-}AlH_3 \xrightarrow{R_2CO}$$

$$(R_2CHO)_2AlH_2 \xrightarrow{R_2CO} (R_2CHO)_3AlH \xrightarrow{R_2CO} (R_2CHO)_4Al^- \xrightarrow{H^+} 4R_2CHOH$$

Experimental work has shown that the rate of reduction is decreased as the size of R increases. One contributing factor is steric hindrance, and this is made use of by employing reducing complexes containing large groups, *e.g.*, when the substrate contain is two carbonyl groups ad one is sterically hindered, lithium tri-t-butoxyaluminium hydride, reduce only the nonhindered carbonyl group. This reagent is generally used in tetrahydrofuran (THF) solution. In addition to steric hindrance, there is also the inductive effect in the anion, *e.g.*,

$$(R-O\leftarrow)_3Al\leftarrow H \quad >C=O$$

The electronegative oxygen atom exerts a strong – I effect and consequently makes hydride ion transfer more difficult. Hence, the reactivity of lithium tri-t-butoxyaluminium hydride is lower than that of LAH.

Many other metallic hydrides are also reducing reagents, and because of their different reactivities, they offer a means of *selective* reduction.

Lithium borohydride (in ether or THF) is more reactive than sodium borohydride but less than LAH.

Lithium tri-t-butoxyaluminium hydride (in THF) useful for reducing acid chlorides to aldehydes at –78 C.

Sodium aluminium hydride (in THF) is similar to LAH, but it reduces esters to aldehydes.

Diborane (in THF) reduces many functional groups.

Aluminium hydride (in ether or THF) is similar to LAH, but is better for reducing α, β-unsaturated carbonyl compounds to unsaturated alcohols.

The reducing power of a *given* metallic hydride is affected by the nature of the solvent and the presence of certain other compounds, *e.g.*, **LAH and aluminium chloride** in ether form AlH_3, AlH_2Cl, or $AlHCl_2$ according to the proportion of reagents used:

$$3LiAlH_4 + AlCl_3 \longrightarrow 4AlH_3 + 3LiCl$$

$$LiAlH_4 + AlCl_3 \longrightarrow 2AlH_2Cl + LiCl$$

$$LiAlH_4 + 3AlCl_3 \longrightarrow 4AlHCl_2 + LiCl$$

Hence, the reducing power of the reagent will depend on which one is actually present; all are milder reducing reagents than LAH, *e.g.*, LAH--$AlCl_3$ does not reduce alkyl halides. It has also been shown that a solution of LAH in pyridine which has been allowed to stand will reduce a carbonyl group but not a carboxyl or carbalkoxyl group. Thus, LAH has been converted into a milder reducing reagent, the structure of which appears to be uncertain.

Sodium borohydride-aluminium chloride (in diethylene glycol dimethyl ether, *i.e.*, diglyme is more reactive than sodium borohydride itself.

Catalytic Hydrogenation

Many functional groups are reduced catalytically the most common catalysts being nickel, platinum, palladium, rhodium, and ruthenium; other catalysts used are copper chromite and copper-barium-chromium oxide. The catalytic activity of a given metal is dependent on its method of preparation, *e.g.*, nickel. It also depends on other factors;

(*a*) the presence of other compounds which may either increase (*promoters*) or decrease or inhibit (*poisons*) catalytic activity;

(*b*) the nature of the solvent: neutral, basic, or acidic.

In addition to having different reactivities, catalysts have different selectivities, *e.g.*, carboxylic acids can be hydrogenated to alcohols with a Ru---C catalyst but not with a Pt catalyst. In general, catalysts are deposited on the surface of a support such as charcoal, alumina, calcium carbonate, etc., and the activity of the catalyst may be changed by changing the support.

5. Primary, secondary and tertiary alcohols may be prepared by means of a Grignard reagent and the appropriate carbonyl compound.

6. A number of alcohols are obtained by fermentation processes.

7. In recent years synthetic methods have become very important for preparing various alcohols commercially:

(i) By the hydration of alkenes .

(ii) By heating a mixture of carbon monoxide and hydrogen under pressure in the presence of a catalyst, *e.g.*, zinc chromite plus small amounts of alkali metal or iron salts. A mixture of alcohols containing methyl, ethyl, n-propyl, isobutyl and higher-branched alcohols is obtained, the individuals being separated by factional distillation (see below).

(iii) The **Oxo process** (*Oxo synthesis, carbonylation* or *hydroformylation reaction*). A mixture of alkene, carbon monoxide and hydrogen, under pressure and elevated temperature, in the presence of a catalyst, forms aldehydes. A common catalyst is cobalt carbonyl hydride, $[CoH(CO)_4]$, and the product is a mixture of isomeric straight-chain and branched-chain aldehydes (the former predominating). These are reduced catalytically to the corresponding alcohols, *e.g.*, propene gives a mixture of n- and isobutanols:

$$2MeCH = CH_2 + 2CO + 2H_2 \longrightarrow Me(CH_2)_2CHO + Me_2CHCHO$$

$$\xrightarrow[Cu-Zn]{H_2} Me(CH_2)_2\ CH_2OH + Me_2CHCH_2OH$$

The aldehydes are first separated by fractional distillation.

(*iv*) Methanol, ethanol, propanols and butanols are prepared industrially by the oxidation of natural gas.

Most of the methods given above can be used for the preparation of any particular class of alcohol: it is only a question of starting with the appropriate compound, *e.g.*, all three types may be prepared via Grignard reagents; *primary* alcohols from formaldehyde ($R^1 = R^2 = H$), *secondary* from aldehydes ($R^1 = R$; $R^2 = H$), and *tertiary* from ketones (R^1 and R^2 = alkyl groups).

$$R^1C(=O)R^2 + R^3MgX \longrightarrow R^1C(OMgX)(R^2)-R^3 \xrightarrow{H^+} R^1R^2R^3COH$$

General Properties of the Alcohols

The alcohols are neutral substances: the lower members are liquids, and have a distinctive smell and a burning taste; the higher members are solids and are almost odourless.

In a group of isomeric alcohols, the primary alcohol has the highest boiling point and the tertiary the lowest, with the secondary having an intermediate value. The lower members are far less volatile than is to be expected from their molecular weight, and this is due to association through hydrogen bonding *extending over a chain of molecules,* thus, giving rise to a large molecule' the volatility of which would be expected to below:

$$\underset{\underset{R}{|}}{O}-H-\underset{\underset{R}{|}}{O}-H-\underset{\underset{R}{|}}{O}-H-\underset{\underset{R}{|}}{O}-H$$

The lower alcohols are very soluble in water, and the solubility diminishes as the molecular weight increases. Their solubility in water is to be expected, since the oxygen atom of the hydroxyl group in alcohols can form hydrogen bonds with the water molecules. In the lower alcohols the hydroxyl group constitutes a large part of the molecule, whereas as the molecular weight of the alcohol increases the hydrocarbon character of the molecule increases, and hence the solubility in water decreases. This, however; is not the complete story; the structure of the carbon chain also plays a part, *e.g.*, n-butanol is fairly soluble in water, but t-butanol is miscible with water in all proportions.

General Reactions of the Alcohols

1. Alcohols react with organic and inorganic acids to form *esters*:

$$R^1CO_2H + R_2OH \longrightarrow R^1CO_2R^2 + H_2O$$

Esters of the halogen acids are, as we have seen, the alkyl halides.

The order of reactivity of an alcohol with a given acid is primary alcohol > secondary > tertiary, provided the mechanism is bimolecular for all cases. With a given alcohol, the order of reactivity of the halogen acids in the bimolecular mechanism is HI > HBr > HCl.

Let us consider the following example:

$$EtOH + H{+} + X \overset{\text{fast}}{\rightleftharpoons} X^- + EtOH_2^+ \rightleftharpoons \overset{\delta-}{X}\cdots Et\cdots \overset{\delta+}{O}H_2 \xrightarrow{\text{fast}} X\text{—}Et + H_2O$$

The order of reactivity of the halogen acids is attributed to the variation in polarisability of the anion; the greater the size of the anion, the greater is its polarisability. What this amounts to is that the larger the anion, the more readily it can donate a lone pair to form a covalent bond. Also, the nature of the alkyl group will influence the mechanism of halide formation (in aqueous solution) in the same way as it does their hydrolysis.

Protonated t-alcohols readily eliminate a molecule of water to form a carbonium ion and so react by the S_N1 mechanism. Straight-chain primary alcohols will favour the S_N2 mechanism, but if branched-chain, then the S_N1 mechanism may operate. Secondary alcohols would be expected to react by both methanisms.

2. Alcohols react with phosphorus halides to form alkyl halides.

3. Alcohols combine with phenyl isocyanate to form phenyl-substituted urethans:

$$C_6H_5NCO + ROH \longrightarrow C_6H_5NHCO_2R$$

4. Strongly electropositive metals (K, Na, Mg, Al, Zn) liberate hydrogen from alcohols to form *alkoxides*, *e.g.*, sodium reacts with ethanol to form sodium ethoxide:

$$2C_2H_5OH + 2Na \longrightarrow 2C_2H_5O^-Na^+ + H_2$$

Alkoxides are white deliquescent solids, readily soluble in water with hydrolysis:

$$RO^-Na^+ + H_2O \rightleftharpoons ROH + NaOH$$

Alkoxides react with carbon disulphide to form xanthates:

$$RONa + S{=}C{=}S \longrightarrow RO{-}C({=}S){-}\bar{S}Na^+$$

The fact that alcohols liberate hydrogen by the action of metals shows that alcohols can behave as acids, but since they do not affect the pH of water, they are weaker acids than water.

This accounts for alkoxides being hydrolysed by water and for the fact that the ethoxide ion (the cB of EtOH) is a stronger base than the hydroxide ion (the cB of H_2O).

Since alkyl groups have a +I effect, there will be an increased electron displacement towards the oxygen atom in going from primary to secondary to tertiary alcohol. This may be represented (qualitatively) as follows:

$$Me \rightarrow CH_2 \rightarrow \overset{\delta-}{O} \rightarrow H \qquad \overset{Me}{\underset{Me}{\searrow\nearrow}} CH_2 \twoheadrightarrow \overset{2\delta-}{O} \twoheadrightarrow H \qquad Me \rightarrow \overset{Me}{\underset{Me}{\searrow\nearrow}} CH_2 \ggg \overset{3\delta-}{O} \ggg H$$

The greater the negative charge on the oxygen atom, the closer is the covalent pair in the O—H bond driven to the hydrogen atom and consequently separation of a proton becomes increasingly difficult. Thus the acid strengths of alcohols will be in the order: prim > s > t. This is the order of reactivity of alcohols towards metals, and with t-alcohols the reaction with sodium is so slow that it is better to use the more electropositive potassium for these alcohols, *e.g.*, potassium t-butoxide is the usual salt prepared from t-butanol.

From what has been said above, it can be seen that the tendency for the C—O bond to break will be the reverse of that for the O—H bond, *i.e.*, reactions involving the breaking of the C—O bond will follow the order of reactivity: t > s > prim. This order would be expected from the consideration of the stabilities of the carbonium ions produced .

If we now consider β-substituted alcohols containing strong –I groups, the acidity of the alcohol is increased. *e.g.*, pK_a of ethanol of ethanol is ~ 18 whereas that of trifluoroethanol is 12.4.

$$Me \rightarrow CH_2 \rightarrow O—H \qquad CF_3 \leftarrow\leftarrow\leftarrow CH_2 \leftarrow\leftarrow\leftarrow O—H$$

The Me group is weakly + I, whereas the CF_3 group is strongly –I.

Thus the covalent pair is pulled away from the oxygen atom in the C—O bond in the latter compound, and consequently the covalent pair i the O—H bond is drawn towards the O atom, thereby facilitating release of the hydrogen as a proton.

A number of alkoxides are important as synthetic reagents; *e.g.*, sodium ethoxide, C_2H_5ONa; aluminium ethoxide, (C_2H_5O) Al; aluminium t-butoxide $[(CH_3)_3\ CO]_3$ Al. The aluminium alkoxides may be conveniently prepared by the action of aluminium amalgam or aluminium shavings o the alcohol.

5. Primary and secondary alcohols may be acetylated with acetyl chloride, *e.g.*, ethanol gives ethyl acetate:

$$CH_3COCl + C_2H_5OH \longrightarrow CH_3CO_2C_2H_5 + HCl$$

With tertiary alcohols the reaction is often accompanied by dehydration of the alcohol to alkene or by the formation of a tertiary alkyl chloride; *e.g.*, t-butanol gives a good yield of t-butyl chloride:

$$(CH_3)_3\ COH + CH_3COCl \longrightarrow (CH_3)_3\ CCl + CH_3CO_2H$$

However, in the presence of a base such as dimethylaniline, the ester is produced in good yield.

$$CH_3COCl + (CH_3)_3COH \xrightarrow{PhNMe_2} CH_3CO_2C(CH_3)_3 + PhNMe_2\ HCl \quad (63–68\%)$$

A possible explanation for this is that t-butyl esters are very readily decomposed by hydrochloric acid as follows:

$$MeCOCl + Me_3COH \longrightarrow MeCOOCMe_3 + HCl \longrightarrow MeCO_2H + Me_3CCl$$

In the presence of a base, the HCl is removed as base· HCl, and consequently the second step cannot occur.

Acetylation can also be carried out with acetic anhydride in the presence of a catalyst, *e.g.*, pyridine. This method is successful for prim and s-alcohols, but fails with t-alcohols, *e.g.*, t-butanol is not acetylated.

However, t-alcohols can be acetylated if *p*-toluenesulphonic acid is used as acetalyst.

6. Alcohols may be oxidised, and the products of oxidation depend on the class of the alcohol and on the nature of the oxidising agent.

Halogens, in *aqueous* solution, oxidise alcohols to carbonyl compounds. According to Swain *et al.* (1961), the mechanism involves hydride ion removal:

$$Br{-}Br \quad H{-}CHMe{-}\ddot{O}{-}H \longrightarrow Br^- + HBr + MeCH{=}\overset{+}{O}H \xrightarrow{-H^+} MeCHO$$

In *alkaline* solution, the mechanism proposed is:

$$MeCH_2O{-}H \quad \bar{O}H \rightleftharpoons H_2O + MeCH_2{-}O^-$$

$$MeCH_2{-}\bar{O} + Br_2 \longrightarrow MeCH{=}O + HBr + Br^-$$

In both cases, the carbonyl compound is then halogenated under the catalytic influence of acid or alkali.

On the other hand, alcohols are oxidised by hydrogen peroxide and ferrous sulphate to diols which, according to Coffman *et al.* (1960), are produced by dimerisation of free-radical intermediates, *e.g.*, t-butanol forms, 2,5-dimethylhexane-2,5-diol.

$$Fe^{2+} + H2O2 \longrightarrow Fe^{3+} + HO^- + HO\cdot$$

$$(CH_3)_2C(OH)CH_3 \xrightarrow[-H_2O]{\cdot OH} (CH_3)_2C(OH)CH_2\cdot \xrightarrow{2} (CH_3)_2C(OH)CH_2CH_2C(OH)(CH_3)_2$$

t-Butanol is particularly useful since all the alkyl hydrogens are equivalent and so only one free radical is produced.

7. Alcohols may be dehyrated to alkenes, the ease of dehydration being t > s > prim. Dehydration may be effected by heat alone (400–800°C), but in the presence of a catalyst, lower temperatures may be used, *e.g.*, when passed over heated alumina, t-alcohols are dehydrated at about 150°C, s-alcohols at 250°C and prim.-alcohols at 350°C. Also, primary alcohols are dehydrated by concentrated sulphuric acid at about 170°C, and secondary and tertiary alcohols by boiling dilute sulphuric acid (this is used to avoid polymerisation of the alkene).

Isomerisation usually occurs with dehydration by acid or alumina, but in the latter case, isomerisation is suppressed by the addition of a small amount of pyridine.

The mechanism of dehydration with acid is described on p. 108 and, on the basis that a carbonium ion is formed as an intermediate, it can be seen why the ease of dehydration is t > s > prim.; this is the order of stabilities of the carbonium ions.

With secondary ad tertiary alcohols, dehydration may occur in two ways, *e.g.*,

$$CH_3CH_2CH(OH)CH_3 \xrightarrow[(1:1H_2SO_4)]{-H_2O} \begin{cases} CH_3CH_2CH{=}CH_2 \\ CH_3CH{=}CHCH_3 \end{cases}$$

Experiment shows that hydrogen attached to the adjacent carbon atom joined to the least number of hydrogen atoms is eliminated most easily. Thus, in the above reaction, the main product is but-2-ene (65–80 per cent). This elimination therefore occurs i accordance with Saytzeff's rule for the dehydrohalogenation of alkyl halides, and thc reason is the same. However, when thorium oxide or lanthanide metal oxides are used as the dehydrating agents, the yield of 1-alkene from 2-alcohols is generally above 98 per cent (Lundeen *et al.*, 1963).

As mentioned above, rearrangement often occurs with acid-catalysed dehydration. All three types of alcohol may behave in this way via a carbonium ion that may undergo methyl and/or hydride ion 1,2-shift. In the following examples, only methyl shifts are shown (all the possible hydride ion shifts lead finally to the same products resulting from some of the competing methyl shifts).

(i) $Me_2CHCH_2OH \xrightarrow[(ii)\ -H_2O]{(i)\ H^+} Me_2CHCH_2^+ \xrightarrow{-H^+} Me_2C{=}CH_2$

$Me_2CHCH_2^+ \xrightarrow{Me^-\ shift} Me\overset{+}{C}HCH_2Me$

$\underset{(major\ product)}{MeCH{=}CHMe} \xleftarrow{-H^+} Me\overset{+}{C}HCH_2Me \xrightarrow{-H^+} CH_2{=}CHCH_2Me$

N.B. **An illustration of hydride ion shift is:**

$$Me_2CHCH_2^+ \longrightarrow Me_2\overset{+}{C}CH_3 \xrightarrow{-H^+} Me_2C{=}CH_2$$

(ii) $Me_3CCHOHMe \xrightarrow[(ii)\ -H_2O]{(i)\ H^+} Me_3C\overset{+}{C}HMe \xrightarrow{-H^+} \underset{(very\ little)}{Me_3CCH{=}CH_2}$

$Me_3C\overset{+}{C}HMe \xrightarrow{Me^-\ shift} Me_2\overset{+}{C}CHMe_2$

$\underset{(major\ product)}{Me_2C{=}CMe_2} \xleftarrow{-H^+} Me_2\overset{+}{C}CHMe_2 \xrightarrow{-H^+} \underset{(moderate\ amount)}{CH_2{=}CMeCHMe_2}$

(iii) $Me_3CCH_2OH \xrightarrow[(ii)\ -H_2O]{(i)\ H^+} Me_3CCH_2^+$

$Me_3CCH_2^+ \xrightarrow{Me^-\ shift} Me_2\overset{+}{C}CH_2Me$

$\underset{(major\ product)}{Me_2C{=}CHMe} \xleftarrow{-H^+} Me_2\overset{+}{C}CH_2Me \xrightarrow{-H^+} \underset{(small\ amount)}{CH_2{=}CMeCH_2Me}$

In each case, the major product is in accordance with Saytzeff's rule. Another point to note is that the rearrangements occur extremely rapidly. In fact, in the examples given, rearrangement is faster than proton elimination. Furthermore, Phelan *et al.* (1967) have shown that ease of anion migration is Ph >> Me > H.

When a double bond is produced in the product and is accompanied by a 1,2-shift, the reaction is said to be a *retropinacol rearrangement.* This type of rearrangement, when occurring in *open chain* compounds, is also sometimes called the *Wagner rearrangement.*

Alcohols may also be converted into alkenes via their esters or xanthates, but in these cases no rearrangements occur, *e.g.*, $Me_3CCH(OAc)Me \longrightarrow Me_3CCH{=}CH_2$ (*cf.* the action of acid on the alcohol, above).

Dehydration may be used as the first step in the conversion of a primary alcohol of suitable structure into a secondary or tertiary alcohol, or a s-alcohol of suitable structure into a t-alcohol, *e.g.* (*N. B.* HI additions are in accordance with Markownikoff's rule):

(*i*) $MeCH_2CH_2OH \xrightarrow[350^\circ C]{Al_2O_3} MeCH{=}CH_2 \xrightarrow[(ii)\ AgOH]{(i)\ HI} MeCHOHMe$

(*ii*) $Me_2CHCH_2OH \xrightarrow[350^\circ C]{Al_2O_3} Me_2C{=}CH_2 \xrightarrow[(ii)\ AgOH]{(i)\ HI} Me_3COH$

(*iii*) $Me_2CHCHOHMe \xrightarrow[350^\circ C]{Al_2O_3} Me_2C{=}CHMe \xrightarrow[(ii)\ AgOH]{(i)\ HI} Me_3C(OH)CH_2Me$

The 'reverse' procedure, *i.e.*, the conversion t → s → prim. alcohol can be carried out via hydroboration, *e.g.*,

$$Me_2C(OH)CH_2Me \xrightarrow{-H_2O} Me_2C{=}CHMe \xrightarrow{B_2H_6} (Me_2CHCHMe{-})_2BH \xrightarrow{heat} (Me_2CHCH_2CH_2{-})_2BH$$

$$(Me_2CHCHMe{-})_2BH \xrightarrow{H_2O_2;\ OH^-} Me_2CHCHOHMe$$

$$(Me_2CHCH_2CH_2{-})_2BH \xrightarrow{H_2O_2;\ OH^-} Me_2CHCH_2CH_2OH$$

It is also possible to step down the alcohol series by first dehydrating to the alkene, which is then subjected to ozonolysis, *e.g.*,

(*a*) $RCH_2CH_2OH \xrightarrow[350^\circ C]{Al_2O_3} RCH{=}CH_2 \xrightarrow[(ii)\ Zn/H^+]{(i)\ O_3} RCHO \xrightarrow[cat.]{H_2} RCH_2OH$

$$(b)\ RCHOHCH_3 \xrightarrow[300°C]{ThO_2} RCH{=}CH_2 \cdots\cdots\rightarrow RCH_2OH$$

8. Alcohols combine with acetylene in the presence of mercury compounds as catalyst to form acetals:

$$2ROH + CH{\equiv}CH \xrightarrow{Hg^{2+}} CH_3CH(OR)_2$$

If, however, the reaction is carried out in the presence of potassium alkoxides at high temperature and under pressure, vinyl ethers are obtained.

METHODS OF DISTINGUISHING BETWEEN THE THREE CLASSES OF ALCOHOLS

1. By means of *oxidation.* The nature of the oxidation products of an alcohol depends on whether the alcohol is primary, secondary of tertiary.

(*i*) A *primary* alcohol on oxidation first gives an *aldehyde*, and this, on further oxidation, gives an acid. *Both aldehyde and acid contain the same number of carbon atoms as the original alcohol, e.g.*,

$$CH_3CH_2OH \xrightarrow{[O]} CH_3CHO \xrightarrow{[O]} CH_3CO_2H$$

(*ii*) A *secondary* alcohol, on oxidation, first gives a *ketone with the same number of carbon atoms as the original alcohol.* Ketones are fairly difficult to oxidise, but prolonged atoms than the original alcohol, *e.g.*, methyl-n-propylmethanol gives first pentan-2-one, and then a mixture of acetic and propionic acids:

$$CH_3CHOHCH_2CH_2CH_3 \xrightarrow{[O]} CH_3COCH_2CH_2CH_2CH_3 \xrightarrow{[O]}$$
$$CH_3CO_2H + CH_3CH_2CO_2H$$

(*iii*) *Tertiary* alcohols are resistant to oxidation in neutral or alkaline solution, but are readily oxidised by acid oxidising agents to a mixture of *ketone and acid, each containing fewer carbon atoms than the original alcohol.*

$$(CH_3)_2C(OH)CH_2CH_3 \xrightarrow{[O]} (CH_3)_2CO + CH_3CO_2H$$

The oxidising agents usually used for oxidising alcohols are: acid dichromate, acid or alkaline potassium permanganate, and dilute nitric acid.

The accepted mechanism for the oxidation of primary and secondary alcohols by dichromate and pemanganate involves an ester intermediate, *e.g.*, with acid dichromate:

$$Cr_2O_7^{2-} + H_2O \rightleftharpoons 2HCrO_4^-$$

$$Me_2CHOH + HCrO_4^- + 2H^+ \rightleftharpoons Me_2CH—O—CrO_3H_2^+ + H_2O$$

$$Me_2C(H)—O—CrO_3H_2^+ \xrightarrow{slow} Me_2CO + H_3O^+ + H_2CrO_3$$

H_2O H

The rate-determining step is the removal of hydrogen from the intermediate ester: This has been demonstrated by Westheimer *et. al.* (1949), who showed that Me_2CDOH is oxidised at about 1/7 the rate of either Me_2CHOH or $(CD_3)_2CHOH$. Thus, the r/d step involves the fission of the C—H bond (this accounts for the kinetic isotope effect).

This mechanism cannot operate with t-alcohols, since there is no H—COH group present. According to Sager (1956), the mechanism involves dehydration, *e.g.*,

$$Me_3COH \xrightarrow[(ii)\ —H_2O]{(i)\ H^+} Me_3C^+ \xrightarrow{—H^+} Me_2C{=}CH_2 \xrightarrow{[O]} Me_2\overset{O}{C—CH_2} \xrightarrow{[O]} Me_2CO + HCO_2H$$

2. The three classes of alcohols differ in their behaviour when the vapour is passed over copper at 300°C:

(*i*) A primary alcohols is dehydrogenated to an aldehyde, *e.g.*,

$$CH_3CH_2OH \xrightarrow[300°C]{Cu} CH_3CHO + H_2$$

(*ii*) A secondary alcohol is dehydrogenated to a ketone, *e.g.*,

$$CH_3CH(OH)CH_3 \xrightarrow[300°C]{Cu} CH_3COCH_3 + H_2$$

(*iii*) A tertiary alcohol is dehydrated to a alkene, *e.g.*,

$$(CH_3)_2C(OH)CH_2CH_3 \xrightarrow[300°C]{Cu} (CH_3)_2C{=}CHCH_3 + H_2O$$

Methyl alcohol, methanol (*carbinol*) is prepared industrially by several methods. The earliest method was the destructive distillation of wood, whereby tar and an aqueous fraction known as *pyroligneous acid* are obtained. Pyroligneous acid contains methanol, acetone and acetic acid, and all three compounds may be obtained by suitable treatment. It was this method which gave rise to the name 'wood spirit' for methanol. The modern methods are synthetic.

(*i*) Synthesis gas is passed at a pressure of 200 atmospheres over a catalyst containing the oxides of copper, zinc and chromium at 300°C:

$$CO + 2H_2 \longrightarrow CH_3OH$$

If the proper precautions are taken, the yield of methanol is almost 100 per cent, an its purity is above 99 per cent. By changing the catalyst and the ratio of carbon monoxide to hydrogen, methanol and a variety of higher alcohols are produced.

(*ii*) By the catalytic oxidation of methane. A mixture of methane and oxygen (ratio by volume of 9 : 1) at a pressure of 100 atmospheres is passed through a copper tube at 200°C:

$$CH_4 + \tfrac{1}{2}O_2 \longrightarrow CH_3OH$$

Methanol is a colourless, inflammable liquid, b.p. 64°C, and is poisonous. It is miscible with water in all proportions, and is also miscible with most organic solvents.

It burns with a faintly luminous flame, and its vapour forms explosive mixtures with air or oxygen when ignited. It combines with calcium chloride to form $CaCl_2{\cdot}4CH_3OH$, and hence cannot be dried this way (*cf.* ethanol).

Methanol is used as a solvent for paints, varnishes, shellac celluloid cements, etc.; in the manufacture of dyes, perfumes, formaldehyde, etc. It is also used for making methylated spirit and automobile antifreeze mixtures.

Structure of Methanol

Analysis and molecular-weight determinations show that the molecular formula of methanol is CH_4O. Assuming that carbon is quadrivalent, oxygen bivalent and hydrogen univalent, only one structure is possible:

```
      H
      |
  H—C—O—H    or   CH3OH
      |
      H
```

This is supported by all the chemical reactions of methanol, *e.g.*, (*i*) only one hydrogen atom is methanol is replaceable by sodium; this suggests that one hydrogen atom is in a different state of combination from the other three.

(*ii*) Methanol is formed from methyl chloride by hydrolysis with sodium hydroxide. Methyl chloride can have only the structure CH_3Cl. It is reasonable to suppose that the methyl group in methyl chloride is unchanged by the action of dilute alkali, and that the reaction takes place by the replacement of the chlorine atom by a hydroxyl group.

(*iii*) The presence of the hydroxyl group is confirmed, for example, by the reaction between methanol and phosphorus pentachloride, when methyl chloride, hydrogen chloride and phosphoryl chloride are formed. Thus one oxygen atom (bivalent) and one hydrogen atom (univalent) have been replaced by one chlorine atom (univalent). This implies that the oxygen and hydrogen atoms exist as a univalent group in methanol: the only possibility is as a hydroxyl group, OH. It is the hydrogen of the hydroxyl group which is displaced by sodium.

Infra-red spectroscopic studies of hyroxy-compounds show that they absorb (O—H stretch) in the region of 3650 - 3580 cm^{-1}(v). This is true only if there is no hydrogen bonding. Intermolecular hydrogen bonding produces absorption)broad band) in the region 3350 – 3230 cm^{-1}(v), whereas intramolecular hydrogen bonding absorbs in the region 3590 – 3420 cm^{-1} (v), Moreover, the former shows changes in intensity and frequency shifts on dilution, whereas the latter is unaffected by dilution; thus the two may be distinguished.

In addition to O—H stretching vibrations, there are also C—O stretching vibrations which are characteristic of the type of the type of alcohol: primary, close to 1050 (s); secondary, close to 1100 (s); tertiary, close to 1150 cm^{-1}(s). Thus the classes of alcohols may be identified in these regions.

Two infra-red spectra of ethanol are shown in Fig. 1.1 (as a film) and in Fig. 1.2 (as a 1 per cent solution in carbon tetrachloride). IN the former, all the OH groups are associated, but in the latter, some OH groups are free.

However, a further point that may be mentioned here is the τ-value for methylene protons in the —CH_2—O—group is between 6.5–6.65 p.p.m., and when the alcohol is acetylated, the τ-value changes to ~ 5.9. This change may be used to identify a primary alcohol. Secondary alcohols can also be identified; > CH—O—(τ = ~ 6.1) and the acetylated alcohol (τ = ~ 5.0). These down-field shifts are due to the increased deshiedling effect of the acetoxyl group ($OOCCH_3$) compared to hydroxyl oxygen.

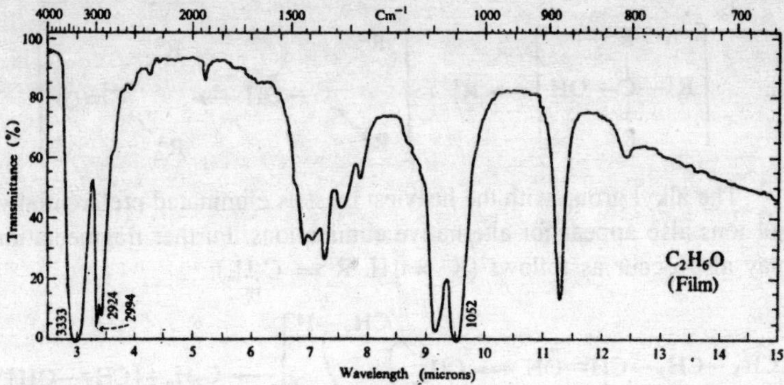

Fig. 1.1

3 333 cm^{-1} : O—H (str.) association (through hydrogen bonding). 2994 and 2 994 and 2 994 and 2 924cm^{-1} : C—H (str.) in Me and/or CH_2. 1052cm^{-1} : C—O (str.); characteristic of primary alcohols.

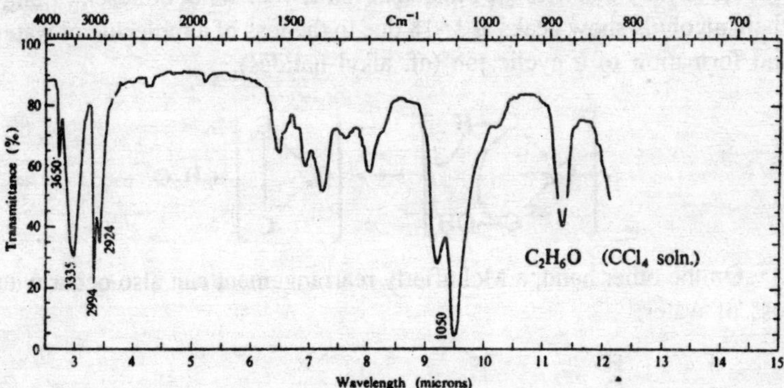

Fig. 1.2

3 650cm^{-1} : O—H (str.); free OH. 3 333, 2 994, 2 924, and 1050 cm^{-1}; As for Fig. 1.1

Mass Spectra

The mass spectrum of ethanol but have we shall describe the mass spectra of alcohols from a general point of view. Usually, strong peaks are shown by fragmentation involving β-cleavage (*i.e.*, the C—COH bond), and the fragment containing the hydroxyl group is stabilised by resonance:

$$\left[R^1-\underset{R^3}{\overset{R^2}{C}}-\overset{\cdot+}{O}H\right] \longrightarrow R^{1\cdot}+\left[\begin{matrix}R^2\\ \\R^3\end{matrix}\!\!>C-\overset{\cdot+}{O}H \longleftrightarrow \begin{matrix}R^2\\ \\R^3\end{matrix}\!\!>C=\overset{+}{O}H\right]$$

The alkyl group with the heaviest mass is eliminated preferentially, but ions also appear for alternative eliminations. Further fragmentation may also occur as follows (R^2 = H, R_3= C_2H_5):

$$\left[CH_3-CH_2-CH=\overset{+}{O}H \longleftrightarrow CH_2\begin{matrix}CH_2-H\\ \\ \overset{+}{C}H-OH\end{matrix}\right] \longrightarrow C_2H_4+[CH_2-OH]^+ \quad m/e\ 31$$

Hence, both *s*- and *t*-alcohols can also show peaks at *m/e* 31 (the most characteristic ion of alcohols).

Primary and secondary alcohols usually give weak molecular ions, and for tertiary alcohols the molecular ion is very weak or absent. Long-chain alcohols show peaks at *M*-18 due to the loss of a molecule of water and formation of a cyclic ion (cf. alkyl halides):

$$\left[C_n\begin{matrix}C-H\\ \\C-OH\end{matrix}\right]^{\dagger} \longrightarrow \left[C_n\begin{matrix}C\\ |\\C\end{matrix}\right]^{\dagger} + H_2O$$

On the other hand, a McLafferty rearrangement can also occur with loss of water:

$$\left[\begin{matrix} & H & \\ & O & \\ CH_2 & & H \\ | & & | \\ CH_2 & & CHR \\ & CH_2 & \end{matrix}\right]^{\dagger}_{M^+} \longrightarrow H_2O+CH_2=CH_2+\underset{M-46}{[RCH=CH_2]}$$

In general, long-chain alcohols also show the fragmentation patterns of alkanes and alkenes.

Ethyl alcohol, ethanol, is prepared industrially by several methods, *e.g.*, ethylene (from cracked petroleum) is absorbed in concentrated sulphuric acid (98 per cent) at 75-80°C, under pressure (17.5–35kg/cm^2). Ethyl hydrogen sulphate and ethyl sulphate are formed:

$$C_2H_4 + (HO)_2SO_2 \longrightarrow C_2H_5OSO_2OH$$

$$C_2H_5OSO_2OH + C_2H_4 \longrightarrow (C_2H_5O)_2SO_2$$

The reaction mixture is then diluted with about an equal volume of water, and warmed. Hydrolysis takes place and ethanol together with some diethyl ether is formed:

$$C_2H_5OSO_2OH + H_2O \longrightarrow C_2H_5OH + H_2SO_2$$

$$(C_2H_5O)_2SO_2 + 2H_2O \longrightarrow 2C_2H_5OH + H_2SO_4$$

$$(C_2H_5O)_2SO_2 + C_2H_5OH \longrightarrow (C_2H_5)_2O + C_2H_5OSO_2OH$$

The ether is kept to a minimum by separating the ethyl sulphate from the reaction products, and hydrolysing it separately. The hydrolysed liquids are distilled, and the aqueous ethanol distillate is concentrated by fractional distillation.

Ethanol is also manufactured by the *direct* hydration of ethylene with steam under pressure in the presence of a suitable catalyst, *e.g.*, phosphoric acid on a support.

$$C_2H_4 + H_2O \longrightarrow C_2H_5OH$$

Some ether (~ 5 per cent) is formed as a by-product.

The earliest method of preparing ethanol is by fermentation and this is still used for the manufacture of beer, wine, brandy, etc., and also a source of ethanol. The starting material is starch, which is obtained from sources depending on the particular country: common sources of starch are wheat, barely, potato, etc. Molasses is also used as the starting material for ethanol.

The grain, *e.g.*, wheat or barley, is mashed with hot water, and then heated with malt (germinated barley) at 50°C for 1 hour. Malt contains the enzyme *diastase* which, by hydrolysis, converts starch into the sugar, maltose (*q.v.*):

$$(C_6H_{10}O_5)_n + \frac{n}{2} H_2O \xrightarrow{\text{maltase}} \frac{n}{2} C_{12}H_{22}O_{11}$$

If molasses is used, then this step is unnecessary, since it contains carbohydrates already present as sugars which can be fermented.

The liquid is cooled to 30°C and fermented with yeast for 1– 3 days. Yeast contains various enzymes, among which are *maltase*, which converts the maltose into glucose, and *zymase*, which converts the glucose into ethanol:

$$C_{12}H_{22}O_{11} + H_2O \xrightarrow{\text{maltase}} 2C_6H_{12}O_6$$

$$C_6H_{12}O_6 \xrightarrow{\text{zymase}} 2C_2H_5OH + 2CO_2$$

The carbon dioxide is recovered and sold as a by-product. The fermented liquor or 'wort', which contains 6-10 per cent ethanol and some other compounds, is fractioned into three fractions:

(*i*) *First runnings*, which consists mainly of acetaldehyde.

(*ii*) *Rectified spirit*, which is 93–95 per cent w/w ethanol.

(*iii*) *Final runnings* or *fusel oil*, which contains n-propyl, n-butyl, isobutyl, n-amyl, isoamyl and 'active] amyl alcohol.

Industrial alcohol is ordinary rectified spirit. *Methylated spirit* is of two kinds: (*a*) *Mineralised methylated spirit* is 90 per cent rectified spirit, 9 per cent methanol and 1 per cent petroleum oil, and a purple dye. (*b*) *Industrial methylated spirit* is 95 per cent rectified spirit and 5 per cent methanol, whose purpose is to '*denature*' the rectified spirit, *i.e.*, make it unfit for drinking purposes.

Absolute alcohol is 99.5 per cent ethanol, and is obtained from rectified spirit. When an aqueous solution of ethanol is fractionated, it forms a constant-boiling mixture containing 96 per cent ethanol from which 100 per cent ethanol may be obtained by adding a small amount of benzene, and then distilling. The first fraction is the ternary azeotrope, *i.e.*, a constant-boiling mixture containing three constituents, b.p. 64.8°C (water, 7.4 per cent; ethanol, 18.5 per cent; benzene, 74.1 per cent). After all the water has been removed, the second fraction that distils over is the binary azeotrope, b.p. 68.2°C (ethanol, 32.4 per cent; benzene, 67.6 per cent). After all the benzene has been removed, pure ethanol, b.p. 78.1°C, distils over.

Ethanol cannot be dried by means of calcium chloride, since a compound (an *alcoholate*) is formed, *e.g.*, $CaCl_2 \cdot 3C_2H_5OH$ (*cf.* methanol). Distillation of rectified spirit over calcium oxide, and then over calcium, gives absolute alcohol. This method is often used in the laboratory, and was formerly used industrially.

Ethanol is a colourless, inflammable liquid, b.p. 78·1°C. It is miscible with water in all proportions, and is also miscible with most organic solvents. Ethanol ad methanol resemble each other very closely, but they may be distinguished :

(*i*) by the fact that ethanol gives the haloform reaction (*q.v.*), whereas methanol does not; and

(*ii*) ethanol gives acetic acid on oxidation; methanol gives formic acid. These two acids are readily distinguished from each other.

Ethanol is used for the preparation of esters, ether, chloral, chloroform, etc. It is also used as a solvent for gums, resins, paints, varnishes, etc., and as a fuel.

Structure of Ethanol

Analysis and molecular-weight determinations show that the molecular formula of ethanol is C_2H_6O. Assuming that carbon is quadrivalent, oxygen bivalent, and hydrogen univalent, two structures are possible.

$$\underset{\text{(I)}}{CH_3-CH_2-OH} \qquad \underset{\text{(II)}}{CH_3-O-CH_3}$$

(*i*) Only one hydrogen atom in ethanol is replaceable by sodium or potassium. This indicates that one hydrogen atom is in a different state of combination from the other five. In (I), one hydrogen atom differs from the other five, but in (II) all the hydrogen atoms are equivalent.

(*ii*) When ethanol is treated with hydrochloric acid or phosphorus pentachloride, one oxygen atom (bivalent) and one hydrogen atom (univalent) are replaced by one chlorine atom (univalent) to give ethyl chloride, C_2H_5Cl. This implies the presence of a hydroxyl group (*cf.* methanol).

(*iii*) When ethyl chloride is hydrolysed with dilute alkali, ethanol is obtained. This reaction also indicates the presence of a hydroxyl group in ethanol.

(*iv*) Ethanol may be prepared as follows:

$$C_2H_6 \xrightarrow{Cl_2} C_2H_5Cl \xrightarrow{NaOH} C_2H_5OH$$

The arrangement of the six hydrogen atoms in ethane is known, and it is reasonable to suppose that five retain their original arrangement in ethyl chloride and ethanol, since these five hydrogen atoms do not enter (presumably) into the above reactions. Thus there is an ethyl group C_2H_5—in ethanol. This is so in (I), but not in (II).

(*v*) Structure (II) is definitely eliminated, since it can be shown that it is the structure of dimethyl ether (*q.v.*), a compound that has very little resemblance, physically or chemically, to ethanol.

Hence, (I) is accepted as the structure of ethanol, and it accounts for all the known properties of ethanol.

It can be seen from the various examples given on structure determination, *e.g.*, methane, butanes, ethylene, methanol and ethanol, that the method of approach follows certain definite lines. First the molecular formula is obtained. Then, if the compound is simple–in the sense that it contains a small number of unlike atoms, and that the total number of atoms is also small–the valencies of the atoms present are assumed, and various possible structures are written down. If the compound is 'simple' the number of possible structures will not be large (four or five at the most). Then by considering the chemical properties of the compound in question, the structure which best fits the observed facts is accepted as the correct one.

If the compound is not 'simple', the procedure is to detect the presence of as many functional groups as possible; to degrade the compound into simpler substances whose structures are already known or which may be determined by further degradation; to build up structural formulae based on the facts obtained; and then to choose the structure which best fits the facts. Finally a synthesis is attempted, and if successful, will usually give proof of the correctness of the structure suggested.

It should be noted that the arguments used for structure determination by purely chemical methods are based on the assumption that no rearrangement have occurred in the series of reactions performed. Thus, the possibility of rearrangement must always be borne in mind when carrying out chemical reactions.

As was pointed out physical methods are being used nowadays for structural determination. Once the molecular formula is obtained, functional groups may be identified, etc. In this way, the structures of many compounds have now been ascertained with a minimum of chemical evidence.

Propyl Alcohols

Two isomeric propyl alcohols are possible, and both are known.

*n-Propyl alcohol, propan-*1*-ol, n-propanol*, was originally obtained from fusel oil (see above), but it is now produced by the hydogenation of carbon monoxide. A more recent method is by the catalytic reduction of propargyl alcohol (from acetylene and formaldehyde).

$$CH{\equiv}CCH_2OH + 2H_2 \xrightarrow{Ni} 2CH_3CH_2CH_2OH$$

n-Propanol, b.p. 97.4°C, is miscible with water, ethanol and ether. It is used in the preparation of propionic acid, toilet preparations such as lotions, etc.

Isopropyl alcohol, propan-2-ol, isoprpanol, is prepared industrially by passing propene (from cracked petroleum) into concentrated sulphuric acid, then diluting with water, and distilling off the isopropanol. Isopropyl ether is obtained as a by-product (*cf.* ethanol):

$$CH_3CH{=}CH_2 + (HO)_2SO_2 \longrightarrow (CH_3)_2CHOSO_2OH$$

$$(CH_3)_2CHOSO_2OH + CH_3CH{=}CH_2 \longrightarrow [(CH_3)_2CHO]_2SO_2$$

$$(CH_3)_2CHOSO_2OH + H_2O \longrightarrow (CH_3)_2CHOH + H_2SO_4$$

$$[(CH_3)_2CHO]_2SO_2 + 2H_2O \longrightarrow 2(CH_3)_2CHOH + H_2SO_4$$

$$[(CH_3)_2CHO]_2SO_2 + (CH_3)_2CHOH \longrightarrow [(CH_3)_2CH]_2O + (CH_3)_2CHOSO_2SO_4$$

Isopropanol is also prepared by *direct* hydration by passing a mixture of propene and steam at 220-250°C, and under pressure (220 atm), over a catalyst of tungsten oxide plus zinc oxide, on a silica carrier.

Isopropanol, b.p. 82.4°C, is soluble in water, ethanol and ether. It is used for preparing esters, acetone, keten, as a solvent, and for high-octane fuel.

Butyl Alcohols

Four isomers are possible, and all are known.

n-Butyl alcohol, butan-1-ol, n-butanol, b.p. 117.4°C, is prepared industrially from propene by the Oxo process. It is widely used as a solvent.

Isobutyl alcohol, 2-methylpropan-1-ol, n-butanol, b.p. 108°C, is also obtained by the Oxo process. It behaves as a primary alcohol, but it readily rearranges due to the presence of a branched chain near the COH group, *e.g.*, when it is treated with hydrochloric acid, isobutyl chloride and t-butyl chloride are obtained:

$$(CH_3)_2CHCH_2OH \xrightarrow{HCl} (CH_3)_2CHCH_2Cl + (CH_3)_3CCl$$

s-Butyl alcohol, butan--2-ol, s-butano, b.p. 100°C, is prepared industrially by the hydration of 1- or 2-butene (from cracked petroleum) by means of concentrated sulphuric acid (*cf.* ethanol and isopropanol). It is used for the preparation of butanone, esters, and as a lacquer solvent.

t-Butyl alcohol, t-butanol (*trimethylmethanol*), m.p. 25.5°C, is prepared synthetically by the hydration of isobutene (from cracked petroleum). It is mainly used as an alkylating agent.

Amyl Alcohols

Eight isomers are possible, and all are known:

1. $CH_3CH_2CH_2CH_2CH_2OH$, n-amyl alcohol, pentan 1-ol, n-pentanol, p.b. 138°C.
2. $(CH_3)_2CHCH_2CH_2OH$, isoamyl alcohol, isopentanol, p.b. 130°C.
3. $CH_3CH_2CH(CH_3)CH_2OH$, 'active' amyl alcohol, 2-methylbutan-1-ol, b.p. 128°C.
4. $(CH_3)_3CCH_2OH$, neopentyl alcohol, t-butylcarbinol, b.p. 113°C.
5. $CH_3CH_2CH_2CH(OH)CH_3$, pentan-2-ol, methyl-n-propylcarbinol, p.b. 120°C.
6. $CH_3CH_2CH(OH)CH_2CH_3$, pentan-3-ol, diethylcarbinol, b.p. 117°C.
7. $(CH_3)_2CHCH(OH)CH_3$, 3-methylbutan-2-ol, isopropylmethylcarbinol, b.p. 114°C.
8. $(CH_3)_2C(OH)CH_2CH_3$, 2-methylbutan-2-oil, t-amyl alcohol, t-pentanol, ethyldimethylcarbinol, b.p. 102°C.

There amyl alcohols, *viz*., n-pentanol, isopentanol and 'active' amyl alcohol, have been isolated from fusel oil (*q.v.*). The last two are the chief constituents of fusel oil, and all three are produced by the fermentation of protein matter associated with the carbohydrates in starch. This mixture of amyl alcohols (from fusel oil) is used for the preparation of esters (for artificial essences), scents, and as a laboratory reducing agent with sodium it is better than ethanol owing to its higher boiling point (this mixture of amyl alcohols will be referred to in future as isopentanol).

A mixture of amyl alcohols, known as *pentasol*, is prepared industrially by chlorinating at 200°C, and in the dark, a mixture of n-

and isopentanes (from petroleum) to the amyl chlorides which are hydrolysed with dilute sodium hydroxide solution plus a little sodium oleate for emulsification, to the amyl alcohols.

Seven isomeric amyl chlorides are theoretically possible, but in practices six are obtained. 2-Chloro-2-methylbutane (the only t-chloride produced) readily eliminates hydrogen chloride to form trimethylethylene. Pentasol (the mixture of six amyl alcohols) finds great use as a solvent in the lacquer industry.

Commercial 's-amyl alcohol' (80 per cent 2- and 20 per cent 3-pentanol) is made by hydrating 1-and 2-pentenes (from cracked petroleum).

A number of higher alcohols occur as esters in waxes, *e.g.*, cetyl (palmityl) alcohol (hexadecan-1-ol), $C_{16}H_{33}OH$, m.p. 49°C, occurs as the palmitate in spermaceti (obtained from the oil of the sperm whale); carnaubyl alcohol (tetracosan-1-ol), $C_{24}H_{49}OH$, m.p. 69°C, as esters in wool-grease; ceryl alcohol (hexacosan-1-ol), $C_{26}H_{53}OH$, m.p. 79°C, as the cerotate in chinese wax; myricyl (melissyl) alcohol (triacontan-1-ol), $C_{30}H_{61}OH$, m.p. 88°C, as esters in bees-wax.

A number of higher alcohols are now prepared industrially by the catalytic reduction of the ethyl (or glyceryl) esters of the higher monocarboxylic acids, particularly the alcohols lauryl (dodecan-1-ol), $C_{12}H_{25}OH$, m.p. 24°C; myristyl (tetradecan-1-ol), C14H29OH, m.p. 38°C; palmityl (hexadecan-1-ol), $C_{16}H_{33}OH$, m.p. 49°C; and stearyl (octaecan-1-ol), $C_{18}H_{37}OH$, m.p. 59°C.

These alcohols are used in the form of sodium alkyl sulphates, $ROSO_2ONa$, as detergents, emulsifying agents, wetting agents, insecticides, fungicides, etc. These sodium salts lather well, and are not affected by hard water, and hence can be used as a soap substitute.

POLYHYDRIC ALCOHOLS

Dihydric Alcohols or Glycols

Dihydric alcohols are compounds containing two hydroxyl groups. They are classified as α, β, γ . . . glycols, according to the relative positions of the two hydroxyl groups: α is the 1,2 glycol; β, 1,3; γ, 1,4 . . . Although, it is unusual to find a compound with two hydroxyl groups attached to the same carbon atom, ether derivatives of these, 1,1 glycols are stable, *e.g.*, acetals. The commonest glycols are the α-glycols.

Nomenclature

The common names of the α-glycols are derived from the corresponding alkene from which they may be prepared by direct hydroxylation, *e.g.*,

$CHCH_2CH_2OH$ ethylene glycol

$(CH_3)_2COHCH_2OH$ isobutene glycols,

β-, γ . . . Glycols are named as the corresponding *polymethlene* glycols, *e.g.*,

$HOCH_2CH_2CH_2OH$ trimethylene glycol

$HOCH_2CH_2CH_2CH_2CH_2OH$ pentamethylene glycol

According to the I.U.P.A.C. system of nomenclature, the class suffix is -*diol*, and numbers are used to indicate the positions of side-chains and two hydroxyl groups *e.g.*,

$$HOCH_2CH_2\underset{|}{\overset{CH_3}{C}}HCH_2\underset{|}{\overset{CH_3}{C}}HCH_2OH$$ 2, 4-dimethylhexane-1,6-diol

Ethylene glycol, glycol (*ethane*-1,2-*diol*), is the simplest glycol, and may be prepared as follows:

(*i*) By passing ethylene into cold dilute alkaline permanganate solution.

(*ii*) By passing ethylene into hypochlorous acid, and then hydrolysing the ethylene chlorohydrin by boiling with aqueous sodium hydrogen carbonate:

$$CH{=}CH_2 + HOCl \longrightarrow ClCH_2CH_2OH \xrightarrow{\text{aq. } NaHCO_3} HOCH_2CH_2OH + NaCl + CO_2$$

(*iii*) By boiling ethylene dibromide with aqueous sodium carbonate:

$$BrCH_2CH_2Br + NaCO_3 + H_2O \longrightarrow HOCH_2CH_2OH + 2NaBr + CO_2 \ (50\%)$$

The low yield in this reaction is due to the conversion of some ethylene dibromide into vinyl bromide:

$$BrCH_2CH_2Br + Na_2CO_2 \longrightarrow CH_2{=}CHBr + NaBr + NaHCO_3$$

If aqueous sodium hydroxide is used instead of sodium carbonate, vinyl bromide is again obtained as by-product. The best yield of glycol is obtained by heating ethylene dibromide with potassium acetate in glacial acetic acid, and subsequently hydrolysing the glycol diacetate with hydrogen chloride in methanolic solution:

$$BrCH_2CH_2Br + 2CH_3CO_2K \longrightarrow$$

$$\underset{(90\%)}{CH_2(OCOCH_3)CH_2(OCOCH_3)} + 2KBr \xrightarrow{HCl} \underset{(83\text{-}84\%)}{HOCH_2CH_2OH}$$

(*iv*) By treating ethylene oxide with dilute hydrochloric acid:

$$\begin{matrix} CH_2 \\ | \\ CH_2 \end{matrix}\!\!>\!O + H_2O \xrightarrow{HCl} HOCH_2CH_2OH$$

Glycol is prepared industrially by method (*iv*), and by the catalytic reduction of methyl glycollate which is produced synthetically.

Glycol is a colourless viscous liquid, p.b. 197°C, and has a sweet taste (the prefix *glyc*-indicates that the compound has a sweet taste: Greek *glukus*, sweet). It is miscible in all proportions with water and ethanol, but is insoluble in ether. It is widely used as a solvent and as an antifreeze agent.

The high boiling points and high solubility i water of polyhydric alcohols are due to hydrogen bonding involving all hydroxyl groups.

The chemical reactions of glycol are those which might have been expected of a monohydric primary alcohol. One hydroxyl group, however, is almost always completely attacked before the other reacts.

(*i*) Sodium at 50°C forms the monosodium salt, and the di-sodium salt only when the temperature is raised to 160°C.

(*ii*) Hydrogen chloride forms ethylene chlorohydrin at 160°C, whereas at 200°C, ethylene dichloride is formed.

Ethylene chlorohydrin (β-chloroethyl alcohol, 2-chloroethanol) is a colourless liquid, p.b. 128.8°C.

It is very useful in organic syntheses since it contains two different reactive groups; *e.g.*, by heating with aqueous sodium cyanide it may be converted into ethylene cyanohydrin, which, on hydrolysis, gives β-hydroxypropionic acid:

$$HOCH_2CH_2Cl + NaCH \longrightarrow HOCH_2CH_2CN + NaCl \ (79\text{-}80\%)$$

$$HOCH_2CH_2CN \xrightarrow{HCl} HOCH_2CH_2CO_2H$$

(*iii*) When glycol is treated with phosphorus trichloride or phosphorus tribromide, the corresponding ethylene dihalide is obtained:

$$\begin{matrix} CH_2OH \\ | \\ CH_2OH \end{matrix} \xrightarrow{PBr_3} \begin{matrix} CH_2Br \\ | \\ CH_2Br \end{matrix}$$

Phosphorus tri-iodide, however, produces ethylene. Ethylene di-iodide is formed as an intermediate, but readily eliminates iodine :

$$\begin{matrix} CH_2OH \\ | \\ CH_2OH \end{matrix} \xrightarrow{PIr_3} \left[\begin{matrix} CH_2I \\ | \\ CH_2I \end{matrix}\right] \longrightarrow I_2 + \begin{matrix} CH_2 \\ \| \\ CH_2 \end{matrix}$$

(*iv*) When glycol is treated with organic acids or inorganic oxygen acids, the mono- or di-esters are obtained, depending of the relative amounts of glycol ad acid; *e.g.*, glycol diacetate is obtained by heating glycol with acetic acid in the presence of a small amount of sulphuric acid as catalyst:

$$\begin{matrix} CH_2OH \\ | \\ CH_2OH \end{matrix} + 2CH_3CO_2H \xrightarrow{PIr_3} \begin{matrix} CH_2OCOCH_3 \\ | \\ CH_2OCOCH_3 \end{matrix} + 2H_2O$$

When glycol is heated with dicarboxylic acids, condensation polymers (polyesters) are obtained:

$$n\text{HOCH}_2\text{CH}_2\text{OH} + n\text{HO}_2\text{C(CH}_2)_x\text{CO}_2\text{H} \longrightarrow$$

$$\text{HOCH}_2(\text{CH}_2)_x\text{COOCH}_2]_n\text{CH}_2\text{OH}$$

(*v*) Glycol condenses with aldehydes or ketones in he presence of *p*-toluenesulponic acid to yield respectively cyclic acetals or cyclic ketals (1,3-*dioxolans*):

$$\begin{matrix} CH_2OH \\ | \\ CH_2OH \end{matrix} + O{=}C\begin{matrix} R^1 \\ R^2 \end{matrix} \longrightarrow \begin{matrix} CH_2O \\ | \\ CH_2O \end{matrix}\!\!>\!C\begin{matrix} R^1 \\ R^2 \end{matrix} + H_2O$$

Dioxolan formation offers a means of protecting a carbonyl group in reactions carried out in alkaline media. The carbonyl compound may be regenerated by the addition of periodic acid to an aqueous dioxan solution of the dioxolan:

$$\begin{matrix} R^1 \\ R^2 \end{matrix}\!\!>\!C\begin{matrix} O{-}CH_2 \\ | \\ O{-}CH_2 \end{matrix} \xrightarrow{HIO_4} R^1COR^2 + 2CH_2O$$

(*vi*) When glycol is oxidised with nitric acid, glycollic and oxalic acids may be readily isolated. All theoretically possible oxidation products have been isolated, but since, except for glycollic and oxalic acids, they are more readily oxidised than glycol itself, they have only been obtained in very poor yields.

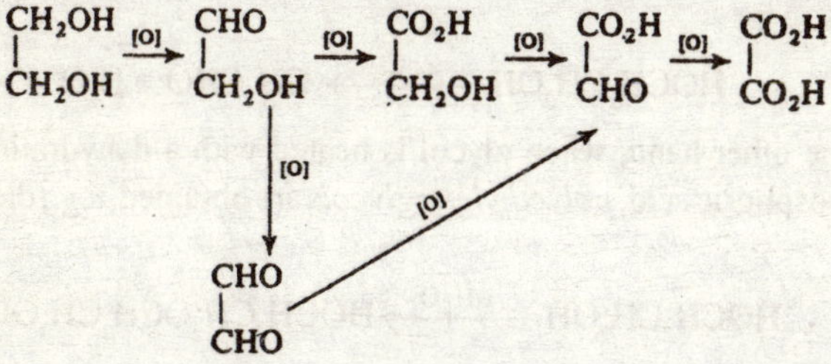

$$\begin{matrix} CH_2OH \\ | \\ CH_2OH \end{matrix} \xrightarrow{[O]} \begin{matrix} CHO \\ | \\ CH_2OH \end{matrix} \xrightarrow{[O]} \begin{matrix} CO_2H \\ | \\ CH_2OH \end{matrix} \xrightarrow{[O]} \begin{matrix} CO_2H \\ | \\ CHO \end{matrix} \xrightarrow{[O]} \begin{matrix} CO_2H \\ | \\ CO_2H \end{matrix}$$

$$\begin{matrix} CHO \\ | \\ CH_2OH \end{matrix} \xrightarrow{[O]} \begin{matrix} CHO \\ | \\ CHO \end{matrix} \xrightarrow{[O]} \begin{matrix} CO_2H \\ | \\ CHO \end{matrix}$$

Fig. 1.3

Oxidation of glycols with acid, alkaline or neutral permanganate, acid dichromate or chromic acid (chromium trioxide in acetic acid), lead tetra-acetate or periodic acid results in fission of the carbon-carbon bond to give aldehydes, acids and/or ketones.

Oxidation of 1,2-glycols occurs more rapidly with *cis*-glycols than with the corresponding *trans*-isomers, and this has been interpreted as evidence for the formation of cyclic intermediates, *e.g.*,

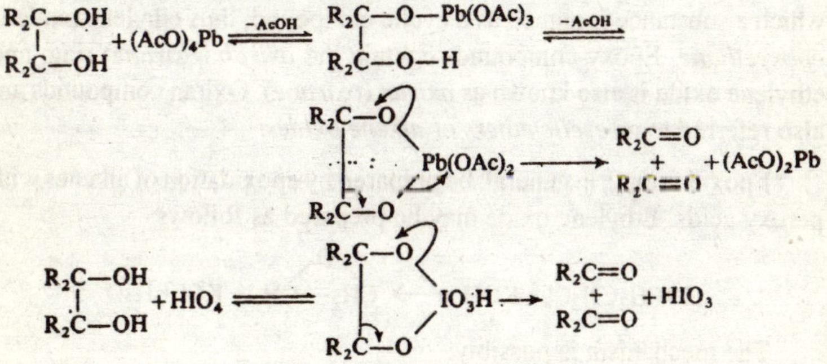

$$\begin{matrix} R_2C{-}OH \\ | \\ R_2C{-}OH \end{matrix} + (AcO)_4Pb \underset{}{\overset{-AcOH}{\rightleftharpoons}} \begin{matrix} R_2C{-}O{-}Pb(OAc)_3 \\ | \\ R_2C{-}O{-}H \end{matrix} \underset{}{\overset{-AcOH}{\rightleftharpoons}} \begin{matrix} R_2C{-}O \\ | \quad\quad \\ R_2C{-}O \end{matrix} Pb(OAc)_2 \longrightarrow \begin{matrix} R_2C{=}O \\ + \\ R_2C{=}O \end{matrix} + (AcO)_2Pb$$

$$\begin{matrix} R_2C{-}OH \\ | \\ R_2C{-}OH \end{matrix} + HIO_4 \rightleftharpoons \begin{matrix} R_2C{-}O \\ | \quad\quad \\ R_2C{-}O \end{matrix} IO_3H \longrightarrow \begin{matrix} R_2C{=}O \\ + \\ R_2C{=}O \end{matrix} + HIO_3$$

Fig. 1.4

The fact he *trans*-isomers *are* oxidise suggests that the reaction proceeds through a noncyclic intermediate for these compounds. This could be the non-cyclic ester formed in the first step.

These oxidations are also subject to steric hindrance, *e.g.*, glycol is oxidised much faster than pinacol by periodic acid. According to Bunton *et al.* (1957), the rate-determining step for glycol oxidation is the fission of the complex, whereas the for pinacol is the formation of the complex.

(*vii*) Glycol is converted in acetaldehyde when heated with dehydrating agents, *e.g.*,

$$HOCH_2CH_2OH \xrightarrow{ZnCl_2} CH_3CHO + H_2O$$

On the other hand, when glycol is heated with a dehydrating agent such as phosphoric acid, *polyethylene glycols* are obtained, *e.g.*, diethylene glycol:

$$2HOCH_2CH_2OH \xrightarrow{-H_2O} HOCH_2CH_2OCH_2CH_2OH$$

By varying the amount of phosphoric acid and the temperature, the polyethylene glycols up to decaethylene glycol can be obtained. These condensation polymers contain both the alcohol and ether functional groups; they are soluble in water (alcohol function) and are very good solvents (ether function). They are soluble in water (alcohol function) and are very good solvents (ether function). They are widely used as solvents for gums, resins, cellulose esters, etc.

Ethylene oxide, b.p. 10.7°C. According to the I.U.P.A.C. system of nomenclature, an oxygen atom linked to two of the carbon atoms in a carbon chain is denoted by the prefix *epoxy* in all cases other than those in which a substance is named as a cyclic compound; thus ethylene oxide is *epoxyethane*. Epoxy-compounds contain the *oxiran* (*oxirane*) ring, and ethylene oxide is also known as *oxiran* (*oxirane*). Oxiran compounds are also referred to as *cyclic ethers* or *alkene oxides.*

Epoxides may, in general, be prepared by epoxidation of alkenes with peroxy acids. Ethylene oxide may be prepared as follows:

$$HOCH_2CH_2Cl + KOH \longrightarrow \overset{\quad O \quad}{CH_2{-}CH_2} + KCl + H_2O$$

The mechanism is possibly:

$$HO^- \; H{-}\underset{|}{O} \qquad CH_2{-}CH_2Cl \xrightarrow{fast} H_2O + \overset{\quad O^- \quad}{CH_2{-}CH_2} \; Cl$$

$$\xrightarrow{slow} \overset{\quad O \quad}{CH_2{-}CH_2} + Cl^-$$

This is an example of neighbouring group participation.

Ethylene oxide is manufactured by passing ethylene and air under pressure over a silver catalyst at 200-400°C:

$$C_2H_4 + 1/2O_2 \longrightarrow \overset{O}{CH_2—CH_2}$$

Ethylene oxide undergoes molecular rearrangement on heating of form acetaldehyde:

$$\overset{O}{CH_2—CH}(H) \xrightarrow[\text{shift}]{H^-} CH_3CHO$$

Epoxides are reduced by lithium aluminium hydride to alcohols, unsymmetrical epoxides giving as main product the more highly substituted alcohol; thus (I) is the main product:

$$\overset{O}{RCH—CH_2} \xrightarrow{LiAlH_4} \underset{(I)}{RCHOHCH_3} + \underset{(II)}{RCH_2CH_2OH}$$

On the other hand (II) is the main product if LAH—$AlCl_3$ is used (Eliel *et al.*, 1958. Brown *et al.* (1968) have shown that (II) is the major product (> 90%) when BH_3—BF_3 is the reducing reagent.

Epoxides are readily reduced to the corresponding alkenes by t-phosphines, *e.g.*,

$$\overset{O}{RCH—CHR} + Ph_3P \longrightarrow RCH{=}CHR + Ph_3PO$$

They are also oxidised by dimethyl sulphoxide to α-hydroxyketones (acyloins).

$$\overset{O}{RCH—CHR} + Me_2SO \xrightarrow{H^+} RCH(OH)—CR(H)(O—\overset{+}{S}Me_2) \longrightarrow RCHOHCOR + H^+ + Me_2S$$

Ethylene oxide is converted into ethylene glycol in *dilute* acid solution and into ethylene halogenohydrins with *concentrated* halogen acid solutions. Ethylene oxide also forms monoethers with alcohols in the presence of a small amount of acid as catalyst. The mechanisms of these reactions in acid media probably follow the same pattern, *e.g.*, with *dilute* acid (with alcohols, replace H_2O by ROH):

$$\begin{array}{l} CH_2 \\ | \quad \diagdown \\ | \quad\; O + H_3O^+ \\ | \quad \diagup \\ CH_2 \end{array} \rightleftharpoons \begin{array}{l} CH_2 \\ | \quad \diagdown \\ | \quad\; \overset{+}{O}H + H_2O \\ | \quad \diagup \\ CH_2 \end{array}$$

$$H_2O \curvearrowright \begin{array}{l} CH_2 \\ | \quad \diagdown \\ | \quad\; \overset{+}{O}H \\ | \quad \diagup \\ CH_2 \end{array} \longrightarrow \begin{array}{l} H_2\overset{+}{O}-CH_2 \\ \qquad\quad | \\ \qquad CH_2-OH_2 \end{array} \xrightarrow{-H^+} \begin{array}{l} CH_2OH \\ | \\ CH_2OH \end{array}$$

Attack occurs from the position remote from the ^+OH, and the product is the *trans* diol. This cannot be demonstrated with ethylene oxide itself but has been, *e.g.*, with 1,2-epoxycyclohexane. With *concentrated* halogen acid.

$$Cl^- \curvearrowright \begin{array}{l} CH_2 \\ | \quad \diagdown \\ | \quad\; \overset{+}{O}H \\ | \quad \diagup \\ CH_2 \end{array} \longrightarrow \begin{array}{l} CH_2Cl \\ | \\ CH_2OH \end{array}$$

Ethylene oxide also reacts with alcohols under the influence of *basic* catalysts. A possible mechanism is:

$$EtO^- \curvearrowright \begin{array}{l} CH_2 \\ | \quad \diagdown \\ | \quad\; O \\ | \quad \diagup \\ CH_2 \end{array} \longrightarrow \begin{array}{l} EtO-CH_2 \\ \qquad\quad | \\ \qquad CH_2O^- \curvearrowright H-OEt \end{array} \longrightarrow \begin{array}{l} CH_2OEt \\ | \\ CH_2OEt \end{array} + EtO^-$$

Of particular interest is the addition of hydrogen cyanide to form ethylene cyanohydrin:

$$\overset{\;\;O}{CH_2-CH_2} + HCN \longrightarrow HOCH_2CH_2CN$$

The structure of ethylene oxide is uncertain; it may contain 'bent' bonds (see cyclopropane).

When ethylene oxide is heated with methanol under pressure, the monomethyl ether of glycol is formed:

$$\overset{\;\;O}{CH_2-CH_2} + CH_3OH \longrightarrow HOCH_2CH_2OCH_3$$

This is known as *methyl cellosolve*; the corresponding ethyl ether is known as *ethyl cellosolve*. Cellosolves are very useful as solvents since they contain both the alcohol and ether functional groups (*cf.* polyethylene glycols).

The further action of ethylene oxide on cellosolves produces *carbitols*, *e.g.*,

$$HOCH_2CH_2OCH_3 + \underset{\diagdown O \diagup}{CH_2{-}CH_2} \longrightarrow HOCH_2CH_2OCH_2CH_2OCH_3$$

When glycol is treated wit ethylene oxide, diethylene glycol is formed:

$$HOCH_2CH_2OH + \underset{\diagdown O \diagup}{CH_2{-}CH_2} \longrightarrow HOCH_2CH_2OCH_2CH_2OH$$

Diethylene glycol dimethyl ether (*diglyme*), p.b. 160°C, is used as solvent.

Ethylene oxide reacts with ammonia to form a mixture of three aminoalcohols which are usually referred to as the *ethanolamines*:

$$C_2H_4O + NH_3 \longrightarrow HOCH_2CH_2NH_2 \xrightarrow{C_2H_4O}$$

$$(HOCH_2CH_2{-})_2NH \xrightarrow{C_2H_4O} (HOCH_2CH_2{-})_3N$$

The ethanolamines are widely used as emulsifying agents.

The mechanism of these reactions is possibly:

$$H_3\ddot{N}\;\; \begin{matrix} CH_2 \\ | \diagdown O \\ | \diagup \\ CH_2 \end{matrix} \longrightarrow \begin{matrix} H_3\overset{+}{N}{-}CH_2 \\ | \\ CH_2O^- \end{matrix} \longrightarrow \begin{matrix} H^+ + CH_2NH_2 \\ | \\ CH_2O^- \end{matrix} \longrightarrow \begin{matrix} CH_2NH_2 \\ | \\ CH_2OH \end{matrix}$$

In the presence of ethylene oxide, the reaction proceeds with the ethanolamine now being the amino reagent, etc.

The C—O group (stretch) in epoxides in he region 1 260–1 240cm^{-1} (s), and so these compounds may be readily distinguished from ethers.

The NMR spectra of epoxy compounds show a τ-value ~ 7·2-7·6 p.p.m. for methylene protons and 7·0 for methine protons. These ranges are sufficiently different from those of CH_2 and CH groups in other environments that epoxides may usually be recognised in this way.

Dixoan (1,4-*dioxan, diethylene dioxide*), p.b. 101·5°C, is manufactured:

(*i*) By distilling glycol with fuming sulphuric acid:

$$2HOCH_2CH_2OH \longrightarrow O\begin{matrix} \diagup CH_2{-}CH_2 \diagdown \\ \\ \diagdown CH_2{-}CH_2 \diagup \end{matrix}O + 2H_2O$$

(*ii*) Bydistilling ethylene oxide with dilute sulphuric acid:

$$2C_2H_4O \longrightarrow O\begin{matrix} CH_2-CH_2 \\ CH_2-CH_2 \end{matrix}O$$

Dioxan is a useful solvent for cryoscopic and ebullioscopic work.

Propene oxide, p.b. 35°C, is manufactured as follows:

$$(i)\ 2CH_3CH{=}CH2 \xrightarrow{2HOCl} CH_3CHOHCH_2Cl + CH_3CHClCH_2OH$$

$$\xrightarrow{Ca(OH)_2} 2CH_3\overset{O}{CH-CH_2}$$

$$(ii)\ CH_3CH{=}CH_2 \xrightarrow[cat.]{t\text{-}BuOOH} CH_3\overset{O}{CH-CH_2}$$

Its reactions are similar to those of ethylene oxide. When heated with dilute hydrochloric acid, propene glycol, p.b. 188°C, which is used as a solvent, as a substitute for glycerol, etc.

General Methods of Preparation of 1,2-Glycols Other Than Glycol Itself

1. By the reduction of ketones with magnesium amalgam, *e.g.*, pinacol.

$$2CH_3COCH_3 \xrightarrow[(ii)\ H_2O]{(i)\ Mg-Hg} (CH_3)_2C(OH)C(OH)C(OH)(CH_3)_2 (43\text{–}50\%)$$

2. By the action of a Grignard reagent of α-diketones, *e.g.*,

$$CH_3COCOCH_3 \xrightarrow{2RMgX} CH_3\underset{XMgO}{\overset{R}{C}}-\underset{OMgX}{\overset{R}{C}}CH_3 \xrightarrow{H_2O} CH_3\underset{HO}{\overset{R}{C}}-\underset{OH}{\overset{R}{C}}CH_3$$

3. By the action of a Grignard reagent on α-ketonic esters, *e.g.*,

$$CH_3COCO_2C_2H_5 \xrightarrow{3RMgX} CH_3\underset{XMgO}{\overset{R}{C}}-\underset{OMgX}{\overset{R}{C}}R \xrightarrow{H_2O} CH_3\underset{HO}{\overset{R}{C}}-\underset{OH}{\overset{R}{C}}R$$

4. By the catalytic reduction of acyloins:

$$RCH(OH)COR \xrightarrow[Ni]{H_2} RCH(OH)CH(OH)R$$

5. By hydroxylation of unsaturated compounds.

Pinacol (*tetramethylethylene glycol, 2,3-dimethylbutane-2, 3-diol*) is most conveniently prepared by reducing acetone with magnesium amalgam (see above). It crystallises out of solution as the hexahydrate. The most important reaction of pinacol is the rearrangement it undergoes with distilled with dilute sulphuric acid:

$$(CH_3)_2C(OH)C(OH)(CH_3)_2 \xrightarrow{H_2SO_4} CH_3COC(CH_3)_3 + H_2O (65\text{–}72\%)$$

The *pinacol-pinacolone rearrangement* is general for pinacols (ditertiary alcohols). Some diene is formed as by-product:

$$CH_3-\underset{OH}{\overset{CH_3}{C}}-\underset{OH}{\overset{CH_3}{C}}-CH_3 \xrightarrow{H_2SO_4} CH_2=\overset{CH_3}{C}-\overset{CH_3}{C}=CH_2 + 2H_2O$$

POLYMETHYLENE GLYCOLS

A general method of preparing polymethylene glycols is to reduce α,ω-dicarboxylic esters with sodium and ethanol or lithium aluminium hydride, or catalytically. Individual polymethylene glycols are usually prepared by special methods.

Trimethylene glycol (propane-1,3-diol), b.p. 214°C (with decomposition), may be prepared by the hydrolysis of trimethylene dibromide.

Tetramethylene glycol (butane-1, 4-diol), b.p. 230°C, and **hexamethylene glycol (hexane-1, 6-diol),** m.p. 42°C, are most conveniently prepared by reducing the corresponding α,ω-dicarboxylic esters (succinic and adipic, respectively). Tetramethylene glycol is prepared industrially by hydrogenating butynediol. It is used for preparing butadiene, γ-butyrolactone and tetrahydrofuran.

Pentamethylene glycol (pentane-1,5-diol), b.p. 239°C, can be prepared from pentamethylene dibromide, which is obtained from piperidine by the *von Braun reaction*, by heating with potassium acetate in glacial acetic acid, and subsequently hydrolysing the diacetate with alkali (cf. glycol):

$$BrCH_2(CH_2)_3CH_2Br \xrightarrow{CH_3CO_2K} CH_3COOCH_2(CH_2)_3CH_2OOCCH_3$$

$$\xrightarrow{NaOH} HOCH_2(CH_2)_3CH_2OH \quad \text{(v.g.)}$$

Pentamethylene glycol may also be prepared by the catalytic hydrogenation (copper chromite) of tetrahydrofurfuryl alcohol under pressure (40 – 47 per cent yield).

The polymethylene glycols are readily converted into the corresponding mono- or di-halogen derivative by halogen acid, according to the amount of halogen acid used. These polymethylene halides are useful reagents in organic syntheses since they contain two reactive halogen atoms, one or both of which may be made to undergo reaction. If the synthesis requires reaction with one halogen atom only, the most satisfactory procedure is to 'protect' the other halogen atom by ether formation and subsequently decompose the ether with concentrated hydrobromic acid, *e.g.*,

$$C_2H_5ONa + BrCH_2CH_2CH_2Br \longrightarrow NaBr + C_2H_5OCH_2CH_2CH_2Br$$

$$\xrightarrow{AgNO_2} C_2H_5OCH_2CH_2CH_2NO_2 \xrightarrow{HBr} BrCH_2CH_2CH_2NO_2$$

Then *e.g.*,

$$BrCH_2CH_2CH_2NO_2 \xrightarrow{KCN} NCCH_2CH_2CH_2NO_2$$

TRIHYDRIC ALCOHOLS

The only important trihydric alcohol is **glycerol (propane-1, 2, 3-triol)**, b.p. 290°C. It occurs in almost all animal and vegetable oils and fats as the glyceryl esters of mainly palmitic, stearic and oleic acids.

Glycerol is obtained in large quantities as a by-product in the manufacture of soap, and this is still a commercial source of glycerol (see below). It is prepared synthetically as follows:

$$CH_3CH{=}CH_2 \xrightarrow[480-500°C]{Cl_2} ClCH_2CH{=}CH_2 \xrightarrow[150°C\ 12\ atm.]{aq.\ Na_2CO_3}$$

$$\underset{\text{allyl alcohol}}{HOCH_2CH{=}CH_2} \xrightarrow{HOCl} \underset{\text{glycerol }\beta\text{-monochlorohydrin}}{HOCH_2CHClCH_2OH}$$

$$\xrightarrow{NaOH} HOCH_2CHOHCH_2OH$$

An alternative route that is used is:

$$ClCH_2CH{=}CH_2 \xrightarrow{HOCl} \left.\begin{matrix} ClCH_2CHOHCH_2Cl \\ + \\ ClCH_2CHClCH_2OH \end{matrix}\right\} \xrightarrow{CaO}$$

$$\underset{\text{epichlorohydrin}}{\overset{\;\;O}{CH_2{-}CHCH_2Cl}} \xrightarrow{NaOH} HOCH_2CHOHCH_2OH$$

Other methods of manufacture are:

(*i*) $CH_2CH{=}CHO \xrightarrow[cat.]{H_2} CH_2{=}CHCH_2OH \xrightarrow{H_2O_2}$

$$HOCH_2CHOHCH_2OH$$

(*ii*) $CH_2{=}CHCH_3 + O_2 \xrightarrow[300-400°C]{Cu_2O} CH_2{=}CHCHO$

$CH_2{=}CHCH_3 + H_2O \xrightarrow[WO_3-ZnO]{220-250°} CH_3CHOHCH_3$

$CH_2{=}CHCHO + (CH_3)_2CHOH \xrightarrow[400°C]{MgO-ZnO}$

$(CH_3)_2CO + CH_2{=}CHCH_2OH$

$(CH_3)_2CHOH + O_2 \xrightarrow{100-140°C} (CH_3)_2CO + H_2O_2$

$CH_2CHCH_2OH + H_2O_2 \xrightarrow[WO_3]{60-70°C} HOCH_2CHOHCH_2OH$

Acetone is obtained as a by-product.

Glycerol is used as an antifreeze, for making explosives, and, because of its hygroscopic properties, as a moistening agent for tobacco, shaving soaps, etc.

Glycerol contains one secondary and two primary alcoholic groups, and it undergoes many of the reactions to be expected of these types of alcohols. The carbon atoms in glycerol are indicated as shown:

$\overset{\alpha}{CH_2}\,OH\overset{\beta}{CH}OH\overset{\alpha' \, or \, \gamma}{CH_2}\,OH \cdot$

(*i*) When glycerol is treated with sodium, one α-hydroxyl group is readily attacked, and the other α-group less readily; the β-hydroxyl groups is not attacked at all:

$HOCH_2CHOHCH_2OH \xrightarrow{Na} NaOCH_2CHOHCH_2OH \xrightarrow{Na}$

$NaOCH_2CHOHCH_2ONa$

(*ii*) On passing hydrogen chloride into glycerol at 110°C until there is the theoretical increase in weight corresponding to the esterification of one hydroxyl group, both α- and β-glycerol monochlorohydrin are found, the former predominating (66 per cent). Continued action of hydrogen chloride at 110°C, using 25 per cent of acid in excess required by theory for the esterification of two hydroxyl groups, produces glycerol α, α′-dichlorohydrin (α-dichlorohydrin) and glycerol α,β-dichlorohydrin (β-dichlorohydrin), the former predominating (55 – 57 per cent); some other products are also formed:

$HOCH_2CHOHCH_2OH \xrightarrow{HCl} ClCH_2CHOHCH_2OH + HOCH_2CHClCH_2OH$

$\xrightarrow{HCl} ClCH_2CHOHCH_2Cl + ClCH_2CHClCH_2OH$

When either of these dichlorohydrins or glycerol itself is treated with phosphorus pentachloride, glycerol trichlorohydrin (1, 2, 3-trichloropropane) is obtained. This is a liquid, b.p. 156 – 158°C.

When glycerol α, α′-dichlorohydrin is oxidised with sodium dichromate-sulphuric acid mixture, 1,3-dichloroacetone (1,3-dichloropropan-2-one), m.p. 45°C, is obtained:

$$ClCH_2CHOHCH_2Cl \xrightarrow{[O]} ClCH_2COCH_2Cl \quad (68-75\%)$$

When glycerol α, α′-dichlorohydrin is treated with powdered sodium hydroxide in ether solution, epichlorohydrin (3-chloro-1,2-epoxypropane), b.p. 117°C, is obtained.

$$ClCH_2CHOHCH_2Cl + NaOH \longrightarrow ClCH_2\overset{O}{CH—CH_2} + NaCl + H_2O \quad (76-81\%)$$

Epichlorohydrin is also obtained by distilling an alkaline solution of the α-dichlorohydrin under reduced pressure (yield: 90 per cent). It is used for the manufacture of epoxy resins, plasticisers, etc.

Hydrogen bromide and the phosphorus bromides react with glycerol in the same way as the corresponding chlorine compounds, but the analogous iodine compounds behave differently, the products depending on the amount of reagent used.

When glycerol is heated with a *small* amount of hydrogen iodide or phosphorus tri-iodide, *allyl iodide* is the main product:

$$HOCH_2CHOHCH_2OH \xrightarrow{PI_3} [ICH_2CHICH_2I] \longrightarrow I_2 + CH_2{=}CHCH_2I$$

When a *large* amount of phosphorus tri-iodide is used, the main product is *isopropyl iodide:*

$$CH_2{=}CHCH_2I \xrightarrow{HI} [CH_3CHICH_2I] \xrightarrow{-I_2} CH_3CH{=}CH_2 \xrightarrow{HI} CH_3CHICH_3 \quad (80\%)$$

(*iii*) When glycerol is treated with monocarboxylic acids, esters are obtained which may be mono-, di- or tri-esters, according to the amount of acid used; high temperature and an excess of acid favour he formation of the tri-ester (see the glycerides, below).

Nitroglycerine is manufactured by adding glycerol in a thin stream to a cold mixture of concentrated nitric and sulphuric acids:

$$\begin{array}{l} CH_2OH \\ | \\ CHOH + 3HNO_3 \\ | \\ CH_2OH \end{array} \longrightarrow \begin{array}{l} CH_2ONO_2 \\ | \\ CHONO_2 + 3H_2O \\ | \\ CH_2ONO_2 \end{array}$$

Nitroglycerine is an ester, not a nitro-compound; it is *glyceryl trinitrate*. The incorrect name appears to have been introduced due to the use of 'mixed acid' which is normally used for nitration.

Nitroglycerine is a poisonous, colourless, oily liquid, insoluble in water. It usually burns quietly when ignited, but when heated, rapidly struck or detonated, it explodes violently. Nobel (1867) found that nitroglycerine could be stabilised by absorbing it in kieselguhr. This was *dynamite*, which is now, however, usually manufactured by using wood pulp as the absorbent, and adding solid ammonium nitrate. *Blasting gelatin* or *gelignite* is made by mixing nitroglycerine with gun-cotton (cellulose nitrate). The smokeless powder, *cordite*, is a mixture of nitroglycerine, gun-cotton and vaseline. Nitroglycerine is also used in the treatment of angina pectoris.

When heated with formic acid or oxalic acid at 260°C, glycerol is converted into allyl alcohol. With dibasic acids, glycerol forms condensation polymers known as *alkyd resins*, the commonest of which is *glyptal*, formed by heating glycerol with phthalic anhydride.

When boric acid is added to an aqueous solution of glycerol, a complex is produced which has a higher electrical conductivity than boric acid itself, *i.e.*, is a stronger acid. This complex is believed to be a borospiranic acid in which the two rings attached to the boron atom are perpendicular to each other (hence name spiran):

$$\left[\begin{array}{ccccc} C-O & & & & O-C \\ | & \diagdown & & \diagup & | \\ & & B & & \\ | & \diagup & & \diagdown & | \\ C-O & & & & O-C \end{array}\right] H^+$$

This borospiranic acid is formed from α-glycols in which the two hydroxyl groups are in the *cis*-configuration. Ethylene glycol apparently does not form this complex; the reason is unknown.

(*iv*) Glycerol can theoretically give rise to a large variety of oxidation products. The actual product obtained in practice depends on the nature of the oxidising agent used:

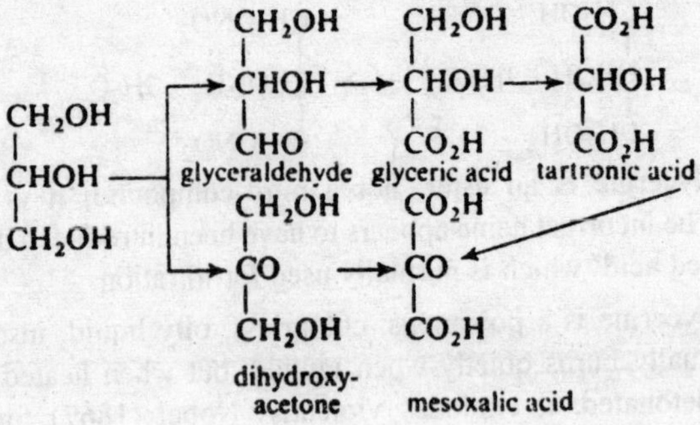

$$
\begin{array}{l}
CH_2OH-CHOH-CH_2OH \longrightarrow CH_2OH-CHOH-CHO \text{ (glyceraldehyde)} \longrightarrow CH_2OH-CHOH-CO_2H \text{ (glyceric acid)} \longrightarrow CO_2H-CHOH-CO_2H \text{ (tartronic acid)} \\
CH_2OH-CHOH-CH_2OH \longrightarrow CH_2OH-CO-CH_2OH \text{ (dihydroxyacetone)} \longrightarrow CO_2H-CO-CO_2H \text{ (mesoxalic acid)} \\
CO_2H-CHOH-CO_2H \text{ (tartronic acid)} \longrightarrow CO_2H-CO-CO_2H \text{ (mesoxalic acid)}
\end{array}
$$

Fig. 1.5

Dilute nitric acid converts glycerol into glyceric and tartronic acids; concentrated nitric acid oxidises it to mainly glyceric acid (80 per cent); bismuth nitrate produces mainly mesoxalic acid. Bromine water, sodium hypobromite and Fenton's reagent (hydrogen peroxide and ferrous sulphate) oxidise glycerol to a mixture of glyceraldehyde (predominantly) and dihydroxyacetone; this mixture is known as *glycerose*. These two compounds are interconvertible in the presence of anhydrous pyridine, this being known as the *Lobry de Bruyn-van Ekenstein rearrangement*.

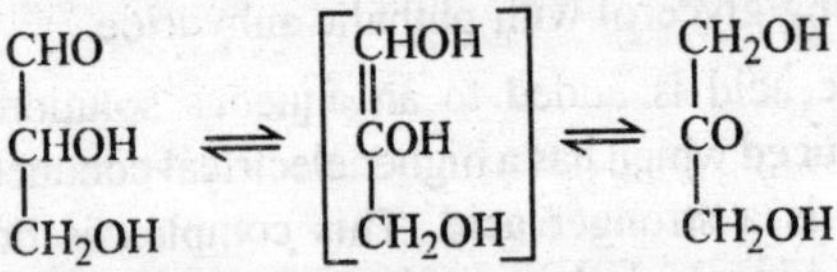

$$
\begin{array}{c} CHO \\ | \\ CHOH \\ | \\ CH_2OH \end{array} \rightleftharpoons \left[\begin{array}{c} CHOH \\ \| \\ COH \\ | \\ CH_2OH \end{array} \right] \rightleftharpoons \begin{array}{c} CH_2OH \\ | \\ CO \\ | \\ CH_2OH \end{array}
$$

(*v*) When heated with potassium hydrogen sulphate, glycerol is dehydrated to acraldehyde:

$$
HOCH_2CHOHCH_2OH \xrightarrow{KHSO_2} CH_2{=}CHCHO + 2H_2O
$$

(*vi*) The mixed ethers of glycerol may be conveniently prepared by the action of sodium alkoxide on a glycerol chlorohydrin, *e.g.*, *triethylin*, the triethyl ether of glycerol, is prepared by heating 1,2,3-trichloropropane with sodium ethoxide:

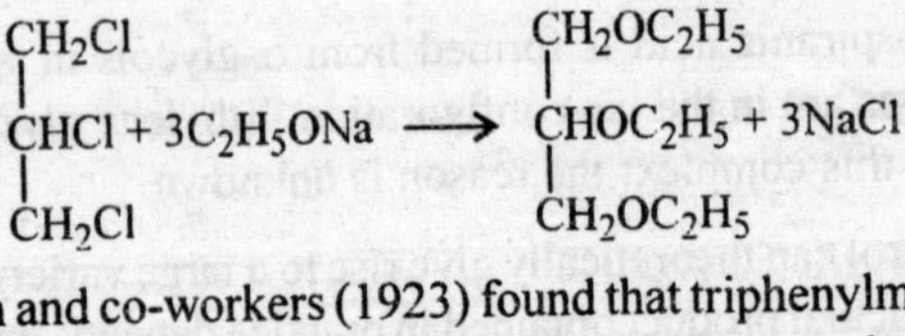

$$
\begin{array}{c} CH_2Cl \\ | \\ CHCl \\ | \\ CH_2Cl \end{array} + 3C_2H_5ONa \longrightarrow \begin{array}{c} CH_2OC_2H_5 \\ | \\ CHOC_2H_5 \\ | \\ CH_2OC_2H_5 \end{array} + 3NaCl
$$

Helferich and co-workers (1923) found that triphenylmethyl chloride (*trityl chloride*), $(C_6H_5)_3CCl$, usually formed ethers only with primary alcoholic groups. Thus, the α-mono- and the α,α′-ditriphenylmethyl (trityl) ethers of glycerol can be prepared, the latter offering a means of preparing β-esters glycerol, *e.g.*,

$$\begin{array}{l} CH_2OH \\ | \\ CHOH \\ | \\ CH_2OH \end{array} \xrightarrow{(C_6H_5)_3CCl} \begin{array}{l} CH_2OC(C_6H_5)_3 \\ | \\ CHOH \\ | \\ CH_2OC(C_6H_5)_3 \end{array} \xrightarrow{RCOCl} \begin{array}{l} CH_2OC(C_6H_5)_3 \\ | \\ CHOCOR \\ | \\ CH_2OC(C_6H_5)_3 \end{array}$$

$$\xrightarrow{HBr} \begin{array}{l} CH_2OH \\ | \\ CHOCOR \\ | \\ CH_2OH \end{array}$$

A better means of obtaining the β-ester is to protect the α,α′-hydroxyl groups by cyclic ether formation with benzaldehyde (Bergmann and Carter, 1930). This cyclic acetal is formed when glycerol and benzaldehyde are heated together, or the cool mixture treated with hydrogen chloride:

$$\begin{array}{l} CH_2OH \\ | \\ CHOH + C_6H_5CHO \\ | \\ CH_2OH \end{array} \xrightarrow{HCl} \begin{array}{l} CH_2O \\ | \quad \diagdown \\ CHOH \quad CHC_6H_5 + H_2O \\ | \quad \diagup \\ CH_2O \end{array}$$

(*vii*) Glycerol can be fermented to produce a variety of compounds, *e.g.*, propionic acid, succinic acid, acetic acid, n-butanol, trimethylene glycol, lactic acid, n-butyric acid, etc. Fermentation by means of a particular micro-organism usually produces more than one compound, *e.g.*, propionic acid bacteria produce propionic acid, succinic acid and acetic acid.

POLYHYDRIC ALCOHOLS

Tetrahydric alcohols. *Erythritol,* $HOCH_2CHOHCHOHCH_2OH$, exists in three forms: dextrorotatory erythritol, m.p. 88°; laevorotatory erythritol, m.p. 88°; *meso*-erythritol, m.p. 121.5°C. The *meso*-form occurs in certain lichens and seaweeds. All three forms may be oxidised to tartaric acid,

$$HO_2CCHOHCHOHCO_2H$$

Pentaerythritol, $C(CH_2OH)_4$, is prepared by the condensation of formaldehyde with acetaldehyde.

Pentahydric alcohols (*pentitols*). There are four pentitols with the structure $HOCH_2(CHOH)_3CH_2OH$, which may be prepared by reducing the corresponding aldopentoses. Two are optically active, forming a pair of enantiomers —D- and L-arabitol—and the other two are optically inactive, existing as *meso* forms—adonitol and xylitol.

Aldopentose	*Pentitol*
D- and L-ribose	adonitol (ribitol)
D- and L-xylose	xylitol
D-arabinose	
D-lyxose	D-arabitol (D-lyxitol)
L-arabinose	
L-lyxose	L-arabitol (L-lyxitol)

Hexahydric alcohols (*hexitols*). **There are ten hexitols with the structure $HOCH_2(CHOH)_4CH_2OH$, which may be prepared by reducing the corresponding aldohexoses. Eight exist as four pairs of enantiomers, and the remaining two as meso forms.**

Aldohexose	*Pentitol*
D-glucose and L-glucose	D-sorbitol (D-glucitol)
L-glucose and D-glucose	L-sorbitol (L-glucitol)
D-mannose	D-mannitol
L-mannose	L-mannitol
D-idose	D-iditol
L-idose	L-iditol
D-talose and D-altrose	D-talitol
L-talose and L-altrose	L-talitol
D- and L-galactose	dulcitol
D- and L-allose	allodulcitol (allitol)

D-*Sorbitol, D-mannitol* and *dulcitol* occur naturally.

LIPIDS

The term 'lipids' embraces a variety of naturally occurring compounds which have in common the property of being soluble in organic

solvents, but sparingly soluble in water. Also, all lipids yield monocarboxylic acids (saturated and unsaturated) on hydrolysis. The lipids include oils, fats, waxes and phospholipids.

Oils and Fats

This group of lipids is often referred to as the 'simple' lipids. They are compounds of glycerol and various organic acids, *i.e.*, they are glyceryl esters of *glycerides*. Oils, which are liquids at ordinary temperatures, contain a larger proportion of unsaturated acids than do the fats, which are solids at ordinary temperatures.

The acids present in the glycerides are almost exclusively straight-chain acids, and almost always contain an even number of carbon atoms. The chief saturated acids are lauric, myristic, palmitic and stearic. The chief unsaturated acids are oleic, linoelic and linoenic, and these usually occur in the *cis*-form.

Palmitic acid is the most abundant of the saturated acids, and acids containing less than eighteen carbon atoms are usually present only as minor constituents, but sometimes they are present in appreciable amounts in insect waxes and marine fats.

Glycerides are named according to the nature of the acids present, the suffix *– ic* of the common name of the acid being changed to *–in*. Glycerides are said to be 'simple' when all the acids are the same, and 'mixed' when the acids are different:

$$\begin{array}{l} CH_2OCOC_{17}H_{35} \\ | \\ CHOCOC_{17}H_{35} \\ | \\ CH_2OCOC_{17}H_{35} \end{array} \qquad \begin{array}{l} CH_2OCOC_{15}H_{31} \\ | \\ CHOCOC_{17}H_{33} \\ | \\ CH_2OCOC_{17}H_{33} \end{array}$$

tristearin (simple glyceride) α-palmito-α′, β-diolein (mixed glyceride)

It is still not certain whether simple glycerides occur naturally; if they do, they are definitely not as common as the mixed glycerides.

According to Desnuelle *et al.* (1959), the structure of natural glycerides is not random; the position of the acid residue depends on the chain-length of the unsaturated acids and on the degree of their unsaturation. In vegetable oils the saturated acids occur mainly at the 1- and 3-positions and unsaturated acids at the 2-position. In animal fats the distribution is not so rigid.

SYNTHESIS OF GLYCERIDES

Simple triglycerides are readily prepared from glycerol and an excess of acyl chloride, or from 1,2,3-tribromopropane and the silver or potassium salts of the acids.

Mixed triglycerides are far more difficult to prepare. A number of methods have been developed, *e.g.*,

(*i*) The sodium salt of an acid is heated with glycerol monochlorohydrin, and the monoester so formed is acylated:

$$\begin{matrix} CH_2Cl \\ | \\ CHOH \\ | \\ CH_2OH \end{matrix} \xrightarrow{R^1CO_2Na} \begin{matrix} CH_2OCOR^1 \\ | \\ CHOH \\ | \\ CH_2OH \end{matrix} \xrightarrow{R^2COCl} \begin{matrix} CH_2OCOR^1 \\ | \\ CHOCOR^2 \\ | \\ CH_2OCOR^2 \end{matrix}$$

The preparation of 1-monoglycerides has been improved by Hartman (1960). The 2- and 3-positions in glycerol are protected by the formation of the isopropylidene derivative, and the protecting group is removed by boric acid in hot 2-methoxyethanol.

(*ii*) 1,3-benzylidene-glycerol is treated with an acid chloride, the benzaldehyde residue is removed by hydrolysis, and the monoester so formed is then further acylated:

$$\begin{matrix} CH_2O \\ | \\ CHOH \\ | \\ CH_2O \end{matrix} \!\!>\! CHC_6H_5 \xrightarrow{R^1COCl} \begin{matrix} CH_2O \\ | \\ CHOCOR^1 \\ | \\ CH_2O \end{matrix} \!\!>\! CHC_6H_5 \xrightarrow{HCl} \begin{matrix} CH_2OH \\ | \\ CHOCOR^1 \\ | \\ CH_2OH \end{matrix}$$

$$\xrightarrow{R^2COCl} \begin{matrix} CH_2OCOR^2 \\ | \\ CHOCOR^1 \\ | \\ CH_2OCOR^2 \end{matrix}$$

In both methods (*i*) and (*ii*) there is, however, still difficulty in knowing with certainty the position of the acyl group introduced first, since the acyl group tends to migrate in the monoester, producing the following equilibrium:

$$\begin{matrix} CH_2OCOR \\ | \\ CHOH \\ | \\ CH_2OH \end{matrix} \rightleftharpoons \begin{matrix} CH_2OH \\ | \\ CHOCOR \\ | \\ CH_2OH \end{matrix}$$

This isomerisation has been shown to be intramolecular.

ANALYSIS OF OILS AND FATS

Oils and fats are characterised by means of physical as well as chemical tests. The usual physical constants that are determined are melting point, solidifying point, density and refractive index. The chemical tests give an indication of the type of fatty acids present in the oil or fat.

The **acid value**, which is the number of milligrams of potassium hydroxide required to neutralise 1 gram of the oil or fat, indicates the amount of free acid present.

The **saponification value** is the number of milligrams of potassium hydroxide required to neutralise the acids resulting from the complete hydrolysis of 1 gram of the oil or fat.

The **iodine value**, which is the number of grams of iodine that combine with 100 grams of oil or fat, gives the degree of unsaturation of the acids in the substance. Several methods are used for determining the iodine value. In *Hubl's method*, a carbon tetrachloride solution of the substance is treated with a solution of iodine and mercuric chloride in ethanol; in *Wijs' method*, iodine monochloride in glacial acetic acid is used. Another method is to use a solution of glacial acetic acid containing pyridine, bromine and concentrated sulphuric acid (*Dam's solution*).

The **Reichert-Meissl value**, which is the number of ml of 0.1 N-potassium hydroxide required to neutralise the distillate of 5 grams of hydrolysed fat, indicates the amount of steam-volatile fatty acids (*i.e.*, acids up to lauric) present in the substance.

The **acetyl value**, which is the number of milligrams of potassium hydroxide required to neutralise the acetic acid obtained when 1 gram of an acetylated oil or fat is hydrolysed, indicates the number of free hydroxyl groups present in the substance.

PREPARATION OF GLYCEROL FROM OILS AND FATS

Glycerol and some acids are prepared for industrial use by the hydrolysis of oils and fats with water under pressure at 220°C. The glycerol is recovered from the aqueous solution; the free fatty acids are used in the manufacture of candles.

$$\begin{array}{l} CH_2OCOR \\ | \\ CHOCOR \\ | \\ CH_2OCOR \end{array} + 3H_2O \longrightarrow \begin{array}{l} CH_2OH \\ | \\ CHOH \\ | \\ CH_2OH \end{array} + 3RCO_2R$$

If an oil is saponified, *i.e.*, hydrolysed with alkali, soaps are obtained. Any metallic salt of a fatty acid is a soap, but the term soap is usually applied to the water-soluble salts since only these have detergent properties. The saturated fats give hard soaps whereas the unsaturated fats, *i.e.*, the oils, give soft soaps. Ordinary soap is a mixture of the sodium salts of the even fatty acids from octanoic to stearic. The sodium salts of a given oil or fat are harder and less soluble than the corresponding potassium salts. Thus soft soaps are usually the potassium salts, particularly when they are derived from oils.

Hardening of Oils

Glycerides of the unsaturated acids are liquid at room temperature and so are unsuitable for edible fats. By converting the unsaturated acids into saturated acids, oils are changed into fats. This introduction of hydrogen is known as the *hardening of oils*. The oil is heated to 150 – 200°C and hydrogen is passed in, under pressure, in the presence of a finely divided nickel catalyst. In the hydrogenation process only a proportion of the unsaturated acids are converted into the saturated acids, otherwise a fat as hard as tallow would be obtained; the hydrogenation is carried out until a fat of the desired consistency is obtained.

Phospholipids

These differ from the simple lipids in that they also contain phosphorus and nitrogen; they are glycerides in which one organic acid residue, the α- or β-, is replaced by a group containing phosphoric acid and a base. One subgroup of the phospholipids is the **phosphatides**. When the base is **cholamine**, $CH_2OCH_2NH_2$, the phosphatide is known as a **kephalin**; when the base is **choline**, $CH_2OHCH_2\overset{+}{N}(CH_3)_3\}O\overset{-}{H}$, the phosphatide is known as a **lecithin**. These compounds have betaine structures, and it appears unlikely that β-kephalins or β-lecithins occur in nature.

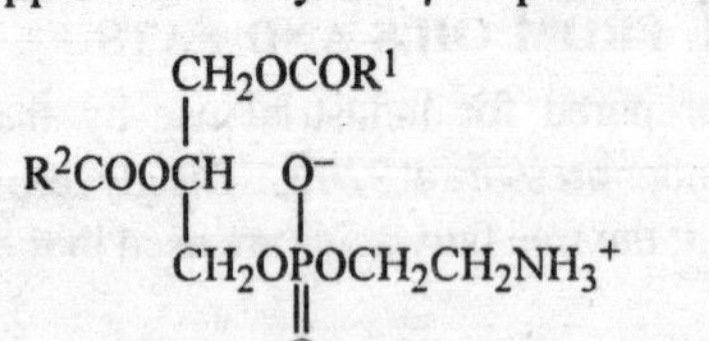

$$\begin{array}{l} CH_2OCOR^1 \\ | \\ R^2COOCH \quad O^- \\ | \qquad\quad | \\ CH_2OPOCH_2CH_2NH_3^+ \\ \qquad \| \\ \qquad O \end{array}$$

α-kephalin

$$\begin{array}{l} CH_2OCOR^1 \\ | \\ R^2COOCH \quad O^- \\ | \qquad\quad | \\ CH_2OPOCH_2CH_2\overset{+}{N}(CH_3)_3 \\ \qquad \| \\ \qquad O \end{array}$$

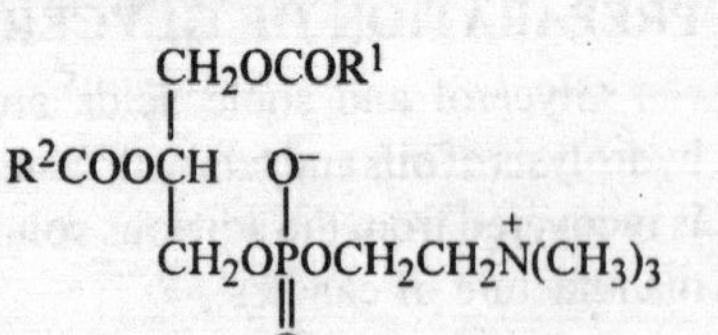

α-lecithin

Kephalins may also contain L-serine instead of aminoethanol. These three types of phospholipids have been called *phosphoglycerides*, in order to distinguish them from another group of phospholipids, the *acetal phospholipids* (*phasmalogens*). The structures of these compounds appear to be uncertain; one proposal is (for an ethanolamine plasmalogen):

$$\begin{array}{l} CH_2OCH{=}CHR^1 \\ | \\ R^2COOCH \quad O^- \\ | \qquad\quad | \\ CH_2OPOCH_2CH_2\overset{+}{N}H_3{}^+ \\ \qquad \| \\ \qquad O \end{array}$$

All of these phosphoglycerides occur in the brain and the spinal chord.

Only a few monocarboxylic acids have been isolated from phosphatides: stearic, oleic, linoelic and arachidonic from kephalins, and plamitic, stearic, oleic, linoeic, linoenic and arachidonic from lecithins. Enzymic studies of phospholipids have shown that saturated acids predominate at the 1-position and unsaturated acids at the 2-position (Hanahan *et al.*, 1960).

Another group of phospholipids is the **sphingolipids**. These are *not* glycerides, and yield, on hydrolysis, fatty acids, phosphoric acid, choline and sphingosine (which has the *trans*-configuration about the double bond). Sphingolipids are also known as **sphingomyelins**, and their structure is believed to be:

$$\begin{array}{l} CH_3(CH_2)_{12}C{-}H \qquad\qquad\qquad O^- \\ \qquad\qquad\quad \| \qquad\qquad\qquad\qquad | \\ \qquad H{-}C{-}CH{-}CHCH_2OPOCH_2CH_2N(CH_3)_3 \\ \qquad\quad | \qquad\qquad | \qquad\quad \| \\ \qquad\; OH \qquad\quad NH \qquad O \\ \qquad\qquad\qquad\quad | \\ \qquad\qquad\qquad RCO \end{array}$$

Glycolipids (cerebrosides) also contain sphingosine, but instead of the phosphocholine group, they have $\cdots\underset{\underset{-NH}{|}}{CH}CH_2O\underbrace{CH(CHOH)_3CH}_{O}CH_2OH$,

i.e., they are glycosides, and apparently only two sugars have been isolated so far, glucose and galactose.

Drying Oils

These are oils which, on exposure to air, change into hard solids, *e.g.*, linseed oil. All drying oils contain a large proportion of the unsaturated acids linoleic and linolenic, and it is the 'drying' property which makes these oils valuable in the paint industry. The mechanism of drying appears to be a complicated process involving oxidation, polymerisation and colloidal gel formation, and it has been found to be catalysed by various metallic oxides, particularly lead monoxide.

These are esters of the higher homologues of both the fatty acids and monohydric alcohols, *e.g.*,

beeswax	myricyl palmitate	$C_{15}H_{31}CO_2C_{30}H_{61}$
spermaceti	cetyl palmitate	$C_{15}H_{31}CO_2C_{16}H_{33}$
carnauba wax	myricyl cerotate	$C_{25}H_{51}CO_2C_{30}H_{61}$

Some waxes are also esters of cholesterol, *e.g.*, cholesteryl esters occur in wool-wax. Also present in waxes are free acids, long-chain alcohols, ketones and alkanes.

Crotyl Alcohol

The crotyl alcohol (*crotonyl alcohol, but-2-en-l-ol*), b.p. 118°C, may be prepared by the reduction of crotonaldehyde with aluminium isopropoxide:

$$CH_3CH = CHCHO \xrightarrow{[(CH_3)_2CHO]_3Al} CH_3CH = CHCH_2OH \text{ (85–90\%)}$$

When treated with hydrogen bromide, crotyl alcohol produces a mixture ofcrotyl and methyl-vinylcarbinyl bromide:

$$CH_3CH = CHCH_2OH \xrightarrow{HBr} CH_3CH = CHCH_2Br + CH_3CHBrCH = CH_2$$

The formation of the rearranged proiduct is an example of anionotropy, and the most widely studied example is the three-carbon system, *i.e.*, the allylic system. Nucleophilic substitutior reactions of allylic halides may occur by the S_N2 mechanism, and when this is operating substitution proceeds normally.

When, however, nucleophilic substitution reactions are carried out under conditions which favour the unimolecular mechanism, then the product may be a mixture of two isomers,

the rearranged product being an example of the *allylic rearrangement*. The mechanism by which this rearrangement occurs is known as SN)', and is believed to occur via a resonance hybrid (ambident) cation:

$$\mathrm{MeC = CHCH_2OH \underset{}{\overset{H^+}{\rightleftharpoons}} MeCH = CHCH_2O\,H_2^+ \underset{+H_2O}{\overset{-H_2O}{\rightleftharpoons}}}$$

$$\mathrm{MeCH{=}CHCH_2^+ \longleftrightarrow Me\overset{+}{C}HCH{=}CH_2}$$

$$\mathrm{MeCH{=}CHCH_2^+ \xrightarrow{Br^-} MeCH{=}CHCH_2Br\ (S_N1)}$$

$$\mathrm{Me\overset{+}{C}HCH{=}CH_2 \xrightarrow{Br^-} MeCHBrCH{=}CH_2\ (S_N1')}$$

There is much evidence to support this S_N1' mechanism, *e.g.*, Hughes *et al.* (1948) examine the conversion of α-methylallyl and crotyl chloride into α-methylallyl and crotyl ether by reaction with ethanolic sodium ethoxide. The concentration of the sodium ethoxide was made so small as to give first-order kinetics, or so large as to give second-order kinetics. The result were:

$$\mathrm{MeCHClCH{=}CH_2 \xrightarrow{2nd\ order} MeCH(OEt)CH{=}CH_2\ (100\%)}$$

$$\mathrm{MeCH{=}CHCH_2Cl \xrightarrow{2nd\ order} MeCH{=}CHCH_2OEt\ (100\%)}$$

$$\mathrm{MeCHClCH{=}CH_2 \xrightarrow{1st\ order} \left\{\begin{array}{l} MeCH(OEt)CH{=}CH_2\ (13\%) \\ MeCH{=}CHCH_2OEt\ (87\%) \end{array}\right. \xleftarrow{1st\ order} MeCH{=}CHCH_2Cl}$$

With second-order substitution, each of chloride gives its own unrearranged ether (thus the S_N2 mechanism). With first-order substitution, however, each chloride gives the *same* mixture isomeric ethers. This latter result may be explained by assuming that *two* mechanisms a operating, S_N1 (without rearrangement) and S_N1' (with rearrangement), both proceeding through a common intermediate ambident cation.

In addition to the S_N1' mechanism, there is also the S_N2', *i.e.*, a bimolecular nucleophilic substitution with allylic rearrangement. This has been difficult to demonstrate, but a good example of it has been given by de la Mare *et al.* (1953). These authors have suppressed the S_N2 reaction by steric factors and reduced the S,1 reaction by means of a reagent of high nucleophilic activity and a solvent of low ionising power. They found that the reaction between α, α-dimethylallyl chloride and sodium thiophenoxide in ethanol gives 62 per cent of rearranged product by a second-order reaction (*i.e.*, S_N2'):

$$Me_2\overset{Cl}{C}—CH=CH_2 \quad SPh \longrightarrow Me_2C=CHCH_2SPh + NaCl$$

One other problem that will be discussed here is the case of ion pairs in the allylic rearrangement (see also). Young *et al.* (1951) have shown that α, α-dimethylallyl chloride in acetic acid undergoes acetolysis accompanied by rearrangement to γ, g-dimethylallyl chloride:

$$Me_2CClCH = CH_2 \xrightarrow{MeCO_2H} \begin{matrix}\text{acetolysis}\\ \text{products}\end{matrix} + Me_2C = CHCH_2Cl$$

The rate of the isomerisation was shown to be proportional to the concentration of the starting chloride only. Thus a possible mechanism is the S_N1'. If this were so, then the rate of isomerisation would be affected by addition of chloride ions (since the first step is ionisation). Experiment showed that the rate was unaffected by added chloride ions, and it was also shown that when an excess of radioactive chloride ion was added, the rate of isomerisation was far greater than the incorporation of Cl^{*-}. It thus appears that the chloride ion is set free during the isomerisation and is mainly the 'first ion' to recombine, even though the solution contains a large excess of Cl^{*-} ions. Young *et al.*, explained this result by assuming that the starting chloride undergoes ionisation to form an ion-pair, and this is followed by internal return:

$$Me_2CClCH=CH_2 \rightleftharpoons \underset{\text{ion pair}}{[Me_2C\cdots CH\cdots CH_2]^+Cl^-}$$

$$\rightleftharpoons Me_2C=CHCH_2Cl$$

Nevertheless, some ion pairs escape from the solvent cage, and these Tree' carbonium ions are capable of undergoing acetolysis and combination with Cl^{*-}

A number of unsaturated alcohols occur in essential oils. The essential oils are volatile oils with pleasant odours, and are obtained from various plants to which they impart their characteristic odours. They are widely used in the perfume industry.

The simplest acetylenic alcohol is *propargyl alcohol* (prop-2-yn-l-ol), b.p. 114°C. This may be prepared from 1,2,3-tribromopropane:

$$BrCH_2CHBrCH_2Br \xrightarrow[KOH]{\text{ethanolic}} CH_2 = CBrCH_2Br$$

$$\xrightarrow[KOH]{\text{aqueous}} CH_2 = CBrCH_2OH \xrightarrow[KOH]{\text{ethanolic}} CH \equiv CCH_2OH$$

Propargyl alcohol (together with butynediol) is prepared by the interaction of acetylene and formaldehyde in the presence of silver or cuprous acetylide as catalyst:

$$C_2H_2 + HCHO \xrightarrow{Ag_2C_2} CH \equiv CCH_2OH$$

$$\xrightarrow[HCHO]{Ag_2C_2} HOCH_2C \equiv CCH_2OH$$

Propargyl alcohol behaves in many ways like acetylene, *e.g.*, it adds on two or four bromine atoms, and forms the silver or cuprous compounds when treated with an ammoniacal solution of silver nitrate or cuprous chloride. It may be catalytically reduced to allyl and n-propyl alcohols.

EXERCISES

1. Write out the structures of all the possible isomeric diols derived from the butanes, and indicate which can be oxidised with periodic acid.
2. Arrange the following in order of increasing b.p., and give your reasons. $(CH_2OH)_2$, $(CH_2OMe)_2$, $HOCH_2CH_2OMe$.
3. Complete the following equations:

(*i*) $HOCH_2CH_2OH + ? \xrightarrow{pyridine}$ [cyclic O–CO–O ring] $\equiv$ $\begin{matrix} CH_2-O \\ | \\ CH_2-O \end{matrix}\!\!>CO$

(*ii*) $Me_2CO + ? \xrightarrow{H^+} Me_2C<\begin{matrix} O- \\ O- \end{matrix}$ (five-membered ring)

(*iii*) $MeCHO + HOCH_2CH_2SH \xrightarrow{H^+} ?$

(*iv*) $R_2C(OH)CH_2R \xleftarrow{?} R_2\overset{O}{C-}CHR \xrightarrow{?} R_2CHCHOHR$

(*v*) [cyclopentene] $\xrightarrow[(ii)\ H^+]{(i)\ HCO_3H} ?$

(*vi*) $CH_2CH(CH_2)_2CHCH_2 \xrightarrow[(ii)\ H_2O_2;\ OH^-]{(i)\ B_2H_6} ?$

(*vii*) $(CH_2)_8(CO_2Et)_2 \xrightarrow{?} (CH_2)_8(CH_2OH)_2$

4. Write out the structures and I.U.P.A.C. names of isomeric secondary heptanols containing one methyl side-chain.

5. The dehydration of butan-2-ol with acid gives a mixture of 1- and 2-butenes. Write out the equations, state which predominates, and give your reasons. Draw the energy profiles of the reactions.

6. Convert ethanol into (i) n-propanol; (ii) n-butanol.

7. Write out the order of rates of oxidation with periodic acid of the following diols, and give your reasons: $HOCH_2CH_2OH$, MeCHOHCHOHMe, Me_2COHCH_2OH, $Me_2C(OH)C(OH)Me_2$, $Me_2COHCHOHMe$.

8. Convert:

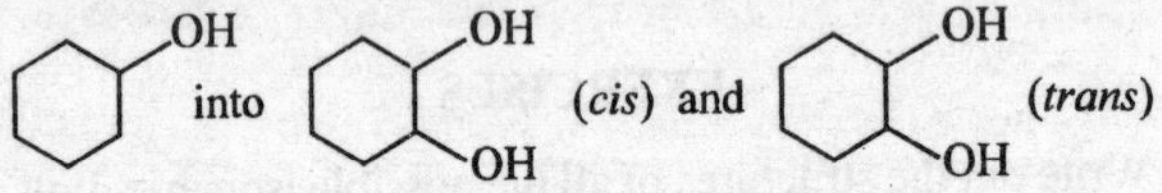

   OH (cyclohexanol) into OH, OH (*cis*) and OH, OH (*trans*)

9. Arrange FtOH, CF_3CH_2OH, and $CC_{13}CH_2OH$ in order of increasing strength as acids and give your reasons.

10. The dehydration of n-BuOH with acid gives 1- and 2-butene. Explain this, state which product predominates, and why.

11. Complete the following equations:

    (i) $n\text{-}C_3H_7CO_2H \xrightarrow{?} n\text{-}C_4H_9OH$

    (ii) $Me_2CO + EtMgI \longrightarrow ? \xrightarrow{H^+} ?$

    (iii) $EtCO_2Et + 2MeMgI \longrightarrow ? \xrightarrow{H_2O} ?$

    (iv) 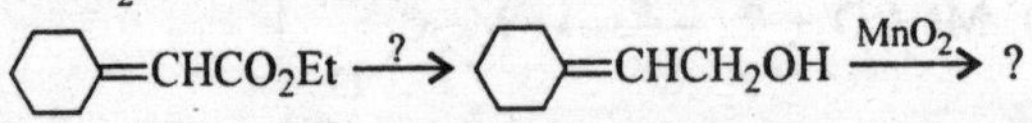 $=CHCO_2Et \xrightarrow{?} \; =CHCH_2OH \xrightarrow{MnO_2} ?$

    (v) $CH_3CH_2CH_2OH \xleftarrow{?} BrCH_2CH_2COCl \xrightarrow{?} BrCH_2CH_2OH$

12. How many NMR signals would be given by the following alcohols in the presence of a trace of acid? Indicate all signals other than singlets by (m).

    (i) $CH_3CH_2CH_2OH$; (ii) Me_2CHCH_2OH;

    (iii) Me_3COH; (iv) Me_3CCH_2OH;

    (v) $MeCH_2CMe_2OH$.

13. Convert propene into (i) n-PrOH; (ii) isoPrOH.

14. How could you show the existence of the following equilibrium?

$$ROH + H^+ \xrightleftharpoons{H_2SO_4-H_2O} R^+ + H_2O$$

15. What are the disadvantages of preparing alkenes by:

(*i*) RX + base; ROH + acid?

16. (1-methylcyclohexyl)CHOHMe $\xrightarrow{H^+}$ A (C_9H_{16})

A, on ozonolysis, gives nonane-2,8-dione. What is **A** and how is it formed?

17. Complete the following equation and comment:

1-methylcyclohexene $\xrightarrow{KMnO_4}$? $\xrightarrow[AcOH]{CrO_3}$ 2-hydroxy-2-methylcyclohexanone (Me, OH, O)

18. Convert (i) n-C4H9OH into MeC≡CMe; (iii) n-PrOH into *cis*-hex-2-ene.

19. A concentrated aqueous solution of HBr reacts with EtOH to give EtBr, but a concentrated aqueous solution of NaBr does not. Explain.

20. What are A to E in the following reactions?

(A) $C_4H_{10}O \xrightarrow{SOCl_2} C_4H_9Cl$ **(B)**

(A) $\xrightarrow{H_2SO_4}$ C_4H_9**(C)**

(D) $C_3H_6O_2 \xleftarrow[(ii)\ H_2O_2]{(i)\ O_3} C_4H_9$**(C)** $\xrightarrow{HCl} C_4H_9Cl$**(E)**

21. Suggest a mechanism for the conversion of ethylene glycol into dioxan by conc. H_2SO_4.

22. How many N.M.R. signals can be expected for:

(*i*) CH_2—CH_2 (epoxide, bridged by O) (*ii*) CH_2—CHCHO (epoxide, bridged by O) (*iii*) O(—CH_2—CH_2—)O(—CH_2—) ring

23. Tetramethylethylene, on oxidation with acid dichromate, gives the compound $C_6H_{12}O$. This compound showed an i.r. band at about 1700 cm^{-1}, and undergoes the haloform reaction. Suggest a structure for this compound, and give your reasons.

2

MONO-DI-CARBOXYLIC ACIDS AND THEIR DERIVATIVES

INTRODUCTION TO MONOCARBOXYLIC ACIDS

Saturated monocarboxylic acids are also referred to as the fatty acids, a name which arose from the fact that some of the higher members, particularly palmitic and stearic acids, occur in natural fats. Their general formula is $C_nH_{2n}O_2$ but, because their functional group is the carboxyl group, —CO_2H, they are more conveniently expressed as $C_nH_{2n+1}CO_2H$, since these show the nature of the functional group. The structure of the carboxyl group is written as $-C\overset{\displaystyle O}{\underset{\displaystyle OH}{\lessgtr}}$ (or —COOH), but, as we shall later, it does not represent accurately the behaviour of this group. Also, since only the carboxyl group hydrogen atom is replaceable by a metal, the monocarboxylic acids are *monobasic*.

Nomenclature

The monocarboxylic acids are commonly known by the trivial names, which have been derived from the source of the particular acid; *e.g.*, formic acid, HCO_2H, was so named because it was first obtained by the distillation of ants; the Latin word for ant is *formica*. Acetic acid, CH_3CO_2H, is the chief constituent of vinegar, the Latin word for which is *acetum*; etc.

Another system of nomenclature considers the acids, except formic acid, as derivatives of acetic acid, *e.g.*,

$CH_3CH_2CO_2H$	methylacetic acid
$(CH_3)_3CCO_2H$	trimethylacetic acid
$(CH_3)_2CHCH_2CO_2H$	isopropylacetic acid

In the above two systems the positions of substituents in the chain are indicated by Greek letters, the α-carbon atom being the one joined to the caboxyl group, *e.g.*,

$CH_3CH(OH)CH_2CH_2CO_2H$ γ-hydroxyvaleric acid

$CH_3CHClCH(CH_3)CO_2H$ β-chloro α-methylbutyric acid.

According to the I.U.P.A.C. system of nomenclature, the suffix of the monocarboxylic acids is *-oic*, which is added to the name of the alkane corresponding to the longest carbon chain containing the carboxyl group, *e.g.*,

HCO_2H methanoic acid

$CH_3CH_2CO_2H$ propanoic acid

The positions of side-chains (or substituents) are indicated by numbers, *the carboxyl group always being given number* 1:

$CH_3CH(CH_3)CH(CH_3)CH_2CO_2H$ 3,4-dimethylpentanoic acid

Alternatively, the carboxyl group is regarded as a substituent, and is denoted by the suffix carboxylic acid, *e.g.*,

$\overset{4}{C}H_3\overset{3}{C}H_2\overset{2}{C}H(CH_3)\overset{1}{C}H_2CO_2H$ 2-methylbutane-1-carboxylic acid

General Methods of Preparation

1. Oxidation of alcohols, aldehydes or ketones with acid dichromate yields acids:

$$RCH_2OH \xrightarrow{[O]} RCHO \xrightarrow{[O]} RCO_2H\ (g.\text{–}v.g.)$$

$$R_2CHOH \xrightarrow{[O]} R_2CO \xrightarrow{[O]} R^1CO^2H + R^2CO_2H\ (g.)$$

Many *primary* alcohols, on oxidation with acid dichromate, form esters in addition to acids, *e.g.*, n-butanol gives n-butyric acid and n-butyl n-butyrate (41–47 per cent).

Several mechanisms have been proposed, but the generally accepted one involves hemiacetal formation.

$$RCH_2OH \xrightarrow{[O]} RCHO \xrightarrow{RCH_2OH} RCH(OH)OCH_2R \xrightarrow{[O]} RCO_2CH_2R$$

2. A very good synthetic method is the hydrolysis of cyanides with or alkali:

$$RC{\equiv}N \xrightarrow{H_2O} \left[R{-}C\begin{matrix}{/\!/}OH \\ \diagdown NH\end{matrix} \right] \longrightarrow R{-}C\begin{matrix}{/\!/}O \\ \diagdown NH_2\end{matrix} \xrightarrow{H_2O} RCO_2H + NH_3 \quad \text{(g.-r.g.)}$$

The amide, $RCONH_2$, may be isolated if suitable precautions are taken.

3. By the reaction between a Grignard reagent and carbon dioxide:

$$RMgX + CO_2 \longrightarrow RCO_2MgX \xrightarrow{H_2O} RCO_2H$$

4. Many monocarboxylic acids may be conveniently synthesised from alkyl halides and ethyl malonate or acetoacetic ester. These syntheses will be discussed in detail later.

5. A number of higher acids may be obtained by the hydrolysis of natural fats or may be prehalf-ester of a dicarboxylic acid (Linstead *et al.*, 1950; *cf.*).

$$RCO_2H + HO_2C(CH_2)_nCO_2CH_3 \longrightarrow H_2 + 2CO_2 + R(CH_2)_nCO_2CH_3$$

The chain of a monocarboxylic acid can be extended by five or six carbon atoms in a synthesis involving enamines.

6. Monocarboxylic acids may be obtained by heating a sodium alkoxide with carbon monoxide under pressure.

$$R\overset{-}{O}\overset{+}{Na} + CO \longrightarrow RCO_2^-\overset{+}{Na}$$

Acids may also be obtained by heating the alcohol with carbon monoxide under pressure in the presence of a catalyst.

The sodium salts of acetic and propionic acids are formed by the interaction of methylsodium or ethylsodium, respectively, and carbon dioxide, *e.g.*,

$$C_2\overline{H}_5\overset{+}{Na} + CO_2 \longrightarrow C_2H_5CO_2Na$$

Monocarboxylic acids may be prepared by heating an alkene with carbon monoxide and steam under pressure at 300 – 400°C in the presence of a catalyst, *e.g.*, phosphoric acid:

$$CH_2{=}CH_2 + CO + H_2O \longrightarrow CH_3CH_2CO_2H$$

$$CH_3CH{=}CH_2 + CO + H_2O \longrightarrow (CH_2)_2CHCO_2H$$

General Reactions of the Monocarboxylic Acids

1. The acids are acted upon by the strongly electropositive metals with the liberation of hydrogen and formation of a salt:

$$RCO_2H + Na \longrightarrow RCO_2^-Na^+ + \frac{1}{2}H_2$$

Salts are also formed by the reaction between an acid and an alkali:

$$RCO_2H + NaOH \longrightarrow RCO_2^-Na^+ + H_2O$$

2. Monocarboxylic acids react with alcohols to form *esters*:

$$R^1CO_2H + R^2OH \rightleftharpoons R^1CO_2R^2 + H_2O$$

3. Phosphorus trichloride, phosphorus pentachloride or thionyl chloride act upon the monocarboxylic acids to from *acid chlorides*:

$$3RCO_2H + PCl_3 \longrightarrow 3RCOCl + H_3PO_3$$
$$RCO_2H + PCl_5 \longrightarrow RCOCl + HCl + POCl_3$$
$$RCO_2H + SOCl_2 \longrightarrow RCOCl + HCl + SO_3$$

The acid series may be 'stepped up' via the acid chloride, but since the process involves many steps, the yield of the higher acid homologue is usually only *f.-f.g.*:

(*a*) $RCO_2H \xrightarrow{PCl_5} RCOCl \xrightarrow[\text{catalyst}]{H_2} RCHO \xrightarrow[\text{catalyst}]{H_2}$

$$RCH_2OH \xrightarrow{PBr_3} RCH_2Br \xrightarrow{KCN} RCH_2CN \xrightarrow{H^+} RCH_2CO_2H$$

(*b*) $RCO_2H \xrightarrow{PCl_5} RCOCl \xrightarrow{KCN} RCOCN$

$$\xrightarrow{H^+} RCOCO_2H \xrightarrow[\text{reduction}]{\text{Clemmensen}} RCH_2CO_2H$$

Although, method (*b*) involves fewer steps than (*a*), the final yield of acid by route (*a*) is higher than that by (*b*), since the yields in each step of (*b*) are generally low. Furthermore, the yield from (*a*) can be improved by direct reduction of acid to alcohol by means of various metallic hydrides.

4. When the ammonium salts of the monocarboxylic acids are strongly heated, the *acid amide* is formed by the elimination of water:

$$RCO_2NH_4 \longrightarrow RCONH_2 + H_2O$$

5. When the anhydrous sodium salt of a monocarboxylic acid is heated with soda-lime, an alkane and other products are formed.

The mechanism of this decarboxylation is uncertain, but there is much evidence to show that *salts* decompose by an S_E1 reaction :

$$\text{R—C}\overset{\text{O}^-}{\underset{\text{O}}{\Large\lt}} \longrightarrow \bar{\text{R}}\text{:} + \text{C}\overset{\text{O}}{\underset{\text{O}}{\Large\lt}} \xrightarrow{H^+} \text{R—H}$$

The thermal decarboxylation of *free* acids may be:

$$\text{R—C}\overset{\text{O—H}}{\underset{\text{O}}{\Large\lt}} \longrightarrow \overset{+}{\text{H}} + \text{R—C}\overset{\text{O}^-}{\underset{\text{O}}{\Large\lt}} \longrightarrow \overset{+}{\text{H}} + \bar{\text{R}}\text{:} + CO_2 \longrightarrow \text{R—H} + CO_2$$

Monocarboxylic acids may be 'stepped down' by heating the calcium salt with calcium acetate, and the methyl ketone produced is oxidised with acid dichromate:

$$RCH_2CO_2Ca/_2 + CH_3CO_2Ca/_2 \longrightarrow RCH_2COCH_3$$
$$\xrightarrow{[O]} RCO_2H + CH_3CO_2H$$

It should be noted that the second step is based on the general rule that when an unsymmetrical ketone is oxidised, the carbonyl group remains chiefly with the smaller alkyl group. Since monocarboxylic acids containing an odd number of carbon atoms occur only in small amounts, this method of descending the series affords a useful means of obtaining 'odd' acids from 'even'.

When a mixture of the calcium salt (or any of the others mentioned above) and calcium formate is heated, an aldehyde is obtained. The pyrolysis of iron salts gives very good yields of symmetrical ketones. The pyrolysis of iron slats gives very good yields of symmetrical ketones.

Metallic salts, particularly the silver slats, are converted by bromine into the alkyl bromide. This reaction may also be used to 'step down' the acid series.

6. When a concentrated aqueous solution of the sodium or potassium salt of a monocarboxylic acid is electrolysed, an alkane is obtained.

$$2RCO_2K + 2H_2O \longrightarrow \text{R—R} + 2CO_2 + 2KOH + H_2$$

7. Monocarboxylic acids react slowly with chlorine or bromine in the cold, but at higher temperatures and in the presence of a small amount of phosphorus, reaction proceeds smoothly to give a-halogeno-acids; *e.g.*,

$$RCH_2CO_2H \xrightarrow{P/Br_2} RCHBrCO_2H \xrightarrow{P/Br_2} RCBr_2CO_2H$$

8. All the acids, except formic acids, are extremely resistant to oxidation, but prolonged heating with oxidising agents ultimately produces carbon dioxide and water.

9. The reduction products of monocarboxylic acids depend on the nature of the reducing agent. Thus, heating with hydrogen iodide-red phosphorous under pressure, or with hydrogen under pressure at elevated temperature in the presence of a nickel catalyst, produces an alkane.

$$RCO_2H + 3H_2 \xrightarrow{Ni} RCH_3 + 2H_2O$$

If the catalyst is Ru—C, the alcohol is formed. Alcohols are also produced by reduction with various metallic hydrides.

10. The acids undergo the *Schmidt reaction* to form a primary amine:

$$RCO_2H + HN_3 \xrightarrow{H_2SO_4} RNH_2 + CO_2 + N_2$$

Schmidt's reaction with acids is a modification of the Curtius reaction. The following mechanism has been proposed :

$$R-C(=O)OH \xrightarrow{H^+} R-\overset{+}{C}(OH)OH \xrightarrow{HN_3} R-C(OH)(OH)-NH-\overset{+}{N}\equiv N \xrightarrow{-H_2O} R-C(OH)=NH-\overset{+}{N}\equiv N \xrightarrow{-N_2}$$

$$O=C(R)-\overset{+}{N}-H \xrightarrow{-H^+} O=C=NR \xrightarrow{H_2O} RNH_2 + CO_2$$

The rearrangement (1,2-shift) has been shown to be intramolecular, *e.g.*, Keynon *et al.* (1939) showed retention of optical activity when α-phenylpropionic acid underwent the Schmidt reaction:

$$C_6H_5CH(CH_3)CO_2H \longrightarrow C_6H_5CH(CH_3)NH_2$$

Formic Acid

Formic acid (*methanoic acid*) is prepared industrially by heating sodium hydroxide with carbon monoxide at 210°C, and at a pressure of 6–10 atmospheres:

$$NaOH + CO \longrightarrow HCO_2Na$$

Anhydrous formic acid is obtained from the aqueous solution (70–77 per cent) by the addition of butyl formate followed by distillation.

The first fraction is an azeotrope of ester and water, and then the excess of ester is removed from the formic acid by fractionation.

The most convenient laboratory preparation of formic acid is to heat glycerol with oxalic acid at 100–100 C. Glyceryl monoxalate is produced, and decomposes into glyceryl monoformate (monoformin) and carbon dioxide. When the evolution of carbon dioxide ceases, more oxalic acid is added, whereupon formic acid is produced:

$$\begin{matrix} CH_2OH \\ | \\ CHOH \\ | \\ CH_2OH \end{matrix} + \begin{matrix} COOH \\ | \\ COOH \end{matrix} \longrightarrow \begin{matrix} CH_2OCOCOOH \\ | \\ CHOH \\ | \\ CH_2OH \end{matrix} + H_2O \xrightarrow{100\text{-}110°C}$$

$$CO_2 + \begin{matrix} CH_2OCHO \\ | \\ CHOH \\ | \\ CH_2OH \end{matrix} \xrightarrow{(COOH)_2} \begin{matrix} CH_2OCOCOOH \\ | \\ CHOH \\ | \\ CH_2OH \end{matrix} + HCO_2H \xrightarrow{100\text{-}110°C}$$

The distillate contains formic acid and water. The aqueous formic acid solution cannot be fractionated to give anhydrous formic acid because the boiling point of the acid is 100.5°C. The procedure adopted is to neutralise the aqueous acid solution with lead carbonate, and concentrate the solution until lead formate crystallises out. The precipitate is then recrystallised, dried, and heated at 100°C in a current of hydrogen sulphide:

$$(HCO_2)_2Pb + H_2S \longrightarrow 2HCO_2H + PbS$$

The anhydrous formic acid which distils over contains a small amount of hydrogen sulphide, and may be freed from the latter by adding some dry lead formate and redistilling. This procedure for obtaining the anhydrous acid from its aqueous solution can only be used for volatile acids.

Formic acid is a pungent corrosive liquid, m.p. 8.4°C, b.p. 100.5°C, miscible in all proportions with water, ethanol and ether. It forms salts which, except for the lead and silver salts, are readily soluble in water. Formic acid is a stronger acid than any of its homologues.

Formic acid is dehydrated to carbon monoxide by concentrated sulphuric acid and when heated under pressure at 160°C, is decomposed into carbon dioxide and hydrogen:

$$CO + H_2O \xleftarrow{H_2SO_4} HCO_2H \xrightarrow{heat} CO_2 + H_2$$

The same decomposition takes place at room temperature in the presence of a catalyst such as iridium, rhodium, etc.

When metallic formates are heated with an alkali, hydrogen is evolve.

$$HCO_2Na + NaOH \longrightarrow H_2 + Na_2CO_3$$

When calcium or zinc formate is strongly heated, formaldehyde is produced:

$$(HCO_2)_2Ca \longrightarrow HCHO + CaCO_3$$

When sodium or potassium formate is rapidly heated to 360°C, hydrogen is evolved and the oxalate is formed.

$$2HCO_2Na \longrightarrow (COONa)_2 + H_2$$

Formic acid forms esters, but since it is a relatively strong acid it is not necessary to use a catalyst: refluxing 90 per cent formic acid with the alcohol is usually sufficient.

Formic acid differs from the rest of the members of the monocarboxylic acid series in being a powerful reducing agent; it reduces ammoniacal silver nitrate and the salts of many of the heavy metals, *e.g.*, it converts mercuric chloride into mercurous chloride.

Acetic acid (*ethanoic acid*) is prepared industrially by various methods, *e.g.*, (*i*) the air oxidation of acetaldehyde in acetic acid in the presence of manganous ion as catalyst: (*ii*) the air oxidation of n-butane in the liquid phase under pressure and at 130–230°C in the presence of a suitable catalyst, *e.g.*, manganous ion; (*iii*) by reaction between carbon monoxide and methanol under pressure in the presence of cobalt octacarbonyl at about 210°C:

$$CO + CH_3OH \longrightarrow CH_3CO_2H.$$

One of the earliest methods for preparing acetic acid was by the destructive distillation of wood to give pyroligneous acid. This contains about 10 per cent acetic acid, and was originally treated by neutralising with lime and then distilling off the volatile compounds (these are mainly methanol and acetone). On distillation with dilute sulphuric acid, the residue gives dilute acetic acid. More recently, the acetic acid is extracted by means of solvents, *e.g.*, isopropyl ether.

Vinegar, which is 6–10 per cent aqueous solution of acetic acid, may be made in several ways. Malt vinegar is prepared by the oxidation of wort by means of the bacteria *Mycoderma aceti*:

$$CH_3CH_2OH + O_2 \longrightarrow CH_3CO_2H + H_2O$$

In the 'quick' vinegar process', beech shavings, contained in barrels, are moistened with strong vinegar containing the bacteria. A 10 per cent aqueous solution of ethanol containing phosphates and inorganic salts (which are necessary for the fermentation) is then poured through the barrels, and the ethanol is thereby oxidised to acetic acid. A plentiful supply of air is necessary, otherwise the oxidation is incomplete and acetaldehyde is produced.

Acetic acid is a pungent corrosive liquid, m.p. 16.6°C, b.p. 118°C, miscible in all proportions with water, ethanol ad ether It is stable towards oxidising agents, and so is a useful solvent for chromium trioxide oxidations. Acetic acid is commonly used as a solvent, and in the preparation of acetates, acetone, acetic anhydride, etc.

Strengths of the monocarboxylic acids. If the structure of the carboxyl group were simply (I), then, because of the –I effect of the carbonyl group (as shown in (II), proton release will be facilitated as compared with alcohols.

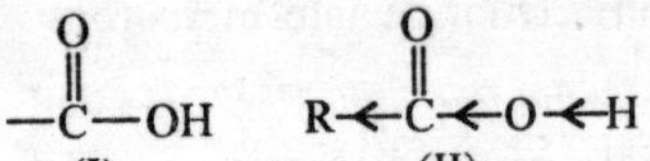

$$-\overset{\overset{\displaystyle O}{\|}}{C}-OH \qquad R\leftarrow\overset{\overset{\displaystyle O}{\|}}{C}\leftarrow O\leftarrow H$$

(I) (II)

Thus carboxylic acids will be stronger acids than alcohols. However, the inductive effect alone cannot account for the very large difference in acid strength. One explanation is that the carboxyl group is a resonance hybrid and because of the positive charge on the oxygen atom of the OH group, proton release is facilitated; this positive charge is absent in alcohols since they are not resonance hybrids.

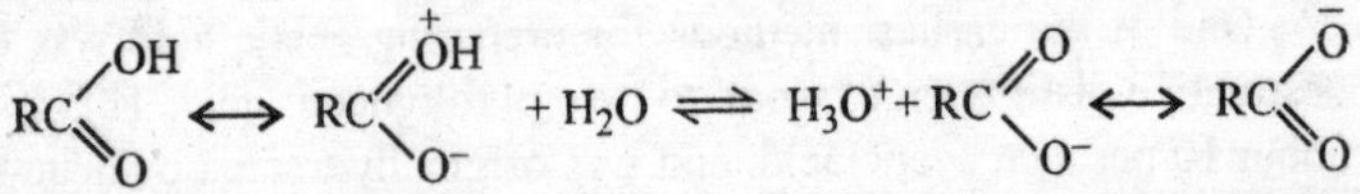

$$RC\begin{matrix}\nearrow OH\\ \searrow\!\!\!\!= O\end{matrix} \longleftrightarrow RC\begin{matrix}=\overset{+}{O}H\\ -O^-\end{matrix} + H_2O \rightleftharpoons H_3O^+ + RC\begin{matrix}=O\\ -O^-\end{matrix} \longleftrightarrow RC\begin{matrix}-\bar{O}\\ =O\end{matrix}$$

This argument can be extended as follows. When the acid ionises, *i.e.*, donates its proton to water (which is the solvent), the conjugate base of the acid. *i.e.*, the carboxylate ion, is a resonance hybrid of two *equivalent* resonating structures. At the same time, since the negative charge is spread, the resonance energy of this anion will be greater than that of the un-ionised acid, which is a resonance hybrid of two different resonating structures, one with separation of *unlike* charges. Therefore the internal energy of the anion is lower than that of the un-ionised acid. In the case of alcohols, thee is no resonance stabilisation in the alkoxide ion. It therefore follows that $\Delta G^{\ominus}$ for the ionisation of monocarboyxlic acids will be more negative than that of alcohols. Hence, the equilibrium constant of the former will be greater than that of the latter, *i.e.*, the pK_a values of the acids will be lower than those of the alcohols. Thus alcohols are weaker acids than the monocaboxylic acids.

X-ray and electron-diffraction measurements of the carboxylate ion have shown that the two C—O bonds are of equal length. This is in keeping with the ion being a resonance hybrid.

Table 2.1

Acid	*pK_a*	*Acid*	*pK_a*
n-Butyric	4.82	Trifluoroacetic	0.23
Isobutyric	4.86	Bromoacetic	2.9
n-Valeric	4.86	Iodoacetic	3.16
Isovaleric	4.78	α-Chloropropionic	2.80
Trimethylacetic	5.05	β-Chloropropionic	4.08
Hexanoic	4.88	Nitroacetic	1.68
Heptanoic	4.89	Hydroxyacetic	3.83
Formic	3.75	Chloroacetic	2.85
Acetic	4.76	Dichloroacetic	1.25
Propionic	4.87	Trichloroacetic	0.66
Octanoic	4.90	Hydrochloric	– 7.00

The treatment of the carboxyl group and the carboxylate ion from the M.O. point of view also leads to the same results as those obtained by the V.B. method, the increased stability now being due to *delocalisation*. Let us first consider the *undissociated* carboxylic acid (Fig. 2.1*a*). The carbon atom is linked to the hydroxyl group by a σ-bond, and to the oxygen of the carbonyl group by one σ- and one π-bond. This is shown in (Fig. 2.1*b*). If the oxygen atom of the hydroxyl group has sp^2 hybridisation, this would leave a *p* orbital perpendicular to the plane of the carboxyl group, and it would contain a pair of electrons (Fig. 2.1*c*). This orbital could now overlap with both the *p* orbital of the carbon atom and that of the oxygen atom of the carbonyl group. The combination of these three orbitals would give rise to M.O.s embracing *three* nuclei (Fig. 2.1*d*). Since *four* electrons are involved, they must be accommodated, in the ground state, in the lowest two M.O.s. The M.O. with the lowest energy level has *one* node (Fig. 2.1*e*), and the next energy level has *two* nodes (Fig. 2.1*f*). Thus, (Fig. 2.1*g*) represents the ground state of the undissociated carboxyl group. Because of delocalisation, this group has become more stable. Since a hydroxyl group is more electron-attracting than an oxygen atom, the delocalisation of the pair of electrons in the *p* orbital is not very large, *i.e.*, although this lone pair embraces three nuclei, this pair is far more likely to be found in the region of the donor atom than anywhere else. Since the oxygen atom of the hydroxyl group has lost 'full-control' of this lone pair, it will acquire a small positive charge, and since the oxygen atom of the CO group has acquired a small share of this lone pair, this oxygen atom acquires a small negative charge (actually this oxygen atom already has a small negative charge due to the inductive effect).

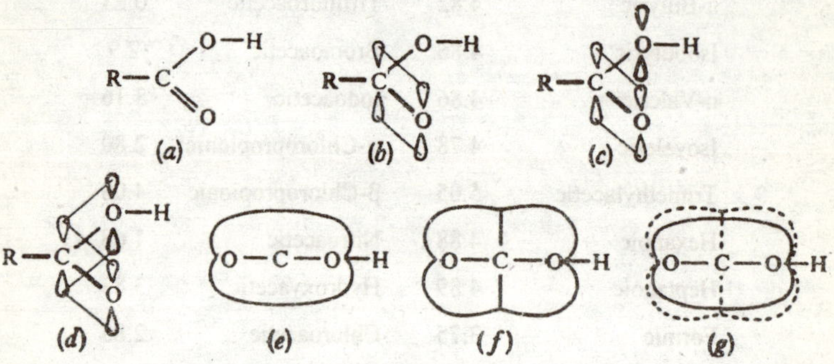

Fig. 2.1

The greater electronegativity of oxygen in a hydroxyl group over that of an oxygen atom alone is due to the hydrogen atom sharing one pair of electrons in forming the bond and thereby decreasing, to some extent, the electron density on that oxygen atom. Now let us consider the *carboxylate ion* (Fig. 2.2*a*). The oxygen atom of the original hydroxyl group retains the σ-electrons when the hydrogen atom is removed as a proton. As in the case of the *carboxy group*, one lone pair enters into conjugation with the π-bond of the C=O group (Fig. 2.2*b*). In the latter case, however, delocalisation is complete, *i.e.*, the lone pair is now likely to be found equally well or either oxygen atom (each atom will have a charge of – 1/2). Since delocalisation is complete, the resonance (delocalisation) energy is much greater than in the case of the undissoicated carboxyl group. Because the carbon atom is sp^2 hybridised in both the carboxyl group and carboxylate ion, the O—C—O angle in both would be expected to be about 120°. This has been observed in practice.

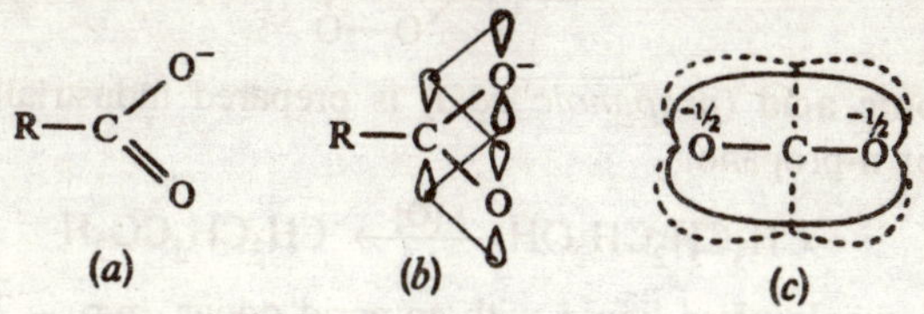

Fig. 2.2

In Figs. 2.1 and 2.2, the orbitals drawn in thicker type indicate that the atoms concerned contain *two* electrons in these orbitals. It should also be noted that lobes and not spheres (with + and – signs) have been used in these diagrams, and that overlapping orbitals have been drawn in an alternative way.

Peracetic (peroxyacetic) acid, $CH_3C(=O){-}O{-}OH$ may be prepared by treating acetic anhydride with concentrated hydrogen peroxide, and then distilling under reduced pressure.

Swern *et al.* (1962) have prepared per-acids (85–97 per cent yield) by treating a solution or slurry of an aliphatic (or aromatic) acid in methanesulphonic acid with 90–95 per cent hydrogen peroxide. This appears to be the best method so far.

Peracetic acid is an unpleasant-smelling liquid, f.p. + 0.1°C, soluble in water, ethanol and ether. It explodes violently when heated above 110°C. It is a powerful oxidising agent; it oxidises the olefinic bond to the oxide.

$$>C{=}C< + CH_3COO_2H \longrightarrow >C\overset{O}{-}C< + CH_3CO_2H$$

It also oxidises primary aromatic amines to nitroso-compounds, *e.g.*, aniline is converted into nitrosobenzene:

$$C_6H_5NH_2 + 2CH_3COO_2H \longrightarrow C_6H_5NO + 2CH_2CO_2H + H_2O$$

Infra-red absorption spectra measurements of per-acids show that these acids exist in solution very largely in the monomeric, intramolecularly hydrogen-bonded form (*inter alia*, Minkoff, 1954).

$$CH_3C\begin{matrix}\nearrow O\cdots H \\ \searrow O - O\end{matrix}$$

Propionic acid (*propanoic acid*) is prepared industrially by the oxidation of n-propanol:

$$CH_3CH_2CH_2OH \xrightarrow{[O]} CH_3CH_2CO_2H$$

It is a colourless liquid with an acrid odour, m.p. – 22°C, b.p. 141°C, miscible with water, ethanol and ether in all proportions.

Butyric acids. There are two isomers possible, and both are known.

n-Butyric acid occurs as the glyceryl ester in butter, and as the free acid in perspiration. It is prepared industrially by the oxidation of n-butanol.

n-Butyric acid is a viscous unpleasant-smelling liquid, m.p. – 4.7°C, b.p. 162°C, miscible with water, ethanol and ether. It is the liberation of free n-butyric acid that gives stale butter its rancid odour.

Isobutyric acid occurs in the free state and as its esters in many plants. It is prepared industrially by the oxidation of isobutanol and may be prepared synthetically as follows:

$$(CH_3)_2CHOH \xrightarrow{PBr_3} (CH_3)_2CHBr \xrightarrow{KCN}$$

$$(CH_3)_2CHCN \xrightarrow[acid]{H_2O} (CH_3)_2CHO_2H$$

Isobutyric acid is a liquid, m.p. – 47°C, b.p. 154°C. Its calcium salt is more soluble in hot water than in cold, whereas calcium butyrate is more soluble in cold water than in hot.

Valeric acids. There are four isomers possible, and all are known:

n-Valeric acid, $CH_3CH_2CH_2CH_2CO_2H$, m.p. – 34.5°C, b.p. 187°C.

Isovaleric acid, $(CH_3)_2CHCH_2CO_2H$, m.p. – 51°, b.p. 175°C.

Ethylmethylacetic acid or *active valeric acid*, CH_3CH_2CH-$(CH_3)CO_2H$, p.b. 175°C.

Trimethylacetic acid or *pivalic acid*, $(CH_3)_3CCO_2H$, m.p. 35.5°, p.b. 164°C.

The higher monocarboxylic acids which occur in nature usually have straight chains, and usually contain an *even* number of carbon atoms. **Caproic** (*hexanoic*), $C_6H_{12}O_2$ (m.p. – 9.5°, b.p. 205°C), **caprylic** (*octanoic*), $C_8H_{16}O_2$ (m.p. 16°, p.b. 237°C) and *capric* (*decanoic*) acid, $C_{10}H_{20}O_2$ (m.p. 31.5° p.b. 270°C), are present as glyceryl esters in goats' butter. *Lauric* (*dodecanoic*), $C_{12}H_{24}O_2$ (m.p. 44°C) and **myristic** (*tetradecanoic*) acid, $C_{14}H_{28}O_2$ (m.p. 58°C), occur as their glyceryl esters in certain vegetable oils. The most important higher acids are *palmitic* (*hexadecanoic*), $C_{16}H_{32}O_2$ (m.p. 64°C), and **stearic** (*octadecanoic*), $C_{18}H_{36}O_2$ (m.p. 72°C), which are very widely distributed as their glyceryl esters (together with oleic acid) in most animal ad vegetable oils and fats. The sodium and potassium salts of palmitic and stearic acids are the constituents of ordinary soaps.

Some still higher acids are found in waxes: **arachidic** (*eicosanoic*), $C_{20}H_{40}O_2$ (m.p. 77°C), **behenic** (*docosanoic*), $C_{22}H_{44}O_2$ (m.p. 82°C), **lignoceric** (*tetracosanoic*), $C_{24}H_{48}O_2$ (m.p. 83.5°C), **cerotic** (*hexacosanoic*), $C_{26}H_{52}O_2$ (m.p. 87.7°C), and **melissic** acid (*triacontanoic*), $C_{30}H_{60}O_2$ (m.p. 90°C).

The odd fatty acids may be obtained by degrading an even acid (see below). Two odd acids which may be prepared readily are **n-heptanoic** or **oenanthylic** acid, $C_7H_{14}O_2$ (m.p. – 10°, p.b. 223.5°C), and **nonanoic** or **pelargonic** acid $C_9H_{18}O_2$ (m.p. 12°, p.b. 254°C). n-Heptanoic acid is prepared by the oxidation of n-heptaldehyde, which is obtained by destructively distilling castor oil which contains ricinoleic acid (*q.v.*):

$$CH_3(CH_2)_5CHO + [O] \xrightarrow[H_2SO_4]{KMnO_4} CH_3(CH_2)_5CO_2H \ (68\text{–}70\%)$$

Nonanoic acid is obtained by the oxidation of oleic acid (*q.v.*).

Margaric acid (*heptadecanoic acid*), $C_{16}H_{33}CO_2H$, m.p. 61°C, is prepared by heating a mixture of the calcium salts of stearic and acetic acids, and then oxidising the heptadecyl methyl ketone so produced.

$$C_{17}H_{35}CO_2ca + CH_3CO_2ca \longrightarrow CaCO_3 + C_{17}H_{35}COCH_3 \xrightarrow{[O]} C_{16}H_{33}CO_2H$$

Odd-numbered monocarboxylic acids occur naturally in very small amounts up to $n = 23$, and their main source is butter fat. Branched-chain acids have also been isolated from butter fat.

General Properties of the Monocarboxylic Acids

The first three acids are colourless, pungent-smelling liquids; the acids from butyric, $C_4H_8O_2$, to nonanoic, $C_9H_{18}O_2$, are oils which smell like goats' butter; and those higher than decanoic acid, $C_{10}H_{20}O_2$, are odourless solids. The lower members are far less volatile than is to be expected from their molecular weight. This can be explained by hydrogen bonding, and electron diffraction studies (Pauling, 1934) have shown that an eight-membered ring is present, *i.e.*, the acids exist as cyclic dimers. The lower members exist as dimers in the vapour phase and in aqueous solution, but in the liquid phase they exist as 'polymers'.

$$R-C\begin{matrix} /\!\!/ O\cdots H-O \backslash \\ \backslash O-H\cdots O /\!\!/ \end{matrix}C-R$$

The melting point of the n-monocarboyxlic acids show alternation or oscillation from one member to the next, the melting point of an 'even' acid being higher than that of the 'odd' acid immediately below and above it in the series (The physical constants of the individuals). A number of homologous series follow this oscillation or 'saw-tooth' rule.

The first four members are very soluble in water, and the solubility decreases as the molecular weight increases. This solubility is due to the acids being capable of forming hydrogen bonds with water.

The infra-red absorption region of the carboxyl group contains a number of bands. The carbonyl group (stretch) absorbs in the same region as that of acyclic ketones. 1725–1700 cm^{-1}(s), but the (free) hydroxyl group (stretch) absorbs at 3560–3500 cm^{-1}(m), which is in the lower frequency region than for a alcohols (3670–3580 cm^{-1}).

Fig. 2.3 is the infra-red spectrum of stearic acid (as a 1 per cent solution in carbon tetrachloride). It should be noted that the band for the bonded O—H (str.) is hidden by the strong C—H (str.))

On the other hand Fig. 2.4 is the spectrum of stearic acid as a Nujol mull. In this case, the stearic acid is in the solid (dispersed) state, and it can be seen that there is an extremely large difference in the region 1300–1180 cm^{-1}.

In this region there are eight maxima, and this pattern is characteristic of the spectra of long-chain n-alkyl compounds (in the solid state), the number of peaks depending on the chain length Nujol (a petroleum fraction), since it is a hydrocarbon mixture, has the bands characteristic of alkanes. The O—H (str.) is hidden (Fig. 2.3). These two spectra illustrate the influence of the physical state of a compound on its infra-red spectrum.

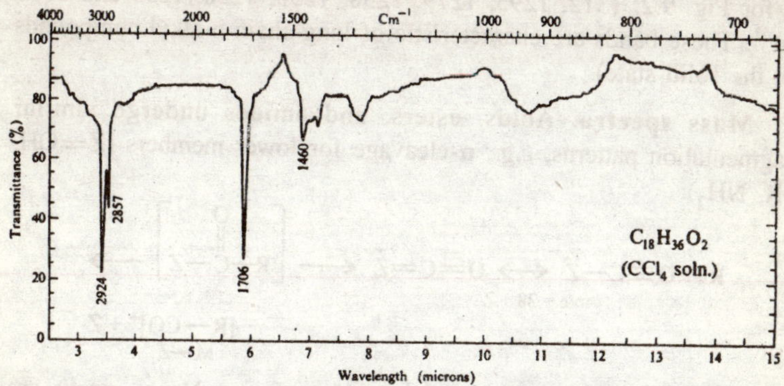

Fig. 2.3

1706 cm^{-1}: C=O (str.) in (saturated ketones and) saturated aliphatic acids. 2942 and 2857 cm^{-1}: C—H (str.) in Me and/or CH_2. 1460 cm^{-1}: C—H (def.) in Me and/or CH_2.

Because of hydrogen bonding (intermolecular) and exchange reactions, the proton peak of the carboxyl group can be found in many different positions in the NMR spectrum.

This applies to hydroxyl protons in general. However, it has been found that in non-polar solvents, the τ-value of the carboxyl proton in aliphatic acids is around – 1.0 p.p.m. (*i.e.*, there is a great shift to lower field due to strong hydrogen bonding).

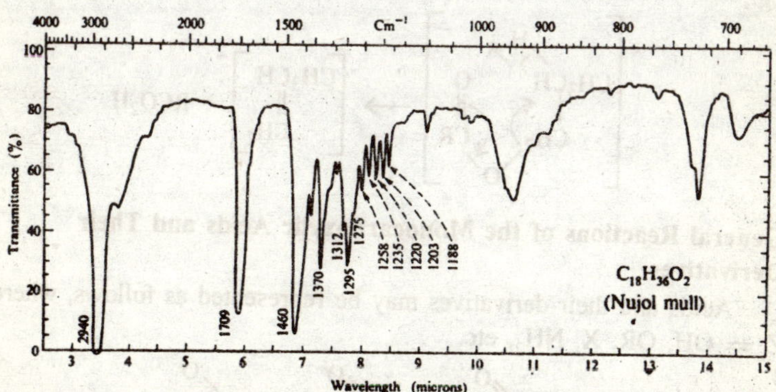

Fig. 2.4

2940, 1460 and 1370 cm^{-1}: Bands for Nujol. 1709 cm^{-1}: C=O (str.) as for Fig. 9.2. 1312, 1295, 1279, 1258, 1235, 1220, 1203 and 1188 cm^{-1}: These bands are characteristic of long-chain n-alkyl compounds (in the solid state).

Mass spectra. Acids, esters, and amides undergo similar fragmentation patterns, *e.g.*, α-cleavage for lower members (Z=OH, OR, NH_2):

$$R\cdot + \overset{+}{O}\equiv C-\ddot{Z} \longleftrightarrow O=C=\overset{+}{Z} \longleftarrow \left[R-\overset{O}{\overset{\|}{C}}-Z\right]^{+} \longrightarrow [R-CO]^{+} + Z\cdot$$

m/e = 28 + Z; M—Z

Also, R and Z may carry the positive charge. Hence, peaks are obtained at $M-17$ and *m/e* 45 (Z = OH), $M-31$, M–45, and *m/e* 59, 73 (Z = OMe and OEt, respectively), and $M-16$ and *m/e* 44 (Z = NH_2). Since they are more volatile than their corresponding acids, esters are used preferentially. All can undergo the McLafferty rearrangement if they contain a γ-hydrogen atom:

$$\text{H–}CH_2CH_2CH_2\text{–}\overset{\cdot+}{O}{=}CZ \longrightarrow CH_2{=}CH_2 + \left[HO\text{–}\overset{|}{C}(\text{–}Z){=}CH_2\right]^{\cdot+}$$

m/e 60 (Z = OH);
74 (Z = OMe);
88 (Z = OEt);
59 (Z = NH_2)

When the alkyl group in O of esters is higher than methyl, esters can also undergo the McLafferty rearrangement to give alkene and acid, *e.g.*,

$$\left[\text{H–}CH_2CH\text{–}CH_2\text{–}O\text{–}C(=O)R\right] \longrightarrow \left[CH_3CH{=}CH_2\right]^{+} + RCO_2H$$

General Reactions of the Monocarboxylic Acids and Their Derivatives

Acids and their derivatives may be represented as follows, where Z = OH, OR, X, NH_2, etc.

$$RC(=O)\ddot{Z} \longleftrightarrow R\overset{+}{C}(O^-)\ddot{Z} \longleftrightarrow RC(O^-){=}Z^+$$

Thus, all are resonance hybrids and the behaviour of the 'carbonyl' group depends on the nature of Z. Z may have a – I and a + R effect and consequently the actual weights of the canonical forms will depend on the relative contributions of these two electronic effects, *e.g.*, for Z $= NH_2$ (acid amide), + R > – I, whereas for Z = Cl (acid chloride), – I > + R. Hence, the rates of reaction of the carbonyl group in these compounds can be expected to be different.

Many of the reactions of the acids and their derivatives may be generalised as:

$$\text{(i)}\quad R{-}\overset{\overset{\displaystyle O}{\|}}{\underset{\underset{\displaystyle Z}{|}}{C}} + Y: \underset{}{\overset{r/d}{\rightleftharpoons}} R{-}\overset{\overset{\displaystyle O^-}{|}}{\underset{\underset{\displaystyle Z}{|}}{C}} + Y \longrightarrow R{-}\overset{\overset{\displaystyle O}{\|}}{C}{-}Y + Z:$$

$$\text{(ii)}\quad R{-}\overset{\overset{\displaystyle O}{\|}}{\underset{\underset{\displaystyle Z}{|}}{C}} + H^+ \rightleftharpoons R{-}\overset{\overset{\displaystyle \overset{+}{O}H}{\|}}{\underset{\underset{\displaystyle Z}{|}}{C}} \longrightarrow R{-}\overset{\overset{\displaystyle OH}{|}}{\underset{\underset{\displaystyle Z}{|}}{C^+}} \xrightarrow[r/d]{Y:} R{-}\overset{\overset{\displaystyle OH}{|}}{\underset{\underset{\displaystyle Z}{|}}{C}}{-}Y$$

$$\longrightarrow R{-}\overset{\overset{\displaystyle O}{\|}}{C}{-}Y + HZ$$

These reactions are typical of the behaviour of the carbonyl group (in aldehydes and ketones), and the greater the positive charge on the carbonyl atom, the more easily is this carbon atom attacked by a nucleophile (in the r/d step). Since the + R effect tends to neutralise the positive charge, the greater the + R effect of Z, the slower will be the reaction. Experimental work has shown the rates of hydrolysis are in the order $RCONH_2 < RCO_2R < RCOCl$, and this is the reverse order of the + R effects of the groups, *i.e.*, NH_2 > OR > Cl.

ESTERS

Esters are compounds which are formed when the hydroxylic hydrogen atom in oxygen acids is replaced by an alkyl group; the acid may be organic or inorganic. The most important esters are derived from the caboxylic acids. The general formula of the carboxylic esters is $C_2H_{2n}O_2$, which is the same as that of the carboxylic acids, and they are named as the alkyl salts of the acid. *e.g.*,

$CH_3COOC_2H_5$ ethyl acetate

$(CH_3)_2CHOOCH(CH_3)_2$ isopropyl isobutyrate

Carboxylic esters are formed by the action of the acid on an alcohol: acid + alcohol:

$$\text{acid} + \text{alcohol} \rightleftharpoons \text{ester} + \text{water}$$

The reaction is reversible, the *forward* reaction being known as *esterification*, and the *backward* reaction as *hydrolysis* (for mechanisms).

General Methods of Preparation of the Carboxylic Esters

1. The usual method is *esterification*. The reaction is always slow, but is speeded up by the presence of small amounts of inorganic acids as catalysts, *e.g.*, when the acid is refluxed with the alcohol in the presence of 5–10 per cent concentrated sulphuric acid:

$$R^1COOH + R^2OH \xrightarrow{H_2SO_4} R^1COOR^2 + H_2O \ (v.g.)$$

Alternatively, hydrogen chloride is passed into the mixture of alcohol and acid until there is a 3 per cent increase in weight, and the mixture is refluxed (the yields are very good). This is known as the *Fische–Speier method* (1895), and is more satisfactory for secondary and tertiary alcohols than the sulphuric acid method, which tends to dehydrate the alcohol to alkene.

Esterification without the use of catalysts, and starting with one mole of acid and one mole of alcohol, gives rise to about 2/3 mole of ester. The yield of ester may be increased by using excess of acid or alcohol, the cheaper usually being the one in excess. Increased yields may also be effected by dehydrating agents, *e.g.*, concentrated sulphuric acid behaves both as a catalyst and a dehydrating agent. The same effect may be obtained by removing the water or ester from the reaction mixture by distillation, which is particularly useful for high-boiling acids and alcohols. On the other hand, the water may be removed from the reaction mixture by the addition of benzene or carbon tetrachloride, each of which forms a binary mixture with water (and may form a ternary mixture with water and the alcohol), the azeotropic mixtures boiling at a lower temperature than any of the components.

2. Acid chlorides or anhydrides react rapidly with alochols to form esters:

$$R^1COCl + R_2OH \longrightarrow R^1CO_2R^2 + HCl \ (v.g.–ex.)$$

$$(R^1CO)_2O + R^2OH \longrightarrow R^1CO_2R^2 + R^1CO_2H \ (v.g.–ex.)$$

The reaction with tertiary alcohols is very show and is often accompanied by side reactions (alkene or alkyl halide formation), but by using the appropriate conditions, esters are produced. Esters of tertiary alochols may also be conveniently prepared by means of a Grignard reagent.

3. Esters may be prepared by refluxing the silver salt of an acid with an alkyl halide in ethanolic solution:

$$R^1CO_2Ag + R^2Br \longrightarrow R^1CO_2R^2 + AgBr \text{ (v.g. –ex.)}$$

This method is useful where direct esterification is difficult.

4. *Methyl* esters are very conveniently prepared by treating an acid with an ethereal solution of diazomethane (*q.v.*):

$$RCO_2H + CH_2N_2 \longrightarrow RCO_2CH_3 + N_2 \text{ (ex.)}$$

According to Morrison *et al.* (1961), the BF_3—MeOH reagent is better than diazomethane for the methylation of carboxylic acids.

5. The interaction between an ether and carbon monoxide at 125–180°C, under 500 atmospheres pressure, in the presence of boron trifluoride plus a little water, produces esters:

$$R_2O + CO \longrightarrow RCO_2R$$

Esters may also be prepared by the Tischenko reaction (with aldehydes); ethyl acetate is made commercially this way.

General Properties of Esters

The carboxylic esters are pleasant-smelling liquids or solids. The boiling points of the straight chain isomers are higher than those of the branched-chain isomers. The boiling points of the methyl and ethyl esters are lower than those of the corresponding acid, and this is due to the fact that the esters are not associated since they cannot form intermolecular hydrogen bonds. The esters of low molecular weight are fairly soluble in water–hydrogen bonding between ester and water is possible–and the solubility decreases as the series is ascended; all esters are soluble in most organic solvents.

The structure of the esters, like their parent acids, is best represented as a resonance hybrid:

$$R^1\text{—}C(=O)\text{—}OR^2 \longleftrightarrow R^1\text{—}C(\text{—}O^-)=\overset{+}{O}R^2$$

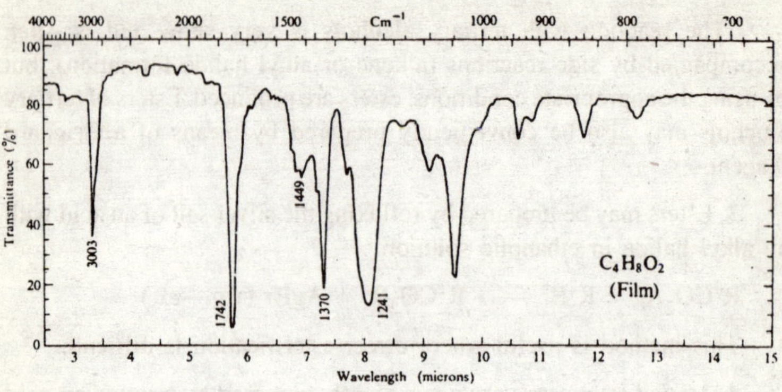

Fig. 2.5

1742 cm^{-1}; C=O (str.) in saturated aliphatic esters. 1241 cm^{-1}: C—O (str.) in *acetates*. 3003 cm^{-1}: C—H (str.) in Me and/or CH_2. 1449 cm^{-1}: C—H (def.) in Me and/or CH_2. 1370 cm^{-1}: C—H (def.) in Me.

The infra-red absorption region of the carbonyl group (stretch) in saturated aliphatic esters is 1750–1735 cm^{-1}(s). This is a higher frequency than that of the carbonyl group in ketones and acids, and so esters may usually be identified.

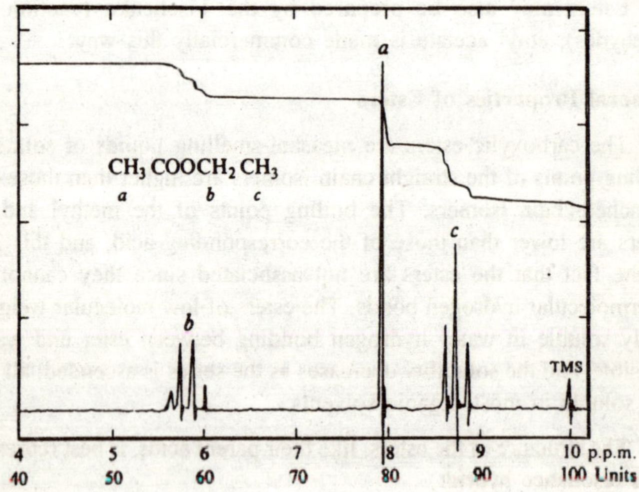

Fig. 2.6

Characteristic bands are also shown by the C—O (stretch) in formates, 1200–1180 cm^{-1}(s); acetates, 1250–1230 cm^{-1}(s); propionates, etc., 1200–1170 cm^{-1}(s).

The infra-red (film) ad NMR (in $CDCL_3$ at 60 MHz) spectra of ethyl acetate are shown in Figs. 2.5 and respectively.

General Reactions of Esters

1. Esters are hydrolysed by acids or alkalis:

$$R^1CO_2H + R^2OH \xleftarrow{H_3O^+} R^1CO_2R^2 \xrightarrow{OH^-} R^1CO_2^- + R^2OH$$

When hydrolysis is carried out with alkali, the salt of the acid is obtained, and since the alkali salts of the higher acids are soaps, alkaline hydrolysis is known as *saponification* (derived from Latin word meaning soap); saponification is far more rapid than acid hydrolysis.

Mechanisms of carboxyl esterification and hydrolysis of carboxylic esters. Mechanistic studies have shown that in alkaline or neutral hydrolysis, it is the neutral ester molecule which undergoes reaction, whereas in acid-catalysed hydrolysis, it is the conjugate acid of the ester which undergoes reaction.

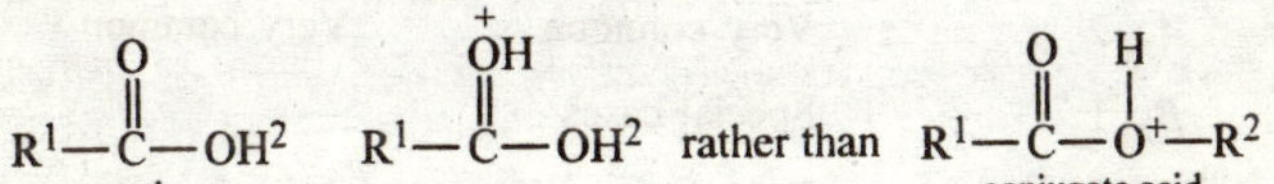

$$\underset{\text{neutral ester}}{R^1-\overset{\overset{\displaystyle O}{\|}}{C}-OH^2} \qquad R^1-\overset{\overset{\displaystyle \overset{+}{O}H}{\|}}{C}-OH^2 \quad \text{rather than} \quad \underset{\text{conjugate acid}}{R^1-\overset{\overset{\displaystyle O}{\|}}{C}-\overset{\overset{\displaystyle H}{|}}{O^+}-R^2}$$

The protonated carbonyl oxygen form of the cA is preferred, and this is based on the work of Fraenkel (1961), who examined the NMR spectrum of methyl formate in, *e.g.*, 100 per cent sulphuric acid and concluded from his results that the ester is protonated chiefly on the carbonyl oxygen.

This formulation is also used for *acids*, partly from analogy with esters and partly from the fact that of the two possible conjugate acids, (I) and (II) , (I) may be written as a resonance hybrid to which two identical canonical forms contribute, whereas for (II) the two contributing structures are different and one has fewer bonds and carries unlike charges.

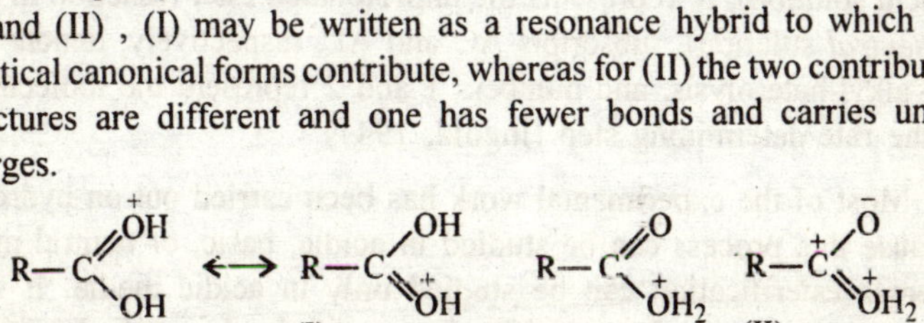

$$R-C\begin{matrix}\nearrow\!\!\!\!/\ \overset{+}{O}H \\ \searrow OH\end{matrix} \longleftrightarrow \underset{\text{(I)}}{R-\overset{+}{C}\begin{matrix}\nearrow OH \\ \searrow\!\!\!\!\backslash\ OH\end{matrix}} \qquad R-C\begin{matrix}\nearrow\!\!\!\!/\ O \\ \searrow \overset{+}{O}H_2\end{matrix} \quad \underset{\text{(II)}}{R-\overset{+}{C}\begin{matrix}\nearrow \overset{-}{O} \\ \searrow \overset{+}{O}H_2\end{matrix}}$$

The hydrolysis of carboxylic esters may be formulated in two ways:

$$R^1—\overset{\overset{O}{\|}}{C}—OR^2 \qquad R^1—\overset{\overset{O}{\|}}{C}—O—R^2$$

acyl-oxygen heterolysis   alkyl-oxygen heterolysis

Both modes of fission have been demonstrated experimentally, and kinetic studies have shown that two types of mechanism may operate for each mode, *unimolecular* and *bimolecular*. Hence, *eight* mechanisms are possible for ester hydrolysis but, so far, *six* have been observed Table 2.2. Also, since in principle, the *acid-catalysed* reactions (esterification and hydrolysis) are reversible, but not *alkaline* hydrolysis, then *four* mechanisms are possible for esterification. So far, *three* have been observed (Table 2.2).

Table 2.2

Type of mechanism	*Hydrolysis*	*Esterification*
$B_{AC}1$	—	—
$B_{AC}2$	Very common	—
$A_{AC}1$	Special cases	Special cases
$A_{AC}2$	Very common	Very common
$B_{AL}1$	Special cases	—
$B_{AL}2$	Rare	—
$A_{AL}1$	Very common for t-alcohols	Very common for t-alcohols
$A_{AL}2$	—	—

The eight possible mechanisms are indicated by the symbols shown in Table 9.2, where *A* represents the cA of the ester or acid (reaction in *acid* solution), *B* represents the unprotonated ester (reaction in *basic* or *neutral* solution), subscripts AC and AL, respectively, denote acyl- and alkyl-heterolysis, and numbers 1 and 2 represent the molecularity of the rate-determining step (Ingold, 1941):

Most of the experimental work has been carried out on hydrolysis because this process can be studied in acidic, basic, or neutral media, whereas esterification can be studied only in acidic media. It might appear at first sight that esterification could also be studied in neutral media, but since carboxylic acids are acids, reaction in the absence of added acid (as catalyst) is still effectively in an acid medium. Because of this, mechanisms of esterification are partly based on the principle of microscopic reversibility.

Bimolecular basic hydrolysis with acyl-oxygen heterolysis (B_{AC} 2 mechanism). It has long been known that the alkaline hydrolysis of esters is a second-order reaction, *i.e.*, rate = *k* [ester] [OH^-]. Evidence for acyl-oxygen heterolysis has been obtained in several ways, *e.g.*, Polanyl *et al.*, (1934) showed that the alkaline hydrolysis of n-pentyl acetate in water enriched with ^{18}O gave n-pentyl alcohol containing *no* ^{18}O. Therefore acyl-oxygen heterolysis must have occurred:

$$\text{CH5CO—OC5H11} + \dot{O}H^- \longrightarrow CH_3CO\dot{O}H + C_5H_{11}O^- \longrightarrow$$

$$CH_3CO\dot{O}^- + C_5H_{11}OH$$

Another method uses optical activity as a diagnostic test for demonstrating acyl-oxygen heterolysis. If the alcohol group is optically active, and if the $B_{AC}2$ mechanism operates, then the ion RO^- will be liberated and will retain its optical activity since the bond R—O is never broken. Various examples of retention are known when the reaction is bimolecular.

A mechanism consistent with the above facts (and with the work of Bender (1951), who used esters labelled with ^{18}O in the carbonyl group) is:

$$HO^- + R^1{-}C(=O){-}OR^2 \underset{\text{fast}}{\overset{\text{slow}}{\rightleftharpoons}} HO{-}C(O^-)(R^1){-}OR^2 \underset{\text{slow}}{\overset{\text{fast}}{\rightleftharpoons}}$$

$$HO{-}C(=O){-}R^1 + {}^-OR^2 \xrightarrow{\text{fast}} {}^-O{-}C(=O){-}R^1 + R^2OH$$

Any R^1 group with – I effect should accelerate hydrolysis by creating a positive charge on the attached carbon atom, thereby facilitating attack by the hydroxide ion. On the other hand, because the r/d step is bimolecular, the larger R^1 is the larger can be expected to be the steric retardation, since the hybridisation of the carbon atom of the carbonyl group changes from sp^2 to sp^3 (with consequent increased crowding). These anticipated results have been observed in practice, *e.g.*,

– I effect

Ester	$MeCO_2Me$	$ClCH_2CO_2Me$	$CHCl_2CO_2Me$
Relative rates	1	761	16000

Steric effect

Ester	$MeCO_2Et$	$EtCO_2Et$	iso$PrCO_2Et$
Relative rates	1.00	0.47	0.10

Bimolecular acid-catalysed hydrolysis and esterification with acyl-oxygen heterolysis ($A_{AC}2$ mechanism). Acid-catalysed hydrolysis and acid-catalysed esterification are the reverse of each other, and so according to the principle of microscopic reversibility the mechanism of acid-catalysed hydrolysis will be that of acid-catalysed esterification, but in reverse.

The rate law for acid-catalysed hydrolysis has been shown to be second-order, first-order in both ester and hydrogen ion concentration, *i.e.*, rate = k [ester] $[H_3O^+]$. Also, evidence for acyl=oxygen heterolysis has been obtained in several ways, *e.g.*, Ingold *et al.* (1939) showed that the acid-catalysed hydrolysis of methyl hydrogen succinate in water enriched with ^{18}O gave methanol with no extra ^{18}O. Thus, acyl-oxygen heterolysis must have occurred:

$$HO_2CCH_2CH_2CO\text{—}OMe + H_2\dot{O} \xrightarrow{H^+} HO_2CCH_2CH_2CO\dot{O}H + MeOH$$

Furthermore, Robers *et al.* (1938) esterified benzoic acid with methanol enriched with 18O and obtained water *not* enriched with 18O. Thus, acyl-oxygen heterolysis must have occurred:

$$PhCO\text{—}OH + Me\dot{O}H \longrightarrow PhCO\dot{O}me + H_2O$$

A mechanism that fits these facts is:

$$R^1\text{—}\overset{O}{\overset{\|}{C}}\text{—}OH \underset{-H^+;\ fast}{\overset{+H^+;\ fast}{\rightleftharpoons}} R^1\text{—}\overset{\overset{+}{O}H}{\overset{\|}{C}}\text{—}OH \underset{-R^2OH;\ fast}{\overset{+R^2OH;\ slow}{\rightleftharpoons}} R^1\text{—}\overset{OH}{\overset{|}{\underset{\underset{H\text{—}O^+\text{—}R^2}{|}}{C}}}\text{—}OH \underset{fast}{\overset{fast}{\rightleftharpoons}}$$

$$R^1\text{—}\overset{\overset{+}{O}H_2}{\overset{|}{\underset{\underset{OR^2}{|}}{C}}}\text{—}OH \underset{+H_2O;\ slow}{\overset{-H_2O;\ fast}{\rightleftharpoons}} R^1\text{—}\underset{\underset{OR^2}{|}}{C}=\overset{+}{O}H \underset{+H^+;\ fast}{\overset{-H^+;\ fast}{\rightleftharpoons}} R^1\text{—}\underset{\underset{OR^2}{|}}{C}=O$$

It should be noted that the r/d step for esterification is the addition of alcohol, whereas that for hydrolysis is the addition of water. In both cases, the r/d step involves change of hybridisation of the carbon atom of the carbonyl group from sp_2 to sp_3, and consequently steric retardation can be expected to increase as R1 increase as size (*cf.* the BAC_2 mechanism), *e.g.* (esterification with methanol):

Acid	$MeCO_2H$	n-$PrCO_2H$	$MeCCO_2H$	Et_3CCO_2H
Relative rates	1.0	0.51	0.037	0.00016

Unimolecular acid-catalysed hydrolysis and esterification with acyl-oxygen heterolysis ($A_{AC}1$ mechanism). Let us consider the following *observed* results (*M* represents the mesityl group, 1,3,5-trimethylphenyl, in *mesitoic acid*, $Me_3C_6H_2CO_2H$):.

$$PhCO_2H + MeOH \xrightarrow{HCl} PhCO_2Me \quad \textit{occurs readily}$$

$$PhCO_2ME + OH^- \dashrightarrow PhCO_2^- + MeOH \quad \textit{occurs readily}$$

$$MCO_2H + MeOH \xrightarrow{HCl} \textit{No reaction}$$

$$MCO_2Me + OH^- \longrightarrow MCO_2^- + MeOH \quad \textit{very slow}$$

Hammett *et al.* (1937) showed that when methyl benzoate is dissolved in concentrated sulphuric acid and the solution then poured into ice/ water, almost all of the ester is recovered. On the other hand, methyl mestioate, on the same treatment, gave a quantitative yield of mestioic acid. Then Newman (1941) showed that when a solution of mesitoic acid in concentrated sulphuric acid is poured into a cold methanol, methyl mesitoate is formed. This different behaviour of mesitoyl derivatives from benzoyl derivatives suggests that different mechanisms are operating in esterification and hydrolysis for these two sets of compounds. Now, Hammett (1937) had shown that when mesitoic acid is dissolved in concentrated sulphuric acid, the van't Hoff factor *i* is 4 (measured by depression of freezing point). Under the same conditions, benzoic acid gives $i = 2$. Hammett therefore proposed that the mesitoyl reactions proceed through an intermediate *acylium* (*oxocarbonium*) ion (*mesitoylium* ion in this case) formed by acyl-oxygen heterolysis:

$$\text{(i)}\quad MCO_2H + H_2SO_4 \underset{}{\overset{\text{fast}}{\rightleftharpoons}} M\overset{+}{C}\begin{smallmatrix}OH\\OH\end{smallmatrix} + HSO_4^-$$

$$\text{(ii)}\quad M\overset{+}{C}\begin{smallmatrix}OH\\OH\end{smallmatrix} + H_2SO_4 \underset{}{\overset{\text{slow}}{\rightleftharpoons}} M\overset{+}{C}\ O + H_3O^+ + HSO_4^-$$

The overall equation may be written:

$$MCO_2H + 2H_2SO_4 \rightleftharpoons M\overset{+}{C}O + H_3O^+ + 2HSO_4^-$$

This accounts for $i = 4$ for mesitoic acid, and for $i = 2$ for benzoic acid if only reaction (i) is involved. Thus, the reactions of mesitoic acid and its ester may be formulated as:

$$\left.\begin{array}{l} MCO_2H \\ MCO_2H \end{array}\right] \xrightarrow[\text{slow}]{H_2SO_4} MC{=}O \xrightarrow{\text{fast}} \left[\begin{array}{l} \xrightarrow{MeOH} MCO_2Me \\ \xrightarrow{H_2O} MCO_2H \end{array}\right.$$

This type of reaction occurs with highly sterically hindered acids and esters, and so steric hindrance must be an important factor.

Further support for this mechanism has been provided by Olah *et al.* (1967), who showed that acids (and esters), when dissolved in FSO_3H—SbF_5—SO_2 (a very strongly acidic medium), at 60°C, lose a molecule of water at 0°C to form the acylium ion:

$$RC(=O)OH \xrightarrow[-60°C]{H^+} RC(=\overset{+}{O}H)OH \longleftrightarrow RC(OH)=\overset{+}{O}H \xrightarrow[0°C]{-H_2O} R\overset{+}{C}{=}O$$

Unimolecular basic hydrolysis with alkyl-oxygen heterolysis (B_{AL}1 mechanism). This mechanism has been shown to occur when the alkyl group of the alcohol is capable of forming a relatively stable carbonium ion, the solvent has high ionising power (thereby encouraging alkyl-oxygen heterolysis), and the hydroxide ion concentration is low (thereby suppressing the bimolecular mechanism). An example of the B_{AL}1 mechanism is the hydrolysis of optically active α-phenylethyl hydrogen phthalate in *faintly* alkaline solution. Kenyon *et al.* (1936) showed that the product was recemissed α-phenylethanol (phenylmethylmethanol); this is in keeping with the formation of an intermediate carbonium ion (sp^2 hybridised) [R = o–$HO_2CC_6H_4$—]:

$$RCO{-}O{-}CHMePh \underset{}{\overset{\text{slow}}{\rightleftharpoons}} RCO_2^- + PhMeCH^+ \xrightarrow[\text{fast}]{H_2O} PhMeCHOH_2^+ \xrightarrow[\text{fast}]{RCO_2^-}$$

$$(\pm)\text{-}PhMeCHOH + RCO_2H$$

Bimolecular basic hydrolysis with alkyl-oxygen heterolysis (B_{AL}2 mechanism). This mechanism is rare. One example is the reaction between methyl benzoate and sodium methoxide in methanol (Bunnett *et al.*, 1950). The products were dimethyl ether and sodium benzoate; these results are readily explained on the basis of the B_{AL}2 mechanism:

$$MeO^- \; Me{-}O{-}OCPh \longrightarrow Me_2O + PhCO_2^-$$

Unimolecular acid-catalysed hydrolysis and esterification with alkyl-oxygen heterolysis ($A_{AL}1$ mechanism). Definite evidence for this mechanism was obtained by Hughes, Ingold *et al.* (1939), who showed that the esterification of acetic acid with optically active octan-2-ol in the presence of sulphuric acid gave a large amount of racemised ester. Racemisation is to be expected if an intermediate carbonium ion is formed, and so the reaction may be formulated as $A_{AL}1$ (R = n-C_6H_{11}CHMe—):

$$ROH + H^+ \underset{fast}{\overset{fast}{\rightleftharpoons}} ROH_2^+ \underset{fast}{\overset{slow}{\rightleftharpoons}} H_2O + R^+$$

$$MeC(=\ddot{O})OH + R^+ \underset{slow}{\overset{fast}{\rightleftharpoons}} MeC(=\overset{+}{O}R)(O{-}H) \underset{fast}{\overset{fast}{\rightleftharpoons}} MeC(OR)(=O) + H^+$$

Since all steps are reversible, the $A_{AL}1$ mechanism should, in principle, be possible for ester hydrolysis, Evidence for its occurrence has been obtained by, *e.g.*, Bunton *et al.* (1951), who carried out the acid-catalysed hydrolysis of t-butyl acetate in water enriched with ^{18}O and obtained t-butanol containing ^{18}O.

$$MeC(=O){-}O{-}Bu^t + H^+ \underset{fast}{\overset{fast}{\rightleftharpoons}} Me{-}C(=\overset{+}{O}H){-}O{-}Bu^t \underset{fast}{\overset{slow}{\rightleftharpoons}} MeC(OH)(=O) + \overset{+}{Bu^t}$$

$$\overset{+}{Bu^t} + H_2\dot{O} \underset{fast}{\overset{fast}{\rightleftharpoons}} Bu^t\dot{O}H_2^+ \underset{slow}{\overset{fast}{\rightleftharpoons}} Bu^t\dot{O}H + H^+$$

The $A_{AL}1$ mechanism is common for t-alcohols; these from relatively stable carbonium ions (*cf.* the $B_{AL}1$ mechanism).

2. Esters are converted into alcohols by the Bouveault-Blanc reduction, catalytic hydrogenation, and by metallic hydrides.

3. Esters react with ammonia to form amides. This reaction is example of *ammonolysis* (which means, literally, splitting by ammonia):

$$R^1COOR^2 + NH_3 \longrightarrow R^1CONH_2 + R^2OH\ (g.)$$

With hydrazine, esters form acid hydrazides:

$$R^1COOR^2 + H_2NNH_2 \longrightarrow R^1CONHNH_2 + R^2OH$$

Esters react with phosphorus pentachloride or thionyl chloride to form acyl chlorides, *e.g.*,

$$R^1CO_2R^2 + PCl_5 \longrightarrow R^1COCl + POCl_3 + R^2Cl$$

4. By means of *alcoholysis* (splitting by alcohol), an alcohol residue in an ester can be replaced by another alcohol residue. Alcoholysis is carried out by refluxing the ester with a large excess of alcohol, preferably in the presence of a small amount of acid or sodium alkoxide as catalyst.

$$CH_3CO_2C_4H_9 + C_2H_5OH \xrightleftharpoons{C_2H_5ONa} CH_3CO_2C_2H_5 + C_4H_9OH$$

Alcoholysis of esters is also known as *transesterification.*

The Mechanism of alcoholysis is uncertain; a possibility is (in basic media):

$$R^1\overset{O}{\overset{\|}{C}}(OR^2) + EtO^- \rightleftharpoons R^1\overset{O^-}{\overset{|}{C}}(OR^2)—OEt \rightleftharpoons R^1\overset{O}{\overset{\|}{C}}—OEt + R^2O^-$$

$$R^2O^- + EtOH \rightleftharpoons R^2OH + EtO^-$$

In *acidolysis,* the acid residue is displaced from its ester by another acid residue, *e.g.*,

$$CH_3CO_2C_2H_5 + C_5H_{11}CO_2H \rightleftharpoons C_5H_{11}CO_2C_2H_5 + CH_3CO_2H$$

Acidolysis is a useful reaction for converting the natural ester of a dibasic acid into its acid ester.

According to Parker *et al.* (1968), who used ^{18}O-labelled toluic acids and their esters, the mechanism of acidolysis is:

$$R^1—\overset{O}{\overset{\|}{C}}—OH + R^2—\overset{O}{\overset{\|}{C}}—OR^3 \xrightleftharpoons{H^+} R^1—C(=O)—O—C(—OR^3)(R^2)—OH \rightleftharpoons R^1—C(OR^3)=O + HO—C(=O)—R^2$$

5. When an ester – preferably the methyl or ethyl ester – is treated with sodium in an inert solvent, *e.g.*, ether, benzene or toluene, and subsequently with acid, an **acyloin** is formed (50–70 per cent yield). It is important that the reaction be carried out *in the absence of any free alcohol*. Acyloins are α,β-keto-alcohols, and according to Kharasch and his co-workers (1939), their formation takes place via a free-redical mechanism, *e.g.*, propionin from ethyl propionate:

Acyloins are reduced by the Clemmensen method to alkanes.

6. Carboxylic esters which contain α-hydrogen atoms react with sodamide in liquid ammonia solution to form the acid amide and a

condensation product involving two molecules of the ester, *e.g.*, ethyl acetate gives acetamide and acetoacetic ester :

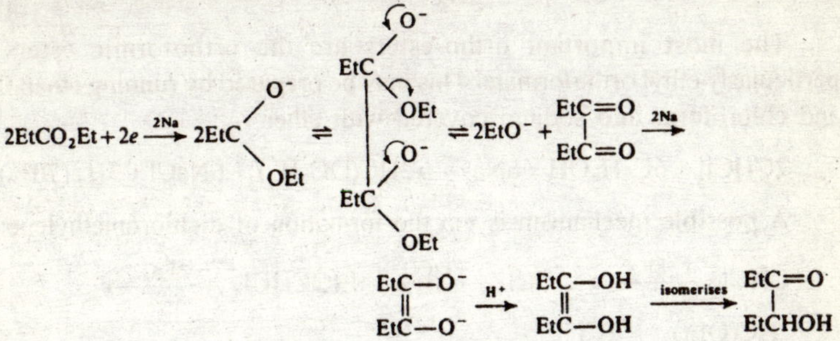

$$2EtCO_2Et + 2e \xrightarrow{2Na} 2EtC\begin{matrix} O^- \\ OEt \end{matrix} \rightleftharpoons \begin{matrix} EtC(O^-)(OEt) \\ | \\ EtC(O^-)(OEt) \end{matrix} \rightleftharpoons 2EtO^- + \begin{matrix} EtC{=}O \\ | \\ EtC{=}O \end{matrix} \xrightarrow{2Na}$$

$$\begin{matrix} EtC{-}O^- \\ \| \\ EtC{-}O^- \end{matrix} \xrightarrow{H^+} \begin{matrix} EtC{-}OH \\ \| \\ EtC{-}OH \end{matrix} \xrightarrow{\text{isomerises}} \begin{matrix} EtC{=}O \\ | \\ EtCHOH \end{matrix}$$

Fig. 2.7

$$CH_3CO_2C_2H_5 + NaNH_2 \longrightarrow CH_3CONH_2 + C_2H_5ONa$$

$$CH_3CO_2C_2H_5 + NaNH_2 \longrightarrow NH_3 + NH_3 + [\ddot{C}H_2CO_2C_2H_5]^- Na^+$$

$$\xrightarrow{CH_3CO_2C_2H_5} CH_3COCH_2CO_2C_2H_5 + C2H_5OH$$

Of particular interest is the *carbonation* of esters, *i.e.*, the introduction of the carboxyl group.

$$R^1CH_2CO_2R^2 + NaNH_2 \xrightarrow{\text{liquid } NH_3}$$

$$NH_3 + [R^1CHCO_2R^2]^- Na^+ \xrightarrow[\text{ether}]{CO_2} \begin{matrix} R^1CHCO_2R^2 \\ | \\ CO_2Na \end{matrix}$$

The yields of these *malonic acid derivatives* are 54–60 per cent with acetates, and becomes progressively lower as the molecular weight of he acid increases.

7. The pyrolysis of esters, particularly acetates, gives alkenes.

Esters are used as solvents for cellulose, oils, gums, resins, etc., and as plasticisers. They are also used for making artificial flavours and essences, *e.g.*, isoamyl acetate – banana oil; amylbutyrate–apricot; isoamyl isovalerate–apple; methyl butyrate–banana oil; amyl butyrate–apicot; isoamyl isovalerate–apple; methyl butyrate–pineapple; etc.

Ortho-esters are compounds of the type $R^1C(OR^2)_3$. They are derived from the ortho-acids, $RC(OH)_3$, which have not yet been isolated, but which are possibly present in aqueous solution:

$$R—C(=O)OH + H_2O \quad RC(OH)_3$$

The most important ortho-esters are the orthoformic esters, particularly ethyl orthoformate. This may be prepared by running ethanol and chloroform into sodium covered with ether:

$$2CHCl_3 + 6C_2H_5OH + 6Na \longrightarrow 2HC(OC_2H_5)_3 + 6NaCl + 3H_3\ (70\%)$$

A possible mechanism is via the formation of dichloromethylene:

$$CHCl_3 \xrightarrow{EtO^-} :CCl_2 \xrightarrow{EtOH} EtOCHCl_2 \xrightarrow{2EtO^-}$$
$$HC(OEt)_3 + 2Cl^-$$

Ethyl orthoformate may be used for preparing ketals and for preparing aldehydes by means of a Grignard reagent.

Esters of the Inorganic Acids

Alkyl sulphates, $(RO)_2SO_2$. Only methanol and ethanol give a good yield of the alkyl sulphate by reaction with concentrated sulphuric acid; the higher alochols give mainly alkenes and ethers. On the other hand, all the alcohols give a fair yield of alkyl hydrogen sulphate when a mixture of alcohol and concentrated sulphuric acid is heated on a steam bath.

Methyl sulphate is prepared industrially:

(*i*) By heating methanol with concentrated sulphuric acid, and then distilling the methyl hydrogen sulphate under reduced pressure:

$$CH_3OH + H_2SO_4 \longrightarrow H_2O + CH_3OSO_3H \xrightarrow{2} (CH_3O)_2SO_2 + H_2SO_4$$

(*ii*) By treating methanol with sulphur trioxide at low temperatures:

$$2SO_3 + 2CH_3OH \longrightarrow (CH_3)_2SO_4 + H_2SO_4$$

Ethyl sulphate may be prepared by the same methods as methyl sulphate, but in addition, there is the industrial preparation by passing ethylene in excess into cold concentrated sulphuric acid:

$$2C_2H_4 + H_2SO_4 \longrightarrow (C_2H_5)_2SO_4$$

Sulphates of the higher alcohols are conveniently prepared by the oxidation of sulphites as follows:

$$2ROH_4 + SOCl_2 \longrightarrow 2HCl + (RO)_2SO \xrightarrow{KMnO_4} (RO)_2SO_2$$

Methyl sulphate, b.p. 188°C, and ethyl sulphate, b.p. 208°C, are heavy poisonous liquids. They are largely used as alkylating agents since the alkyl group will replace the hydrogen atom of the groups —OH, —NH— or —SH. The alkylation may be carried out by treating the compound with the alkyl sulphate in sodium hydroxide solution. Usually only one of the alkyl groups takes part in the reaction, *e.g.*, methylation of a primary amine:

$$RNH_2 + (CH_3)_2SO_4 + NaOH \longrightarrow RNHCH_3 + CH_3NaSO_4 + H_2O$$

The methyl and ethyl esters of the carboxylic acids maybe conveniently prepared by treating the sodium salt of the acid with respectively methyl or ethyl sulphate:

$$RCO_2Na + (C_2H_5)_2SO_4 \longrightarrow RCO_2C_2H_5 + C_2H_5NaSO_4$$

The sodium alkyl sulphates of the higher alcohols are used as detergents, *e.g.*, sodium lauryl sulphate:

Alkyl nitrates, $RONO_2$. The only important alkyl nitrate is ethyl nitrate, $C_2H_5ONO_2$. This may be prepared by heating ethyl iodide with silver nitrate in ethanolic solution:

$$C2H_5I + AgNO3 \longrightarrow C_2H_5ONO_2 + AgI$$

When concentrated nitric acid is added to ethanol, the reaction is usually violent; part of the ethanol is oxidised, and some of the nitric acid is reduced to nitrous acid.

Apparently it is the presence of the nitrous acid which produces the violent oxidation of the alcohol by the nitric acid. This danger may be avoided by first boiling the nitric acid with urea, which destroys any nitrous acid present, and then adding this mixture to cool ethanol, any nitrous acid produced being immediately destroyed by the urea:

$$CO(NH_2)_2 + 2HNO_2 \longrightarrow CO_2 + 2N_2 + 3H_2O$$

Ethyl nitrate is a pleasant-smelling liquid, b.p. 87.5°C. When reduced with tin and hydrochloric acid, it forms hydroxylamine and ethanol:

$$C_2H_2ONO_2 \xrightarrow{Sn/HCl} C_2H_5OH + NH_2OH + H_2O$$

Alkyl nitrites, RONO. Alkyl nitrites are isomeric with the nitroalkanes. The only important alkyl nitrites are the ethyl and amyl nitrites; the latter is actually mainly isoamyl nitrite, since the amyl alcohol used is the isopentanol from fusel oil. Ethyl and amyl nitrites are prepared by adding concentrated hydrochloric acid or sulphuric acid to aqueous sodium nitrite and the alcohol:

$$R—O—H + NO^+ \longrightarrow R—O\begin{matrix}H \\ NO\end{matrix} \xrightarrow{-H^+} RONO$$

Ethyl nitrite, p.b. 17°C, and amyl nitrite, p.b. 99°C, are pleasant-smelling liquids, and are used as a means of preparing nitrous acid in anhydrous media; *e.g.*, an ethanolic solution of nitrous acid may be prepared by passing dry hydrogen chloride into amyl nitrite dissolved in ethanol.

ACID ANHYDRIDES

The acid anhydrides may be regarded theoretically as being derived from an acid by the removal of one molecule of water form two molecules of the acid:

$$2RCO_2H \xrightarrow{-H_2O} R\overset{O}{\overset{\|}{C}}O\overset{O}{\overset{\|}{C}}R$$

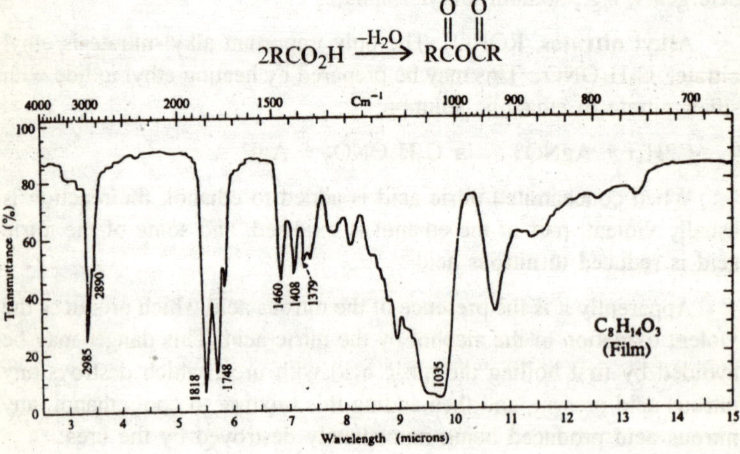

Fig. 2.8

1818 and 1748 cm^{-1}: C=O (str.) in acyclic anhydrides. 1035^{-1}: C—O (str.) in —C—C— (acyclic anhydrides). 2985 and 2890 cm^{-1}: C—H (str.) in Me and/or CH_2. 1460 and 1408 cm^{-1}: C—H (def.) in Me and/or CH_2. 1379 cm^{-1}: C—H (def.) in Me.

It is possible in practice to prepare many acid anhydrides by direct dehydration, but this method is usually confined to the anhydrides of the higher members of the acid series.

The infra-red absorption regions of the carbonyl group (stretch) in acyclic anhydrides are 1840–1800 (s) and 1780–1740 cm^{-1} (s). There is also a band in the region 11801030 cm^{-1} (s) due to the C—O (stretch) in the grouping —C—O—C—. Figure 2.8 is the infra-red spectrum of n-butyric anhydride (as a film).

Nomenclature

The acid anhydrides are named as the anhydride of the acid groups present, the *trivial* name of the acid being used, *e.g.*,

$(CH_3CO)_2O$ acetic anhydride

$(CH_3CH_2CO)_2O$ propionic anhydride

Acetic anhydride may be conveniently prepared by distilling a mixture of anhydrous sodium acetate and acetyl chloride:

$$CH_3\overset{O}{\overset{\|}{C}}Cl + CH_3\overset{O}{\overset{\|}{C}}—O^- \rightleftharpoons CH_3\overset{O^-}{C}(Cl)O\overset{O}{\overset{\|}{C}}CH_3 \longrightarrow CH_3\overset{O}{\overset{\|}{C}}O\overset{O}{\overset{\|}{C}}CH_3 + Cl^-$$

Aceti anhydride is prepared industrially in a number of ways, *e.g.*,

(*i*) By passing acetylene into glacial acetic acid in the presence of mercuric ions as catalyst, and distilling the resulting ethylidene diacetate:

$$C_2H_2 + 2CH_3CO_2H \xrightarrow{Hg^{2+}} CH_3CH(OCOCH_3)_2 \xrightarrow{heat}$$
$$(CH_3CO)_2O + CH_3CHO$$

(*ii*) By passing keten into glacial acetic acid:

$$CH_2{=}C{=}O + CH_3CO_2H \longrightarrow (CH_3CO)_2O$$

Anhydrides of the higher acids may be prepared by heating, and then fractionating, a mixture of the acid and acetic anhydride:

$$2RCO_2H + (CH_3CO)_2O \rightleftharpoons (RCO)_2O + 2CH_3CO_2H$$

This method is only satisfactory for anhydrides which have higher boiling points than acetic acid. A much better method of preparation is to the acid chloride with pyridine and benzene, and the acid, and then heat:

$$RCOCl + RCO_2H + C_5H_5N \longrightarrow (RCO)_2O + C_5H_5\overset{+}{N}HCl^- \quad (g.–ex.)$$

Suschitzky *et al.* (1964) have shown that acid anhydrides which are unaffected by cold water may be readily prepared in high yield and purity by shaking an aqueous solution of a metal salt of the carboxylic acid with an acyl halide at room temperature in the presence of a t-amine, *e.g.*, pyridine; *e.g.*, heptanoic anhydride:

$$CH_3(CH_2)_5CO_2Na + CH_3(CH_2)_5COCl \xrightarrow{\text{pyridine}} [CH_3(CH_2)_5CO]_2O$$

Properties of Acetic Anhydride

Acetic anhydride is a colourless liquid, b.p. 139.5°C, with an irritating smell. It is neutral when pure, and is slightly soluble in water, but readily soluble in ether and benzene. It is hydrolysed slowly in water, but rapidly by alkali:

$$(CH_3CO)_2O + H_2O \longrightarrow 2CH_3CO_2H$$

In undergoes reactions similar to those of acetyl chloride, but with less vigour; only half of the acetic anhydride molecule is used in acetylation, the other half being converted into acetic acid:

$$(CH_3CO)_2O + C_2H_5OH \longrightarrow CH_3CO_2C_2H_5 + CH_3CO_2H$$

$$(CH_3CO)_2O + NH_3 \longrightarrow CH_3CONH_2 + CH_3CO_2H$$

Acetylation with aceti anhydride is usually best carried out in pyridine solution or in the presence of a small amount of sodium acetate or concentrated sulphuric acid as catalyst. Acetic anhydride reacts with dry hydrogen chloride to form acetyl chloride:

$$(CH_3CO)_2O + HCl \longrightarrow CH_3COCl + CH_3CO_2H$$

It is readily halogenated, and may be used in the Friedel–Crafts reaction. It reacts with aldehydes to form alkylidene diacetates; *e.g.*, with acetaldehyde it forms ethylidene diacetate:

$$(CH_3CO)_2O + CH_3CHO \longrightarrow CH_3CH(OCOCH_3)2$$

Acetic anhydride is reduced to ethanol by some metallic hydrides.

Acetic anhydride reacts with nitrogen pentoxide to form **acetyl nitrate**, b.p. 221°C/70 mm, which explodes violently if heated suddenly. It is hydrolysed by water to acetic and nitric acid:

$$CH_3COONO_2 + H_2O \longrightarrow CH_3CO_2H + HNO_3$$

Acetyl nitrate is very useful for preparing certain aromatic nitro-compounds, but is dangerous to handle.

Acylation of a hydroxy-compound to give an ester is generally a better procedure than direct esterification of the acid. It has already been pointed out that sterically hindered acids are difficult to esterify, but Parish *et al.* (1965) have shown that such acids may be readily converted into their esters by dissolving the aid in trifluoroacetic anhydride and adding the hydroxy-compound (yield: up to 97 per cent). According to the authors, the active agent is the mixed anhydride, $RCOOCOCF_3$, that is formed.

Acetyl peroxide, $CH_3COOOCOCH_3$, may be prepared by the action of barium peroxide on acetic anhydride. It is colourless, pungent-smelling liquid, p.b. 63°C/21 mm, and tends to explode on warming. It is powerful oxidising agent.

Propionic anhydride, p.b. 169°C, is manufactured, *e.g.*,

$$CH_3CH_2CO_2H + C_2H_4 + CO \xrightarrow[\text{pressure}]{\text{heat;}} (CH_3CH_2CO)_2O$$

It is mainly used for the preparation of cellulose esters.

Formic anhydride may be prepared (about 90 per cent purity) by the distillation of a mixture of diethoxymethyl formate and n-butyric formic anhydride (Schijf *et al.*, 1965).

$$HCO_2CH(OEt)_2 + PrCOOOCH \longrightarrow (HCO)_2O + PrCO_2CH(OEt)_2$$

Mixed anhydrides containing the formic acid group have been known for some time and may be prepared as follows, *e.g.*,

$$CH_3COCl + HCO_2Na \longrightarrow CH_3COOOCH + NaCl$$

This behaves as a *formylating* reagent.

ACID AMIDES

Acid amides are compounds in which the hydroxyl of the carboxyl group has been replaced by the amino-group, NH_2, to form the *amido-group*, $—CONH_2$.

There are three classes of amides: primary, $RCONH_2$; secondary, $(RCO)_2NH$; and tertiary, $(RCO)_2N$. Only the primary amides are important, and it an be seen from their formulae that all three classes may be regarded as the acyl derivatives of ammonia.

Nomenclature

According to the I.U.P.A.C. system of nomenclature, the suffix -*oic* of the parent acid is replaced by *amide,* but they are commonly named by replacing the suffix *-ic* of the trivial name of the parent acid by *amide*.

$HCONH_2$ methanamide; formamide

CH3CONH$_2$ ethanamide *or* methylcarbonamide; acetamide

General Methods of Preparation of the Amides

1. By heating the ammonium salt of the acid:

$$RCOONH_4 \longrightarrow RCONH_2 + H_2O \quad (g.\text{–}v.g.)$$

Since the ammonium salts tend to dissociate on heating, the reaction is best carried out in the presence of some free acid, which represses the hydrolysis and the dissociation of the ammonium salt.

Amides may also be prepared by heating an acid with urea (Cherbuliez *et al.*, 1946):

$$RCO_2H + CO(NH_2)_2 \longrightarrow RCONH_2 + CO_2 + NH_3 \quad (g.\text{–}v.g.)$$

2. By *ammonolysis*, *i.e.*, the action of concentrated ammonia solution on acid chlorides, acid anhydrides or esters:

$$RCOCl + 2NH_3 \longrightarrow RCONH_2 + NH_4Cl \quad (v.g.)$$

$$(RCO)_2O + 2NH_3 \longrightarrow RCONH_2 + RCO_2NH_4 \quad (v.g.)$$

$$R^1COOR_2 + NH_3 \longrightarrow R^1CONH_2 + R_2OH \quad (g.)$$

The mechanisms of these reactions are uncertain, but a possibility is:

$$R-\overset{\overset{\displaystyle O}{\|}}{\underset{\underset{\displaystyle Z}{|}}{C}}-:NH_3 \rightleftharpoons R-\overset{\overset{\displaystyle O^-}{|}}{\underset{\underset{\displaystyle Z}{|}}{C}}-\overset{+}{N}H_3 \xrightarrow{NH_3} R-\overset{\overset{\displaystyle O^-}{|}}{\underset{\underset{\displaystyle Z}{|}}{C}}-NH_2 + NH_4^+$$

$$\longrightarrow R-\overset{\overset{\displaystyle O}{\|}}{C}-NH_3 + ZH + NH_3$$

N-substituted amides may be prepared by using primary or secondary amines instead of ammonia, *e.g.*,

$$R^1COCl + R^2NH_2 \longrightarrow R^1CONHR^2 + HCl$$

3. By the *graded* hydrolysis of alkyl cyanides. This may be carried out satisfactorily by dissolving the alkyl cyanide in concentrated sulphuric acid and then pouring the solution into cold water, or by shaking the alkyl cyanide with cold concentrated hydrochloric acid:

$$RC{\equiv}N + H_2O \xrightarrow{HCl} RCONH_2 \qquad (g.)$$

Synder *et al.* (1954) have shown that the hydrolysis of cyanides with polyphosphoric acid gives very good yields of amide.

A possible mechanism for this hydrolysis is:

$$RC{\equiv}N: + \overset{+}{H} \longrightarrow RC{\equiv}\overset{+}{N}{-}H \longleftrightarrow R\overset{+}{C}{=}\ddot{N}{-}H \xrightarrow{H_2O} RC(\overset{+}{O}H_2){=}\ddot{N}\ H$$

$$\xrightarrow{-H^+} RC(OH){=}\ddot{N}H \rightleftharpoons RC(=O){-}NH_2$$

Sometimes the conversion of an alkyl cyanide into the amide may be effected by means of alkaline hydrogen peroxide. A possible mechanism is (Wiberg 1953):

$$HOOH + OH^- \longrightarrow H_2O + {}^-OOH$$

$$RC{\equiv}N + {}^-OOH \longrightarrow RC(OOH){=}N^- \xrightarrow{H_2O} RC(O{-}OH){=}NH \xrightarrow{H_2O_2} RCONH_2 + O_2 + H_2O$$

Other methods for preparing amides are the Ritter reaction, the action of ammonia on diazoketones, the Willgerodt reaction, and the Beckmann rearrangement.

General Properties and Reaction of the Amides

Except for formamide, which is a liquid, all the amides are colourless crystalline solids, and those of low molecular weight are soluble in water (with which they can from hydrogen bonds). The lower amides have much higher melting points and boiling points than arc to be expected from their molecular weights; this indicates association (through hydrogen bonding):

$$H_2N{-}CR{=}O\cdots H{-}NH{-}CR{=}O\cdots H{-}NH{-}CR{=}O\cdots$$

This association has been demonstrated from infra-red spectroscopic studies on amides in solution and in the solid state (Richards *et al.*, 1947).

Infra-red spectroscopic evidence has also indicated that amides are resonance hybrids, and this is supported by N.M.R. studies, *e.g.*, the two methyl groups in dimethylformaide give two separate signals at room temperature. This can be explained on the basis that, because of the partial double bond

$$\mathrm{Me_2\ddot{N}{-}CH{=}O \longleftrightarrow Me_2\overset{+}{N}{=}CH{-}O^-}$$

character of the C—N bond, the two methyl groups have *different* environments, one methyl group being *cis* and the other *trans* to the hydrogen atom. At room temperature the rate of rotation about the C—N bond is slow enough for the two methyl groups to be non-equivalent. As the temperature is raised, the signals broaden and finally collapse to a single line.

At these higher temperatures, the molecule has absorbed sufficient energy to overcome the energy barrier to rotation to make the average environment of the two methyl groups the same *i.e.*, the two methyl groups are *now* equivalent.

The infra-red absorption regions of the carbonyl group (stretch) in primary amides are near 1690 (s) and 1650 cm^{-1}(s), and those of the free NH group (stretch) are near 3500 (s) and 3400 cm^{-1} (s). For a hydrogen-bonded NH group, the regions are near 3350 (m) and 3200 cm^{-1} (m).

Dimethylformamide, p.b. 152°C, is a very good solvent for polar and non-polar compounds, and is used as s formylating agent in conjunction with phosphoryl chloride.

It is manufactured by reaction between methyl formate and dimethylamine:

$$HCO_2Me + Me_2NH \longrightarrow HCONMe_2 + MeOH$$

1. Amides are hydrolysed slowly by water, rapidly by acids, and far more rapidly by alkalis:

$$RCONH_2 + H_2O \longrightarrow RCO_2H + NH_3$$

Possible mechanisms are:

Base hydrolysis

$$H\bar{O} + R{-}\overset{O}{\overset{\|}{C}}{-}NH_2 \rightleftharpoons HO{-}\overset{\bar{O}}{\underset{R}{C}}{-}NH_2 \rightleftharpoons HO{-}\overset{O}{\overset{\|}{\underset{R}{C}}} + \bar{N}H_2 \longrightarrow \bar{O}{-}\overset{O}{\overset{\|}{\underset{R}{C}}}{-}NH_3$$

Acid hydrolysis

$$\underset{R}{\overset{O}{\overset{\|}{C}}}-NH_2 + \overset{+}{H} \rightleftharpoons \underset{R}{\overset{O}{\overset{\|}{C}}}-\overset{+}{N}H_3 \xrightleftharpoons{H_2O} H_2\overset{+}{O}-\underset{R}{\overset{O^-}{\overset{|}{C}}}-\overset{+}{N}H_3 \rightleftharpoons$$

$$H_2\overset{+}{O}-\underset{R}{\overset{O}{\overset{\|}{C}}} + NH_3 \longrightarrow HO-\underset{R}{\overset{O}{\overset{\|}{C}}} + NH_4^+$$

2. Amides are very feebly basic and from unstable salts with strong inorganic acids, *e.g.*, $RCONH_2{\cdot}HCl$. The structure of these salts may be (I) or (II):

$$RC(=\overset{+}{O}-H)NH_2 \longleftrightarrow RC(-O-H)=\overset{+}{N}H_2 \quad ; \quad RC(=O)\overset{+}{N}H_3$$

(I) (II)

Kutzennigg *et al.* (1962), from their proton magnetic resonance studies, have provided evidence that the salts of both acetamide and the *N,N*-disubstituted acetamides are protonated on the oxygen atom. This would be in agreement with the fact that amides are resonance hybrids, the O carrying a negative charge and the N a positive charge.

3. Amides are also feebly acidic; *e.g.*, they dissolve mercuric oxide to form covalent mercury compounds:

$$2RCONH_2 + HgO \longrightarrow (RCONH)_2Hg + H_2O$$

When amides are treated with sodium or sodamide in ethereal solution, the sodium *salt* is formed, the structure of which is probably a resonance hybrid (ambident cation):

$$RC(=O)NH^- \longleftrightarrow RC(-O^-)=NH$$

Treatment of the *sodium* salt with iodide gives the *N*-alkyl derivative, whereas with the silver salt, the *O*-alkyl derivative (*imido ester*) is obtained. Imido esters are obtained by alkylation of amides with alkyl sulphates or oxonium salts, *e.g.*, $R_3O^+BF_4^-$

4. Amides are reduced by sodium and ethanol, catalytically, etc.

$$RCONH_2 \xrightarrow{Na/C_2H_5OH} RCH_2NH_2 + H_2O$$

5. When heated with phosphorus pentoxide, or better, with PCl_5, $POCl_3$, or $SOCl_2$, amides are dehydrated to alkyl cyanides:

$$RCONH_2 \xrightarrow{-H_2O} RCN$$

The mechanisms of these dehydrations are not certain, but according to Kirsanov (1954) and Lapidot *et al.* (1962), the mechanism with phosphorus pentachloride is:

$$RCONH_2 \xrightarrow{PCl_5} [RCONHPCl_4] \xrightarrow{-HCl} RC(=O)-N=PCl_3 \longleftrightarrow$$

$$RC(-O^-)=N-\overset{+}{P}Cl_3 \longrightarrow RC\equiv N + POCl_3$$

Alkyl cyanides are also formed when the amides of the higher acids are heated to a high temperature:

$$2RCONH_2 \longrightarrow RCN + RCO_2H + NH_3$$

If this reaction is carried out in the presence of excess of ammonia, then all the amide is converted into alkyl cyanide:

$$RCO_2H + NH_3 \longrightarrow RCO_2NH_4 \xrightarrow{-H_2O} RCONH_2 \xrightarrow{-H_2O} RCN$$

6. When amides are treated with nitrous acid, nitrogen is evolved and the acid is formed:

$$RCONH_2 + HNO_2 \longrightarrow RCO_2H + N_2 + H_2O$$

This reaction is usually slow; a better method is use butyl nitrite:

$$RCONH_2 + BuONO \longrightarrow RCO_2H + N_2 + BuOH$$

7. **Hofmann preparation of primary amines** (1881). This is the conversion of an amide into a primary amine with one carbon atom less by means of bromine (or chlorine) and alkali. The overall equation for the reaction may be written:

$$RCONH_2 + Br_2 + 4KOH \longrightarrow RNH_2 + 2KBr + K_2CO_3 + 2H_2O$$

The mechanism of this reaction has been the subject of a great deal of work. A number of intermediates have been isolated: *N*-bromamides, RCONHBR; salts of these bromamides, $[RCONBr]^-K^+$; isocyanates, RNCO. On this basis, a possible mechanism is the following, which is an example of the 1,2-shift:

The uncertainty of this mechanism lies mainly in the formation of the acyl nitrene. According to Wright (1968), who used ^{14}C- and ^{15}N-

lableed compounds, the Hofmann, Curtius and Lossen rearrangements may not involve nitrenes as intermediates; he has proposed the following *concerted* mechanism:

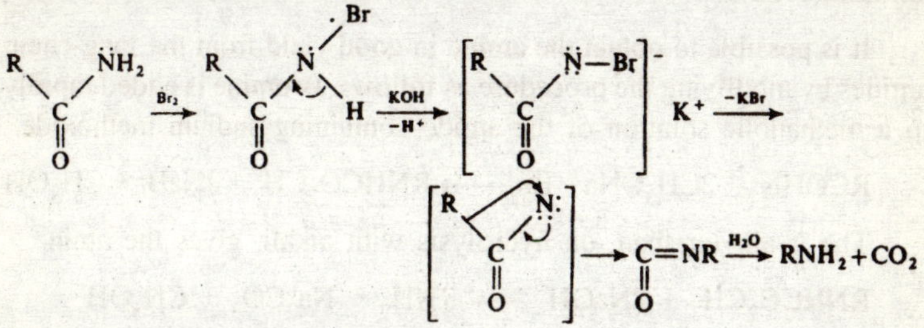

$$RCONH_2 \xrightarrow{Br_2} RCONHBr \xrightarrow[-H^+]{KOH} [RCON^{-}Br]\,K^+ \xrightarrow{-KBr} [RCON:] \longrightarrow C{=}NR \xrightarrow{H_2O} RNH_2 + CO_2$$

(I)

$$[RCON^{-}Br] \xrightarrow{slow} Br^- + R{-}\bar{N}{-}\overset{+}{C}{=}O \longleftrightarrow R{-}\ddot{N}{=}C{=}O$$

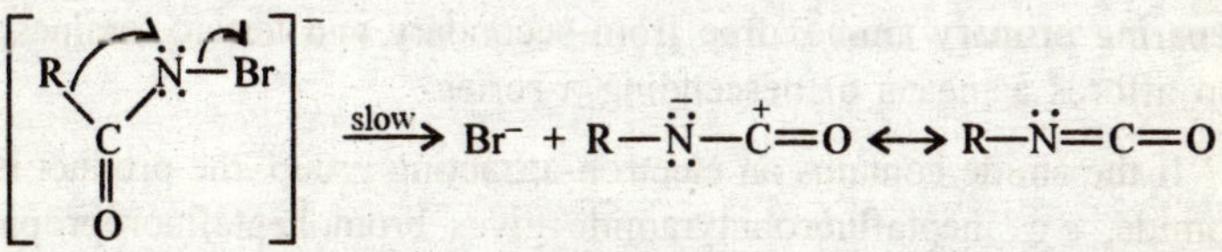

Hauser *et al.* (1937) have shown that electron-releasing groups in the *para*-position of a migrating aryl group accelerate the reaction and electron-withdrawing groups retard the reaction. This supports the formation of a bridged ion (a phenonium ion), (I*a*), where the migrating group is aromatic, and by inference, a bridged ion (I*b*) when the migrating group is

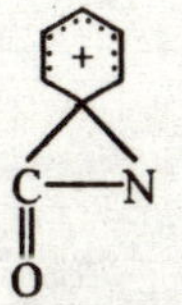

(I*a*) 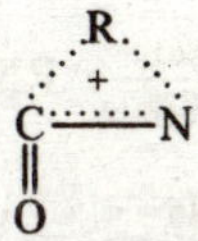 (I*b*)

alkyl. This is further supported by the fact that the rearrangement is intramolecular, *e.g.*, Wallis *et al.* (1931), using the amide of optically active α-benzylpropionic acid, found that there was retention of configuration in the product, α-bonzylethylamine; thus the migrating group is never free:

$$C_6H_5CH_2CHMeCONH_2 \longrightarrow C_6H_5CH_2CHMeNH_3$$

The Hofmann amine preparation can be carried out on all monocarboxylic acids, and the yields are 70–90 per cent for amides containing up to seven carbon atoms; amides with more than seven carbon atoms in the chain give mainly alkyl cyanide:

$$RCH_2CONH_2 \xrightarrow{Br_2KOH} RCN$$

The alkyl cyanide, however, may readily be reduced to the primary amine RCH_2NH_2.

It is possible to obtain the amine in good yield from the long-chain amides by modifying the procedure as follows. Bromine is added rapidly to a methanolic solution of the amide containing sodium methoxide:

$$RCOHN_2 + 2CH_3ONa + Br_2 \longrightarrow RNHCO_2CH_3 + 2NaBr + CH_3OH$$

The *N*-alkylurethan, on hydrolysis with alkali, gives the amine:

$$RNHCO_2CH_3 + 2NaOH \longrightarrow RNH_2 + Na_2CO_3 + CH_3OH$$

The Hofmann amine preparation offers an excellent method of preparing primary amines free from secondary and tertiary amines; it also affords a means of descending a series.

If the amide contains an electron-attracting group, the product is a bromide, *e.g.*, heptafluorobutyramide gives bromoheptafluoropropane (Haszeldine *et al.*, 1956):

$$C_3F_7CONH_2 \xrightarrow[NaOH]{Br_2} C_3F_7Br \ (85\%)$$

Another method for converting primary amides into amines is as follows (Beckwith *et al.*, 1965). On treatment with lead tetra-acetate, the amide gives an acylamine (40–81 per cent); this is readily hydrolysed to the amine:

$$RCONH_2 \xrightarrow[\text{benzene; reflux}]{Pb(OAc)_4} RNHAc$$

Hydroxamic acids

The hydroxamic acids exhibit tautomerism; the keto form (I) is known as the *hydroxamic* form, and the enol form (II) as the *hydroximic* form:

$$\text{(I)} \quad R-\overset{\overset{\displaystyle O}{\|}}{C}-NHOH \rightleftharpoons R-\overset{\overset{\displaystyle O}{|}}{C}=NOH \quad \text{(II)}$$

Hydroxamic acids may be prepared by the action of hydroxylamine on esters or acid chlorides:

$$R-C\begin{matrix}\nearrow O \\ \searrow OC_2H_5\end{matrix} + NH_2OH \longrightarrow R-\overset{\overset{\displaystyle O}{\|}}{C}-NHOH + C_2H_5OH$$

$$RCOCl + NH_2OH \longrightarrow R\text{—}\overset{O}{\overset{\|}{C}}\text{—}NHOH + C_2H_5OH$$

Hydroxamic acids give a coloration with ferric chloride solution, a reaction which is characteristic of enols. When treated with a strong inorganic acid, hydroxamic acids undergo the **Lossen rearrangement** (1875), which results in the formation of a primary amine. The mechanism of the Lossen rearrangement is closely related to that of the Hofmann amine preparation, and so may be formulated:

$$R\text{—}\underset{\|}{\overset{}{C}}(O)\text{—}NH\text{—}OH \xrightarrow{HCl} R\text{—}C(=O)\text{—}NH\text{—}OH_2^+ \xrightarrow{-H_2O} R\text{—}C(=O)\text{—}\overset{+}{N}H \xrightarrow{-H^+}$$

$$R\text{—}C(=O)\text{—}\ddot{N}: \longrightarrow O{=}C{=}NR \xrightarrow{H_2O} RNH_2 + CO_2$$

This rearrangement (1,2-shift) has been shown to be intramolecular, and electron-releasing groups in a migrating aryl group accelerate the reaction, thereby supporting the formation of a phenonium ion.

The amides of the hydroxamic acids are known as **amidoximes**. These are tautomeric substances:

$$RC\begin{matrix} \diagup NH_2 \\ \diagdown\!\!\diagdown NOH \end{matrix} \rightleftharpoons RC\begin{matrix} \diagup\!\!\diagup NH \\ \diagdown NHOH \end{matrix}$$

Aliphatic amidoximes are best obtained by the action of hydroxylamine on an alkyl cyanide:

$$RCN + NH_2OH \longrightarrow RC\begin{matrix} \diagup NH_2 \\ \diagdown\!\!\diagdown NOH \end{matrix}$$

ACID OR ACYL CHLORIDES

The general formula of the *acyl* groups is $R\text{—}\overset{O}{\overset{\|}{C}}\text{—}$. Acid chlorides, which may be prepared by the replacement of the hydroxyl in the carboxyl group by chlorine, are also known as *acyl chlorides* because they contain the acyl group.

Nomenclature : According to the I.U.P.A.C. system the class suffix of the acyl chlorides is *-oyl*; the common names are formed by changing the suffix *-ic* of the *trivial* name of the acid into *-yl, e.g.*,

CH_3COCl — ethanoyl chloride, acetyl chloride

$(CH_3)_2CHCOCl$ — 2-methylpropanoyl chloride, isobutyryl chloride

If the carboxyl group is considered as a substituent, then according to the I.U.P.AC. system, the nomenclature of all substances containing acyl groups is in all cases based on the name 'carbonyl' for the CO group, *e.g.*,

$CH_3CH_2CH_2COCl$ propane-1-carbonyl chloride

When naming amides, the 'yl' is elided before the suffix amide, *e.g.*,

$CH_3CH_2CH_2CONH_2$ propane-1-carbonamide

General Methods of Preparation

1. The acid is heated with phosphorus trichloride or pentachloride, *e.g.*,

$$3RCOOH + PCl_3 \longrightarrow 3RCOCl + H_3PO_3\ (g.)$$

$$RCOOH + PCl_5 \longrightarrow RCOCl + HCl + POCl_3\ (v.g.)$$

The reaction with phosphorus trichloride is accompanied by the formation of small amounts of volatile phosphorus compounds, *e.g.*,

$$RCO_2H + PCl_3 \longrightarrow RCO_2PCl_2 + HCl$$

Thionyl chloride may be used instead of the phosphorus chlorides:

$$ROC_2H + SOCl_2 \longrightarrow RCOCl + SO_2 + HCl\ (v.g.)$$

The inorganic chloride is chosen according to the boiling point of the acyl chloride formed. Phosphorous acid decomposes at 200°C; the boiling point of phosphoryl chloride is 107°C, and that of thionyl chloride is 76°C.

Since acetyl chloride boils at 52°C, any of the three inorganic halides may be used, but it is difficult to separate acetyl chloride from thionyl chloride (which is generally used in excess) by fractionation. n Butyryl chloride boils at 102°C, and so phosphorus pentachoride cannot be used. Usually thionyl chloride is the most convenient.

The mechanism of the formation of acid chlorides is uncertain; it may be similar to that when an alcohol is the substrate.

Acyl halides undergo the following exchange reaction with acids (probably via the acid anhydride):

$$R^1COCl + R^2CO_2H \rightleftharpoons [R^1COOOCR^2 + HCl] \rightleftharpoons R^1CO_2H + R^2COCl$$

This therefore offers a means of preparing acyl halides, but the success depends on displacement of the equilibrium to the right. This is readily achieved by using oxalyl chloride (or bromide) as the acyl halide.

The half-acid chloride of oxalic acid produced is decomposed to carbon monoxide and carbon dioxide under the conditions of the experiment.

2. By distilling the salt of the acid with either phosphorus trichloride, phosphoryl chloride or sulphuryl chloride, *e.g.*,

$$3CH_3CO_2Na + PCl_3 \longrightarrow 3CH_3COCl + Na_3PO_3$$

$$2CH_3CO_2Na + POCl_3 \longrightarrow 2CH_3COCl + NaCl + NaPO_3$$

$$(CH_3CO_2)_2Ca + SO_2Cl_2 \longrightarrow 2CH_3COCl + CaSO_4$$

This method is used industrially since the salts are cheaper than the acid.

Acyl bromides may be prepared by the action of phosphorus tribromide or pentabromide (red phosphorus and bromine) on the acid, or by the action of excess of hydrogen bromide on the acyl chloride:

$$RCOCl + HBr \longrightarrow RCOBr + HCl$$

Acyl iodides may be prepared by the action of phosphorus tri-iodide on acid anhydrides. *Acyl fluorides* are also known, and may be prepared by the action of hydrogen fluoride on acid anhydrides:

$$(RCO)_2O + HF \longrightarrow RCOF + RCO_2H$$

General properties and reactions of the acyl chlorides. The lower acyl chlorides are colourless liquids with irritating odours; the higher members are colourless solids. The chlorine atom is very reactive and so the acid chlorides are important reagents.

The infra-red absorption region of the carbonyl group (stretch) in acyl chlorides is 1815–1785 cm^{-1}(s).

1. The acyl chlorides are readily hydrolysed by water, the lower members reacting vigorously:

$$RCOCl + H_2O \longrightarrow RCO_2H + HCl$$

Acyl chlorides usually react rapidly with compounds containing 'active' hydrogen atoms, *i.e.* hydrogen attached to oxygen, nitrogen or sulphur; *e.g.*, esters are formed with alcohols:

$$R^1COCl + R^2OH \longrightarrow R^1COOR^2 + HCl$$

Amides are formed with ammonia, and *N*-substituted amides with primary and secondary amines:

$$RCOCl + 2NH_3 \longrightarrow RCONH_2 + NH_4Cl$$

$$R^1COCl + R^2NH_2 \longrightarrow R^1CONHR^2 + HCl$$

Hydrazides are formed with hydrazine, and hydroxamic acids with hydroxylamine:

$$RCOCl + H_2NNH_2 \longrightarrow RCONHNH_2 + HCl$$

$$RCOCl + NH_2OH \longrightarrow RCONHOH + HCl$$

The mechanisms of these reactions are uncertain; a possibility is:

$$R^1{-}\overset{\overset{\displaystyle O}{\|}}{C}{-}Cl + R^2OH \xrightleftharpoons{\text{slow}} R^1{-}\underset{\underset{\displaystyle H{-}O^+{-}R^2}{|}}{\overset{\overset{\displaystyle O^-}{|}}{C}}{-}Cl \xrightarrow{\text{fast}} R^1{-}\underset{\underset{\displaystyle OR^2}{|}}{\overset{\overset{\displaystyle O}{\|}}{C}} + HCl$$

2. Acyl chlorides may be reduced catalytically to aldehydes or to alcohols:

$$RCOCl \xrightarrow[Pd]{H_2} RCHO \xrightarrow{H_2} RCH_2OH$$

Acid chlorides are reduced to esters of an enediol by sodium amalgam in an inert solvent, *e.g.*, ether :

$$4RCOCl + 4Na \longrightarrow \begin{matrix} R \quad\quad R \\ C{=}C \\ RCOO \quad\quad OOCR \end{matrix} + 4NaCl$$

Under suitable conditions, small yields of α-diketones have been obtained. Acid chlorides are reduced to alcohols by lithium aluminium hydride or sodium borohydride, and to aldehydes by lithium hydride or lithium tri-t-butoxyaluminium hydride.

3. Acyl chlorides react with salts of the acids to form *acid anhydrides* (*q.v.*):

$$R^1COCl + R^2COONa \longrightarrow R^1CO{\cdot}O{\cdot}OCR^2 + NaCl$$

4. Acyl chlorides may be used in the Friedel–Crafts reaction to produce an aromatic ketone: *e.g.*, acetyl chloride reacts with benzene in the presence of anhydrous aluminium chloride to form acetophenone:

$$C_6H_6 + CH_3COCl \xrightarrow{AlCl_3} C_6H_5COCH_3 + HCl$$

5. Acyl chlorides react with Grignard reagents to produce ketones or tertiary alcohols, according to the conditions; *e.g.*, acetyl chloride forms butanone with ethylmagnesium iodide:

$$CH_3COCl + C_2H_5MgI \longrightarrow CH_3COC_2H_5 + MgClI$$

6. Acyl chlorides are readily halogenated in the α-position.

7. Acyl chlorides from esters (and other products) when heated with an ether in the presence of anhydrous zinc chloride as catalyst:

$$R^1COCl + R_2{}^2O \xrightarrow{ZnCl_2} R^1CO_2R^2 + R^2Cl$$

8. Acyl chlorides add on to the double bond of an alkene in the presence of a catalyst, *e.g.*, zinc chloride or aluminium chloride, to form a chloroketone which, on heating, eliminates a molecule of hydrogen chloride to form an unsaturated ketone:

$$(CH_3)_2C{=}CH_2 + CH_3COCl \xrightarrow{ZnCl_2} (CH_3)_2CClCH_2COCH_3$$
$$\xrightarrow{heat} (CH_3)_2C{=}CHCOCH_2 + HCl$$

Formyl fluoride, HCOF, has been isolated by Olah *et al.* (1961). Formyl chloride, HCOCl, has been obtained in solution by the action of hydrogen chloride on a solution of 1-formylimidazole in chloroform at – 10°C (Staab *et al.*, 1964). A mixture of carbon monoxide and hydrogen chloride behaves as if it were formyl chloride in the Gattermann–Koch aldehyde synthesis.

Acetyl chloride is the most important acyl halide. It is a colourless fuming liquid, b.p. 52°C, and is soluble in ether and chloroform. It is largely used as an *acetylating* agent, *i.e.*, as a means of introducing the acetyl group, into compounds containing 'active' hydrogen atoms. It is also used to detect the presence of hydroxyl groups in organic compounds, and to estimate their number.

The accepted abbreviation for the acetyl group is Ac; *e.g.*, AcOH is acetic acid; Ac_2O, acetic anhydride; AcCl, acetyl chloride; C_6H_5NHAc, acetanilide; etc.

IMIDIC ESTERS AND AMIDINES

Imidic esters, which are also known as *imino-ethers*, are best prepared by passing dry hydrogen chloride into a solution of an alkyl cyanide in anhydrous alcohol; the imidic ester hydrochloride is slowly deposited (Pinner, 1877):

$$R^1C{\equiv}N + R^2OH + HCl \longrightarrow R^1C(=NH_2^+Cl^-)OR^2$$

Imidic ester hydrochlorides form ortho-esters when allowed to stand in an alcohol, and are readily hydrolysed to esters by water:

$$R^1C(OR^2)_3 + NH_4Cl \xleftarrow{2R^2OH} R^1C(=NH_2^+Cl^-)OR^2 \xrightarrow{H_2O} R^1C(=O)OR^2 + NH_4Cl$$

Imidic ester hydrochlorides are decomposed on heating:

$$R^1C(=NH_2^+Cl^-)OR^2 \longrightarrow R^1CONH_2 + R^2Cl$$

When an ethereal solution of an imidic ester hydrochloride is treated with potassium carbonate, the free ester is obtained. When treated with an ethanolic solution of ammonia, the imidic ester hydrochloride forms the **amidine**:

$$R^1C(=NH_2^+Cl^-)OR^2 + 2NH_3 \longrightarrow R^1C(=NH)NH_2 + NH_4Cl + R^2OH$$

Formamidine acetate and acetamidine acetate may be prepared from ortho-esters (R = H or Me) as follows (Taylor *et al.*, 1960):

$$RC(OEt)_3 + NH_3 + CH_3CO_2H \longrightarrow N_2NCR{=}NH_2^+\ CH_3CO_2^-$$

These acetates are better precursors of their respective amidines than are the corresponding hydrochlorides.

Amidines are strong monoacid bases, forming salts with strong acids. Their basic strength may be explained by resonance. In the amidine, the resonating structures are different and one carries *unlike* charges.

$$R{-}C(=NH)NH_2 \longleftrightarrow R{-}C(=\bar{N}H)\bar{N}H_2$$

In the amidine *ion*, the gain of a proton results in *both* contributing structures being equivalent and carrying only a positive charge.

$$R-C(=\overset{+}{N}H_2)NH_2 \longleftrightarrow R-C(NH_2)=\overset{+}{N}H_2$$

Amidines are readily hydrolysed, hydrolysis being catalysed by acids or alkalis:

$$R-C(=NH)NH_2 \quad H_2O \longrightarrow R-C(=O)NH_2 \quad NH_3 \xrightarrow{H_2O} RCO_2NH_4$$

ACID HYDRAZIDES AND ACID AZIDES

Acid hydrazides may be prepared by the action of hydrazine on esters or acyl chlorides:

$$RCOOC_2H_5 + N_2H_4 \longrightarrow RCONHNH_2 + C_2H_5OH$$

$$RCOCl + N_2H_4 \longrightarrow RCONHNH_2 + HCl$$

Girard's reagent 'T' is trimethylaminoacetohydrazide, which may be prepared by interaction between trimethylamine, ethyl chloroacetate and hydrazine:

$$(CH_3)_3N + ClCH_2CO_2C_2H_5 + N_2H_4 \longrightarrow [(CH_3)_3NCH_2CONHNH_2]^+$$

$$Cl^- + C_2H_5OH$$

Girard's reagent 'P' is formed in a similar manner, except that pyridine is used instead of trimethylamine.

The acid hydrazides resemble the acid amides, but differ in the following ways:

(*i*) They are less readily hydrolysed than the amides:

$$RCONHNH_2 + H_2O \longrightarrow RCO_2H + N_2H_4$$

(*ii*) They are reducing agents–hydrazine is a powerful reducing agent.

(*iii*) They from acid azides when treated with nitrous acid; no nitrogen is evolved:

$$RCONHNH_2 + HNO_2 \longrightarrow RCON_3 + 2H_2O$$

Acid azides may also be prepared by the reaction between an acyl chloride and sodium azide:

$$RCOCl + NaN_3 \longrightarrow RCON_3 + NACl$$

The structure of the acid azides is best represented as a resonance hybrid:

$$R-\overset{O}{\overset{\|}{C}}-\ddot{N}=\overset{+}{N}=\overset{-}{\ddot{N}}: \longleftrightarrow R-\overset{O}{\overset{\|}{C}}-\overset{-}{\ddot{N}}-\overset{+}{N}\equiv N: \longleftrightarrow R-\overset{O^-}{\overset{|}{C}}=\ddot{N}-\overset{+}{N}\equiv N:$$

When boiled with an alcohol, acid azides undergo rearrangement to form *N*-alkyl-substituted urethans:

$$R^1CON_3 + R^2OH \longrightarrow RNHCO_2R^2 + N_2 \text{ (v.g.-ex.)}$$

The mechanism of this rearrangement – the **Curtius rearrangement** (1894)– is similar to that of the Hofmann and Lossen rearrangements, and so may be formulated:

$$\underset{\substack{\|\\ O}}{\overset{R\diagdown}{C}}\!\!\diagup\overset{-}{\ddot{N}}-\overset{+}{N}\equiv N: \xrightarrow{-N_2} \left[\underset{\substack{\|\\ O}}{\overset{R\diagdown}{C}}\!\!\diagup\ddot{N}\right] \longrightarrow R-N=C=O \xrightarrow{ROH} RNHCO_2R$$

This mechanism is supported by the fact that when the terminal nitrogen atom is ^{15}N in 3,5-dinitrobenzazide, all of this tracer is found in the nitrogen eliminated in the reaction (Bothner–By *et al.*, 1951).

The Curtius reaction may be used to step down a series (*cf.* the Hofmann amine preparation):

$$RCO_2H \xrightarrow{C_2H_5OH} RCO_2C_2H_5 \xrightarrow{N_2H_4} RCONHNH_2 \xrightarrow{HNO_2}$$

$$RCON_3 \xrightarrow{CH_3OH} RNHCO_2CH_3 \xrightarrow{NaOH} RNH_3$$

The Curtius reactions offers a very good method for preparing primary amines free from secondary and tertiary amines, and may also be used for preparing isocyanates and urethans; *e.g.*, when heated in benzene or chloroform solution, an acid azide rearranges to the alkyl isocyanate which can be isolated. If the isocyanate is warmed with an alcohol, an *N*-substituted urethan is formed:

$$R^1CON_3 \xrightarrow[\text{solution}]{\text{benzene}} R^1NCO \xrightarrow{R^2OH} R^1NHCO_2R^2$$

HALOGEN DERIVATIVES OF THE MONOCARBOXYLIC ACIDS

The halogen derivatives of the monocarboxylic acids are compounds in which one or more hydrogen atoms in the *carbon chain* have been replaced by halogen.

Nomenclature

The following examples illustrate the methods:

CH_2ClCO_2H	monochloroacetic acid
$CH_3CHClCO_2H$	α-chloropropionic acid or 2-chloropropanoic acid
$(CH_3)_2CClCHBrCO_2H$	α-bromo-β-chloroisovaleric acid or 2-bromo-3-chloro-3-methylbutanoic acid

α-Halogeno-acids. Although the monocarboxylic acids themselves are not readily halogenated, their acid chlorides and anhydrides may be halogenated very easily. Bromination takes place only at the α-position, but chlorination, although it occurs mainly at the α-position, may also take place further in the chain; *e.g.*, chlorination of propionic acid results in the formation of the α- and β-chloro-derivatives (see also below):

$$CH_3CH_2CO_2H \xrightarrow{Cl_2} CH_3CHClCO_2H + ClCH_2CH_2CO_2H$$

Fluorination of the monocarboxylic acids has not yet been studied in great detail, but the work done so far seems to indicate that fluorine enters the chain somewhat indiscriminately; *e.g.*, fluorination of butyric acid in carbon tetrachloride solution gives the β-and γ-fluorc-derivatives.

Iodo-derivatives are prepared from the chloro- or bromo-compound by the action of potassium iodide in methanolic or acetaone solution.

The usual method for preparing α-chloro- or bromo-acids is by the **Hell-Volhard-Zelinsky reaction (H.V.Z. reaction),** which is carried out by treating the acid with chlorine or bromine in the presence of a small amount of red phosphorus.

The mechanism of the H.V.Z. reaction is uncertain, but there is some evidence to show that it may proceed through an enol intermediate (Zimmerman *et al.*, 1959):

$$RCH_2CO_2H \xrightarrow{P + Br_2} RCH_2C(=O)Br \rightleftharpoons RCH{=}C(OH)Br \xrightarrow{Br_2}$$

$$HBr + RCHBrCObr \xrightleftharpoons{RCH_2CO_2H} RCHBrCO_2H + RCH_2COBr, \text{ etc.}$$

Enolisation of the acid bromide is far more rapid than that of the acid.

The second α-hydrogen atom may be replaced by chlorine or bromine by using excess of the halogen, but whereas bromination ceases when both α-hydrogen atoms have been replaced, chlorination proceeds further in the chain (see above). Since the H.V.Z. reaction with bromine is specific for α-hydrogen atoms, it can be used to detect the presence of α-hydrogen in an acid; *e.g.*, trimethylacetic acid does not undergo the H.V.Z reaction with bromine.

The H.V.Z. reaction with bromine is applicable to di- and polycarboxylic acids, all the α-positions being substituted if sufficient bromine is used.

Another convenient method of preparing a-bromo-acids is by brominating an alkylmalonic acid and heating the bromo-acid, whereupon it is decarboxylated to the monobasic acid:

$$RCH(CO_2H)_2 \xrightarrow{Br_2} RCBr(CO_2H)_2 \xrightarrow{heat} RCHBrCO_2H + CO_2$$

Sulphuryl chloride, in the presence of small amount of iodine, chlorinates monocarboxylic acids in the α-position, but in the presence of organic peroxides (to give a free-radical reaction), the α-, β-, γ-, etc. positions are substituted. On the other hand, Louw (1966) has shown that cupric chloride (plus LiCl) chlorinates acid chlorides, anhydrides, and carboxylic acid (containing the acid chloride or anhydride as 'catalyst') at the α-position only (yields > 95 per cent). The reaction is carried out in a polar inert solvent, *e.g.*, sulpholane, and only the α-monochloro derivative is obtained regardless of the number of α-hydrogen atoms in the compound.

β-Halogeno-acids may be prepared by treating in α,β-unsaturated acid with halogen acid; *e.g.*, acrylic acid forms β-bromopropionic acid when treated with hydrogen bromide in acetic acid solution:

$$CH_2{=}CHCO_2H + HBr \longrightarrow BrCH_2CH_2CO_2H$$

β-Halogeno-acids may also be prepared by treating an α,β-unsaturated aldehyde with halogen acid, and oxidising the β-chloroaldehyde produced, *e.g.*, acraldehyde gives b-chloropropionic acid when treated with concentrated hydrochloric acid, and the product then oxidised with concentrated nitric acid:

$$CH_2{=}CHCHO + HCl \longrightarrow ClCH_2CH_2CHO \xrightarrow{[O]} ClCH_2CH_2CO_2H \ (60\text{-}65\%)$$

The addition of all the halogen acids to α,β-unsaturated carbonyl compounds takes place very readily, and in a direction contrary to Markownikoff's rule.

When an alkene cyanohydrin is refluxed with 40 per cent hydrobromic acid, the β-bromoacid is obtained; *e.g.*, ethylene cyanohydrin gives β-bromopropionic acid:

$$CH_2(OH)CH_2CN + 2HBr + H_2O \longrightarrow BrCH_2CH_2CO_2H + NH_4Br \ (82\text{-}83\%)$$

Alternatively, when an alkene halohydrin is oxidised with concentrated nitric acid, the halogeno-acid is produced; *e.g.*, trimethylene chlorohydrin gives β-chloropropionic acid:

$$ClCH_2CH_2CH_2OH \longrightarrow ClCH_2CH_2CO_2H \ (78\text{-}79\%)$$

γ-Halogeno-acids may be prepared by the addition of halogen acid to a β,γ-unsaturated acid; the addition occurs contrary to Markownikoff's rule (*cf.* above); *e.g.*, pent-3-enoic acid gives γ-bromo-n-valeric acid when treated with hydrobromic acid:

$$CH_3CH{=}CHCH_2CO_2H + HBr \longrightarrow CH_3CHBrCH_2CH_2CO_2H$$

γ-Halogeno-acids may also be prepared by heating a γ-hydroxyacid with oncentrated halogen acid solution *e.g.*, γ-hydroxybutyric acid gives γ-chlorobutyric acid when treated with concentrated hydrochloric acid:

$$HOCH_3CH_2CH_2CO_2H + HCl \longrightarrow ClCH_3CH_2CH_2CO_2H + H_2O$$

The methods described above apply to particular types of halogeno-acids. However, the following method may be used for preparing any halogeno-acid; *e.g.*, a δ-halogeno-acid:

$$MeO_2C(CH_2)_4CO_2H \xrightarrow[KOH]{AgNO_3} MeO_2(CH_2)_4CO_2Ag \xrightarrow[CCl_4]{Br_2} MeO_2C(CH_2)_3CH_2Br \ (65\text{-}68\%)$$

Properties of Halogeno-acids

The α-halogeno-acids undergo most of the reactions of the alkyl halides, but the halogen atom in the acid is far more reactive than that in the alkyl halide; the enhanced reactivity is due to the – I effect of the adjacent carbonyl group. The reactions of the carboxyl group are unchanged. β-, γ- and δ-halogeno-acids undergo some of the reactions of the alkyl halides, but tend to eliminate a molecule of hydrogen halide to form an unsaturated acid or a lactone. They do not form Grignard reagents, but the halogeno-acids esters react with Grignard reagents to form halogen-substituted tertiary alcohols. The halogeno-acids are reduced to the corresponding acids by sodium amalgam, but the reduction of acids of the type $R^1R^2CClCO_2H$ with lithium aluminium hydride gives chlorohydrins (I), alcohols (II), aldehydes (III) and glycols (IV).

$$\underset{(I)}{R^1R^2CClCH_2OH} \qquad \underset{(II)}{R^1R^2CHCH_2OH} \qquad \underset{(III)}{R^1R^2CHCHO}$$

$$\underset{(IV)}{R^1R^2C(OH)CH_2OH}$$

The amount of each depends on the nature of R^1 and R^2, *e.g.*, very little glycol is obtained unless at least one of these groups is phenyl (Eliel *et al.*, 1956).

The behaviour of an halogeno-acid with alkali depends on the position of the halogen atom relative to the carboxyl group.

(*a*) α-Halogeno-acids are converted into the corresponding α-hydroxyacid:

$$RCHXCO_2H + H_2O \xrightarrow[\text{(ii) } H^+]{\text{(i) NaOH}} RCH(OH)CO_2H$$

On the other hand, if the α-halogeno-acid ester is heated with a tertiary amine, the α,β-unsaturated acid ester is formed by the elimination of a molecule of hydrogen halide; *e.g.*, ethyl α-bromobutyrate gives ethyl crotonate when heated with dimethylaniline, $C_6H_5N(CH_3)_2$:

$$CH_3CH_2CHBrCO_2C_2H_5 \xrightarrow{-HBr} CH_3CH{=}CHCO_2C_2H_5$$

(*b*) β-Halogeno-acids are converted into the corresponding β-hydroxyacid, which, on continued reflux with alkali, eliminates a molecule of water to form the α,β-unsaturated acid:

$$RCHBrCH_2CO_2H \xrightarrow{NaOH} RCH(OH)CH_2CO_2^- \xrightarrow[-H_2O]{NaOH} RCH{=}CHCO_2^-$$

(*c*) γ- and δ-Halogeno-acids are converted into lactones; *e.g.*, γ-chlorobutyric acid gives γ-butyrolactone:

$$\begin{matrix} CH_2-CH_2 \\ | \qquad\quad | \\ ClCH_2 \quad CO_2H \end{matrix} \xrightarrow{OH^-} \begin{matrix} CH_2-CH_2 \\ | \qquad\quad | \\ Cl-CH_2 \quad CO \\ \quad\; ^-O \end{matrix} \longrightarrow Cl^- + \begin{matrix} CH_2-CH_2 \\ | \qquad\quad | \\ CH_2 \quad CO \\ \diagdown O \diagup \end{matrix}$$

This is an example of neighbouring group participation.

(*d*) ε-Halogeno-acids, etc., give the corresponding hydroxyacid; *e.g.*; ε-bromocaproic acid gives ε-hydroxycaproic acid:

$$BrCH_2CH_2CH_2CH_2CH_2CO_2H \xrightarrow[\text{(ii) } H^+]{\text{(i) NaOH}} HOCH_2CH_2CH_2CH_2CH_2CO_2H$$

The most characteristic reaction of the α-halogeno-acid esters is the *Reformatsky reaction.*

The halogeno-acids are all stronger acids than the parent acid, and for a group of isomeric acids, the further the halogen is removed from the carboxyl group the weaker is the acid. The increase in acid strength may be explained by high electronegativity of the halogen atom which exerts a strong inductive effects, thereby facilitating the release of proton from the carboxyl group.

$$X \leftarrow CH_2 \leftarrow \overset{\overset{\displaystyle O}{\|}}{C} \leftarrow O \leftarrow H \qquad\qquad CH_3 \rightarrow \overset{\overset{\displaystyle O}{\|}}{C} \rightarrow O \rightarrow H$$

$$X_2 \nwarrow\swarrow CH \twoheadleftarrow \overset{\overset{\displaystyle O}{\|}}{C} \twoheadleftarrow O \twoheadleftarrow H \qquad\qquad (CH_3)_2 \searrow\nearrow CH \twoheadrightarrow \overset{\overset{\displaystyle O}{\|}}{C} \twoheadrightarrow O \twoheadrightarrow H$$

$$X_3 \rightarrow C \Lleftarrow \overset{\overset{\displaystyle O}{\|}}{C} \Lleftarrow O \Lleftarrow H \qquad\qquad (CH_3)_3 \rightarrow C \Rrightarrow \overset{\overset{\displaystyle O}{\|}}{C} \Rrightarrow O \Rrightarrow H$$

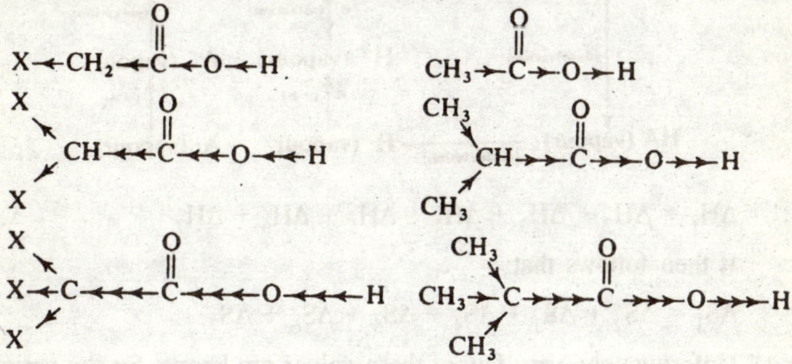

Fig. 2.9

The larger the number of halogen atoms on the α-carbon atom, the stronger is the inductive effect, and consequently the stronger is the acid; also, the further removed the halogen atom is from the carboxyl group, the weaker is the inductive effect at the carboxyl group, and consequently the weaker is the acid. In the same way, since an alkyl group is electron-repelling, increasing their number on the α-carbon atom should *decrease* the strength of the acid since release of the proton will be hindered. Since – I groups in the α-position, *e.g.*, X, NO_2, OH, increase the strength of the acid, and + I groups, *e.g.*, Me, weaken the acid, the measurement of the pK_a's of various substituted acetic acids affords a means of measuring the *relative* inductive effects of different groups.

Relative inductive effects (or electronegativities) may also be measured by NMR spectroscopy. The results, however, must be accepted with reserve, since the applied field itself may affect the electron distribution in the molecule.

To study the strengths of the different members of the carboxylic acids quantitatively, we must use the equations:

(*i*) HA (solvated) $\rightleftharpoons$ H^+ (solvated) + A^- (solvated)

(*ii*) —RT ln K_a = ΔG^θ (ionisation) = ΔH^θ (ionisation − $T\Delta S^\theta$) (ionisation)

This in theory, may be done by using the following thermochemical cycle:

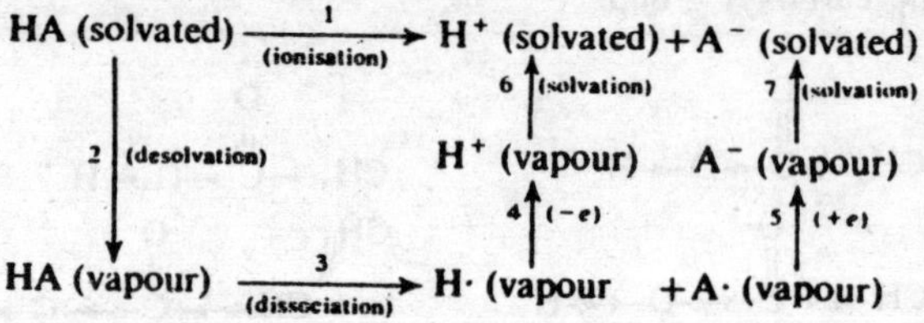

$$\begin{array}{ccccc}
\text{HA (solvated)} & \xrightarrow[\text{(ionisation)}]{1} & \text{H}^+ \text{ (solvated)} & + & \text{A}^- \text{ (solvated)} \\
\Big\downarrow 2 \text{ (desolvation)} & & \Big\uparrow 6 \text{ (solvation)} & & \Big\uparrow 7 \text{ (solvation)} \\
 & & \text{H}^+ \text{ (vapour)} & & \text{A}^- \text{ (vapour)} \\
 & & \Big\uparrow 4 \ (-e) & & \Big\uparrow 5 \ (+e) \\
\text{HA (vapour)} & \xrightarrow[\text{(dissociation)}]{3} & \text{H}\cdot \text{ (vapour} & + & \text{A}\cdot \text{ (vapour)}
\end{array}$$

$$\Delta H_1 = \Delta H_2 + \Delta H_3 + \Delta H_4 + \Delta H_5 + \Delta H_6 + \Delta H_7$$

It then follows that:

$$\Delta S_1 = \Delta S_2 + \Delta S_3 + \Delta S_4 + \Delta S_5 + \Delta S_6 + \Delta S_7$$

Unfortunately, very few of these values are known for the various organic acids. However, it is possible to evaluate ΔH_1 from experimental values of K_a over a range of temperatures, and since ΔG is also, obtained, ΔS_1 can thus be calculated. Some values of ΔH_1 (kJ) and ΔS_1 (JK^{-1}), respectively, are in water: formic aicd, − 0·08368, − 73·64; acetic, − 0·3766, − 92·47; propionic, − 0·6694, − 95·40; chloroacetic, − 4·853, − 71·13.

It can be seen that ΔH_1 is quite small, but ΔS_1 can vary a great deal. Thus the difference in acid strengths must be due to entropy effects. An interesting point to note is the smaller entropy loss for chloroacetic acid makes this a stronger acid than acetic. ΔS_1 for trichloroacetic acid is − 8·368 JK^{-1}, and this acid is a very strong acid. These entropy changes may be explained as follows. Acetic acid is far less polar than the acetate ion, and so the latter is far more solvated than the former. Since solvation produces some sort of arrangement of the solvent molecules around the anion, this loss of randomness of solvent molecules means a decrease in entropy (*i.e.*, ΔS is negative). Because trichloroacetic acid is highly polar, it is highly solvated both in the un-ionised form and in the anion, and so the loss in entropy in this case is very much smaller. It should also be noted that when ionisation occurs, the liberated proton is also solvated. This solvation causes the 'freezing' of four molecules of water (to form $H^+(H_2O)_4$), and so this contributes to the decrease in entropy when the acid ionises. However, since the solvated proton is the same for all acids, only the change in entropy of the anion need be considered.

Chloroacetic acid may be prepared by the H.V.Z. reaction; the reaction can be carried out in direct sunlight, and in the absence of phosphorus. Chloroacetic acid is prepared industrially:

(*i*) by heating trichloroethylene with 74 per cent sulphuric acid:

$$CHCl{=}CCl_2 + 2H_2O \xrightarrow[140°C]{H_2SO_4} ClCH_2CO_2H + 2HCl$$

(ii) By the chlorination of acetic acid in the presence of acetic anhydride-sulphuric acid mixture.

Chloroacetic acid is a deliquescent solid, m.p. 61°C, soluble in water and ethanol. It finds many uses in organic syntheses, and is used in the industrial preparation of indigotin (*q.v.*). It is readily esterified with ethanol in the presence of sulphuric acid.

Ethyl chloroacetate is converted into chloroacetamide when shaken with aqueous ammonia:

$$ClCH_2COOC_2H_5 + NH_3 \longrightarrow ClCH_2CONH_2 + C_2H_5OH \ (78\text{-}84\%)$$

Chloroacetamide is dehydrated to chloroacetonitrile (p.b. 123°C) by phosphorus pentoxide:

$$ClCH_2CONH_2 \xrightarrow[-H_2O]{P205} ClCH_2CN \ (61\text{-}70\%)$$

Dichloroacetic acid may be prepared in the laboratory and industrially by adding calcium carbonate to a warm aqueous solution of chloral hydrate, then adding an aqueous solution of sodium cyanide, and finally heating the mixture:

$$2CCl_3CH(OH)_2 + 2CaCO_3 \xrightarrow{NaCN} (CHCl_2CO_2)_2Ca + 2CO_2 +$$

$$CACl_2 + 2H_2O \xrightarrow{HCl} CHCl_2CO_2H \ (88\text{-}92\%)$$

The action of the sodium cyanide is not understood.

Dichloroacetic acid, p.b. 194°C, gives glyoxylic acid (*q.v.*) on hydrolysis:

$$CHCl_2CO_2H + H_2O \xrightarrow[\text{(ii) } H^+]{\text{(i) NaOH}} OCHCO_2H + 2HCl$$

Vigorous hydrolysis with concentrated alkali gives oxalate and glycollate, due to the glyoxylic acid undergoing he Cannizzaro reaction:

$$2OCHCO_2H + 3NaOH \longrightarrow (CO_2Na)_2 + CH_2(OH)CO_2Na + 2H_2O$$

Trifluoroacetic acid is conveniently prepared by the oxidation of *p*-trifluoromethyltoluidine with chromic acid:

$$CF_3\text{—}C_6H_4\text{—}NH_2 \xrightarrow{[O]} CF_3CO_2H$$

Trifluoroacetic acid is a liquid fumes in the air, and is one of the strongest organic acids known. It does not form fluoroform when heated with alkali, and is reduced to trifluroacetaldehyde by lithium a aluminium hydride. Peroxytrifluoroacetic acid, CF_3CO_3H, is a useful oxidising agent (Emmons, 1954; *cf.* peroxyacetic acid).

Fluoroacetic acid may be prepared by heating a mixture of carbon monoxide, hydrogen fluoride and formaldehyde under pressure.

$$CO + HF + HCHO \longrightarrow FCH_2CO_2H$$

The simplest chloro-acid would be chloroformic acid, $ClCO_2H$, but it is unknown; chlorination of formic acid results in the formation of hydrogen chloride and carbon dioxide:

$$HCO_2H \xrightarrow{Cl_2} 2HCl + CO_2$$

On the other hand, esters of chloroformic acid are known.

Trichloroacetic acid is best prepared by oxidising chloral hydrate with concentrated nitric acid:

$$CCL_3CH(OH)_2 + [O] \xrightarrow{HNO_3} CCl_3CO_2H + H_2O$$

Trichloroacetic acid is a deliquescent solid, m.p. 58°C, and is one of the strongest organic acids.

The presence of three chlorine atoms on a carbon atom adjacent to a carbonyl group causes the C—C bond to break very easily. Thus, when trichloroacetic acid is boiled with dilute sodium hydroxide, or even water, chloroform is obtained:

$$CCL_3CO_2H \longrightarrow CHCl_3 + CO_2$$

The ready fission of the C—C bond may be attributed to the strong inductive effect of the chlorine atoms, and a possible mechanism is:

$$Cl_3C{-}CO_2^- \longrightarrow CO_2 + CCl_3^- \xrightarrow{H_2O} CHCl_3$$

With *concentrated* alkali, formates are produced, and these probably result via the formation of dichloromethylene.

SATURATED DICABOXYLIC ACIDS

The general formula of the saturated dicarboxylic acids is $C_nH_{2n}(CO_2H)_2$ (n = 0 for oxalic acid), and the best-known examples are those which have the two carboxyl groups at the opposite ends of the carbon chain.

Nomenclature

The dicarboxylic acids are commonly known by names which indicate their source, *e.g.*, HO_2CCO_2H oxalic acid; this occurs in plants of the *oxalis* group (for further examples, see the individual acids).

In this trivial system of nomenclature, the positions of side-chains or substituents are indicated by Greek letters, *e.g.*,

$HO_2CCH(CH_3)CH_2CH_2CHClCO_2H$ α-chloro-α′-methyladipic acid

According to the I.U.P.A.C. system of nomenclature, the class suffix is *-dioic, e.g.*,

HO_2CCH,CO,H propanedioic acid

$HO_2CCH(CH_3)CH(CH_3)CH_2CO_2H$ 2,3-dimethylpentanedioic acid

When this method leads to cumbrous names, the alternative' scheme is. to regard the carboxyl group as a substituent, and the name of the acid is then obtained by adding the suffix carboxylic acid, *e.g.*,

$HO_2CCH_2CH_2CO_2H$ 1,2-ethanedicarboxylic acid or ethane-1,2-dicarboxylic acid

General Methods of Preparation

1. The *cyanide synthesis* of dicarboxylic acids is a very useful method: the starting material may be either a halogen-acid or a polymethylene dibromide:

(i) $ClCH_2CO_2H \xrightarrow{KCN} CH_2(CN)CO_2H \xrightarrow{H_3O^+} CH_2(CO_2H)_2$

(ii) $Br(CH_2)_2Br \xrightarrow{KCN} NC(CH_2)_2CN \xrightarrow{H_3O^+} HO_2C(CH_2)_2CO_2H$

2. The **Crum-Brown** and **Walker electrolytic method** (1891, 1893). This is the electrolysis of an aqueous solution of the potassium alkyl esters of the dicarboxylic acids (cf. Kolbe's method):

$$\begin{matrix} K—O_2C—(CH_2)_n—CO_2C_2H_5 \\ K—O_2C—(CH_2)_n—CO_2C_2H_5 \end{matrix} + 2H_2O \longrightarrow \begin{matrix} (CH_2)_nCO_2C_2H_5 \\ | \\ (CH_2)_nCO_2C_2H_5 \end{matrix} + 2CO_2 + 2KOH + H_2 \quad (f.g.–g.)$$

It is obvious that this method can be used to prepare only *even* homologues.

Dicarboxylic acids may also be prepared by the electrochemical hydrodimerisation of suitable α, β-unsaturated compounds.

3. Dicarboxylic acids may be prepared by the acetoacetic ester synthesis and better, by the malonic ester synthesis.

4. Cyclic ketones may be oxidised to dicarboxylic acids, *e.g.*, cyclohexanone gives adipic acid:

$$\text{cyclohexanone (}C_6H_{10}{=}O\text{)} \xrightarrow{HNO_3} HO_2C(CH_2)_4CO_2H$$

Alternatively, lactones can be converted into dicarboxylic acids.

5. The preparation of higher homologues from lower homologues may be carried out in several ways; which method is used depends on the homologue desired. Many *even* higher homologues can be prepared by the Crum-Brown and Walker method (method 2). Any dicarboxylic acid can be stepped up by two carbon atoms as follows:

$$(CH_2)_n(CO_2H)_2 \xrightarrow{LAH} (CH_2)_n(CH_2OH)_2 \xrightarrow{HBr}$$

$$(CH_2)_n(CH_2Br)_2 \xrightarrow{KCN} (CH_2)_n(CH_2CN)_2$$

$$\xrightarrow{H_3O^+} (CH_2)_n(CH_2CO_2H)_2$$

By using malonic ester instead of potassium cyanide, the acid may be stepped up by four carbon atoms:

$$(CH_2)_n(CH_2Br)_n + 2[CH(CO_2C_2H_5)_2]^- Na^+ \longrightarrow$$

$$2NaBr + (CH_2)_n\ [CH_2CH(CO_2C_2H_5)_2]_2 \xrightarrow[(ii)\ HCl]{(i)\ KOH}$$

$$(CH_2)_n[CH_2CH(CO_2H)_2]_2 \xrightarrow{150-200°C} (CH_2)_n(CH_2CH_2CO_2H)_2 + 2CO_2$$

Dicarboxylic acids can be stepped up by ten carbon atoms, by starting with a di-acyl chloride and an examine.

Dicarboxylic acids may also be prepared from oc-bromo-esters as follows.

$$\text{cyclohexane-1,3-dione} \xrightarrow[(ii)\ BrCH_2CO_2Et]{(i)\ KOH-MeOH} \text{2-}(CH_2CO_2Et)\text{-cyclohexane-1,3-dione} \xrightarrow[(ii)\ HCl]{(i)\ N_2H_4,\ OH^-,\ glycol} HO_2C(CH_2)_6CO_2H$$

General Properties

All the dicarboxylic acids are crystalline solids, the lower members being soluble in water, the solubility decreasing with increase in molecular weight: the odd acids are more soluble than the even. Nonc is steam volatile, and except for oxalic acid, the dicarboxylic acids are stable towards oxidising agents. Their melting points follow the oscillation rule, the even acids having higher melting points than the odd. This alternation is due to the relative positions of the carboxyl groups in the crystal structures. These acids dissociate in two steps, the dissociation constant of the first being much greater than that of the second. Furthermore, the acid strength of the dicarboxylic acids decreases as the series is ascended.

Since these acids possess two carboxyl groups, ionisation pair occur in two steps:

$$HO_2C(CH_2)_nCO_2H + H_2O \rightleftharpoons HO_2C(CH_2)_n \quad ...(1)$$

$$HO_2C(CH_2)_n CO_2^- + H_2O \rightleftharpoons {}^-O_2C(CH_2)_n CO_2^- + H_3O^+ \quad ...(2)$$

Acid	pK_a^1	pK_a^2	K_a^1/K_a^2
Oxalic	1.271	4.266	989
Malonic	2.86	5.70	692
Succinic	4.21	5.64	26.9
Glutaric	4.34	5.27	8.51
Adipic	4.41	5.28	7.41

In the-first-stage ionisation (1), the acid can lose a proton from two positions, but the 'half-ion' (conjugate base, cB_1) can add a proton to only one position. Hence, if no other factors operated, it would be anticipated that the acid in equilibrium (1) is twice as strong as a monocarboxylic acid, RCO_2H, *i.e.*, $K_a^1 = 2 \times K_a$, of RCO_2H. In the second-stage ionisation (2), the half-ion can lose a proton from only one position, but the dianion (cB_2) can add a proton to two positions. Therefore for equilibrium (2) $K_a^2 = 1/2\ K_a$, of RCO_2H. Thus, equilibrium (1) is statistically four times as favoured as equilibrium (2). Hence, on the assumption above, *i.e.*, that no other factors operate, the value of K_a^1 would be expected to be four times K_a^2. In practice, the two constants differ from this factor of 4.

An explanation for the experimental results may be as follows. Bjerrum (1923) assumed that the electrostatic effect of the negatively charged carboxylate ion, cB_1, would require additional energy to separate the proton of the carboxyl group, *i.e.*, there is a field effect through the solvent and consequently increases the ratio K_a^1/K_a^2.

At the same time, the carboxylate anion exerts a + I effect with resultant weakening of the acid. It would therefore follow that as the chain increases in length, *i.e.*, as n increases, both the field and + I effects would decrease, and so the ratio would approach 4. This is borne out in practice, Table 4.1: see also conformational effects.

The reactions of the dicarboxylic acids depend, to a large extent, on the length of the carbon chain. The dicarboxylic acids, therefore, will be described individually.

Oxalic acid (*ethanedioic acid*) is one of the most important dicarboxylic acids. It occurs in rhubarb, in sorrel and other plants of the *oxalis* group (hence its name). Oxalic acid is one of the final products of oxidation of many organic compounds, *e.g.*, sugars, starch, etc., give oxalic acid when oxidised with concentrated nitric acid.

Preparation : Oxalic acid is prepared industrially by heating sodium formate rapidly to 360°C.

$$2HOC_2Na \longrightarrow (CO_2Na)_2 + H_2$$

The free acid is obtained from its sodium salt by adding calcium hydroxide solution, then treating the precipitated calcium oxalate with the calculated amount of dilute sulphuric acid, removing the precipitated calcium sulphate, and finally evaporating the filtrate to crystallisation.

The usual laboratory method for preparing oxalic acid is to oxidise sucrose or molasses with concentrated nitric acid in the presence of vanadium pentoxide as catalyst:

$$C1_2H_{22}O_{11} + 18\ [O] \xrightarrow[V_2O_5]{HNO_3} 6(CO_2H)_2 + 5H_2O\ (25\%)$$

Properties and Reactions

Oxalic acid crystallises from water as colourless crystals with two molecules of water of crystallisation; the melting point of the hydrate is 101.5°C; that of the anhydrous acid is 189.5°C. Oxalic acid is poisonous, soluble in water and ethanol but almost insoluble in ether. The dihydrate loses water when heated at 100-105°C, and when heated at about 200°C, oxalic acid decomposes into carbon dioxide, carbon monoxide, formic acid and water:

$$(CO_2H)_2 \longrightarrow CO_2 + HCO_2H \longrightarrow CO + H_2O$$

The anhydrous acid is conveniently obtained by heating the hydrate with carbon tetrachloride.

When heated with concentrated sulphuric acid at 90°C, oxalic acid is decomposed:

$$(CO_2H)_2 \xrightarrow{H_2SO_4} CO + CO_2 + H_2O$$

It is oxidised by permanganate to carbon dioxide:

$$(CO_2H)_2 + [O] \longrightarrow 2CO_2 + H_2O$$

It is only very slowly oxidised by concentrated nitric acid. When fused with potassium hydroxide, it evolves hydrogen:

$$(CO_2H)_2 + 2C_2H_5OH \longrightarrow (CO_2C_2H_5)_2 + 2H_2O \ (80\text{–}90\%)$$

The anhydride of oxalic acid is unknown. When anhydrous oxalic acid is refluxed with ethanol, ethyl oxalate is formed:

$$(CO_2)_2 + 2C_2H_5OH \longrightarrow (CO_2C_2H_5)_2 + 2H_2O \ (80\text{–}90\%)$$

A more general method for preparing di-esters is to heat a mixture of the dicarboxylic acid, alcohol, toluene and concentrated sulphuric acid. Sulphuric acid is the catalyst, and the toluene removes the water by forming a ternary azeotrope of alcohol, water and toluene; the yield of di-ester is 94-98 per cent from oxalic to sebacic acid.

The esterification may be carried out in the absence of toluene: a larger amount of sulphuric acid is required and the yields are lower (70-90 per cent).

The ethyl and propyl esters of oxalic acid are liquid; the methyl ester is solid, and hence has been used to prepare pure methanol by hydrolysis with sodium hydroxide solution. Ethyl oxalate undergoes the Claisen condensation with esters containing two α-hydrogen atoms to form keto-esters, *e.g.*, with ethyl acetate, oxalacetic ester is formed:

$$C_2H_5O_2CCO_2C_2H_5 + CH_3CO_2C_2H_5 \xrightarrow[\text{ethanol}]{C_2H_5ONa \text{ in}}$$

$$\begin{matrix} CHCO_2C_2H_5 \\ \| \\ C(ONa)CO_2C_2H_5 \end{matrix} \xrightarrow{CH_3CO_2H} \begin{matrix} CH_2CO_2C_2H_5 \\ | \\ COCO_2C_2H_5 \end{matrix} \quad (60\text{–}70\%)$$

When oxalic acid is heated with ethylene glycol, the cyclic compound, ethylene oxalate, is obtained:

$$HOCH_2CH_2OH + HO_2CCO_2H \longrightarrow O\begin{matrix} CH_2\text{—}CH_2 \\ \\ CO\text{—}CO \end{matrix}O + 2H_2O$$

This reaction is characteristic of oxalic acid; the other dicarboxylic acids usually react with glycol to form polyesters. When oxalic acid is heated with glycerol, formic acid or allyl alcohol is obtained according to the conditions.

Oxalic acid forms the diamide, *oxamide*. This may be prepared by shaking ethyl oxalate with concentrated ammonium hydroxide solution:

$$C_2H_5O_2CCO_2C_2H_5 + 2NH_3 \longrightarrow NH_2COCONH_2 + 2C_2H_5OH$$

It may also be prepared by passing cyanogen into *cold* concentrated hydrochloric acid:

$$(CN)_2 + 2H_2O \xrightarrow{HCl} (CO_2NH_4)_2$$

Oxamide (decomp. 350°C). in aqueous solution slowly changes into ammonium oxalate, a change which is brought about rapidly by hydrochloric acid (on warming):

$$(CONH_2)_2 \xrightarrow{H_3O^+} (CO_2NH_4)_2$$

When heated with phosphorus pentoxide, oxamide is dehydrated to cyanogen:

$$(CONH_2)_2 \xrightarrow{P_2O_5} (CN)_2 + 2H_2O$$

The monoamide of oxalic acid is also known. It is called oxamic acid, the suffix of the names of all the monoamides of the dicarboxylic acids is -amide, Oxamic acid, m.p. 210°C, may be prepared by the action of concentrated ammonium hydroxide solution on ethyl hydrogen oxalate, or by heating ammonium hydrogen oxalate.

When reduced with zinc and sulphuric acid, oxalic acid forms *glycollic acid*, $HOCH_2CO_2H$. Electrolytic reduction with a lead cathode gives glycollic and glyoxylic acids, the latter being obtained in better yield by reducing oxalic acid with magnesium and sulphuric acid.

When oxalic acid is treated with *excess* of phosphorus pentachloride, *oxalyl chloride*, b.p. 64°C, is formed:

$$(CO_2H)_2 \xrightarrow{PCl_5} (COCl)_2 \qquad \text{(f.–g)}$$

If an excess of phosphorus pentachloride is not used, oxalic acid is decomposed, possibly via the intermediate formation of the half-acid chloride.

$$(CO_2H)_2 \xrightarrow{PCl_5} [HO_2CCOCl] \longrightarrow CO_2 + CO + HCl$$

Alkanes can be carboxylated by oxalyl chloride under the influence of light (Kharasch and Brown, 1942):

$$RH + (COCl)_2 \longrightarrow RCOCl + CO + HCl$$

Oxalic acid is used for the manufacture of ink and for bleaching straw. Its antimony salts are used as mordants in printing and dyeing.

Malonic acid (*propanedioic acid*), m.p. 136°C, was first obtained by the oxidation of malic acid (hence its name):

$$HO_2CCHOHCH_2CO_2H + [O] \xrightarrow{K_2Cr_2O_7/H_2SO_4} \underset{\text{oxalacetic acid}}{HO_2CCOCH_2CO_2H} \xrightarrow{[O]} HO_2CCH_2CO_2H + CO_2$$

Malonic acid may be prepared by heating potassium chloroacetate with aqueous potassium cyanide and hydrolysing the product, potassium cyanoacetate, with hydrochloric acid:

$$ClCH_2CO_2K \xrightarrow{KCN} CH_2(CN)CO_2K \xrightarrow{H_3O^+} CH_2(CO_2H)_2 \ (84\%)$$

When heated to 140-150°C, or when refluxed in sulphuric acid solution, it eliminates carbon dioxide.

$$HO_2CCH_2CO_2H \longrightarrow C_3CO_2 + CO_2$$

All dicarboxylic acids which have both carboxyl groups attached to the *same* carbon atom are decomposed in a similar manner. When malonic acid is heated with phosphorus pentoxide, a small amount of *carbon suboxide* is obtained:

$$HO_2CCH_2CO_2H \xrightarrow{P_2O_5} O = C = C = C = O + 2H_2O$$

Carbon suboxide, b.p. 7°C, may be regarded as a diketen; it combines with water to form malonic acid. Malonic acid does not form a cyclic anhydride, but dimethylmalonic acid does.

When malonic acid is treated with nitrous acid and the product hydrolysed, **mesoxalic acid** *(keto-malonic acid)* is obtained:

$$(HO_2C)_2CH_2 \xrightarrow{HNO_2} (HO_2C)_2C = NOH \xrightarrow{H_3O^+} HO_2CCOCO_2H$$

Mesoxalic acid crystallises from water with one molecule of water which is held firmly, and hence is believed to be water of constitution: $(HO)_2C(CO_2H)_2$ (cf. glyoxylic acid). Ethyl malonate may be oxidised directly to ethyl mesoxalate by selenium dioxide (yield: 23 per cent), or with dinitrogen trioxide (HNO_3 + AS_2O_3; yield: 74-76 per cent).

Ethyl malonate is far more important than the acid because of its synthetic uses; the acid contains an active methylene group, the reactivity of which is more pronounced in the ester. Both the acid and ester may be readily brominated, *e.g.*, monobromomalonic acid, CHBr(CO,H), is formed when a suspension of malonic acid in ether is treated with bromine; owing to the high reactivity of the methylene group, no red phosphorus is required as catalyst.

Ethyl malonate does not form the diamide, *malonamide,* when shaken with concentrated ammonium hydroxide solution; the dimethyl ester, however, gives a good yield of malonamide. Ethyl malonate forms *barbituric acid* when heated with urea in the presence of sodium ethoxide.

CARBONIC ACID AND ITS DERIVATIVES

Orthocarbonic acid, C(OH), is unknown in the free state, but its esters have been prepared. They may be obtained by the reaction between sodium alkoxide and nitrochloroform:

$$4RONa + CCl_3NO_2 \longrightarrow C(OR)_4 + 3NaCl + NaNO_2 \text{ (f.g.)}$$

The alkyl orthocarbonates are ethereal smelling liquids.

Carbonic acid (*metacarbonic acid*), $0 = C(OH)_2$, is unknown in the free state, but its salts and esters have been prepared. Alkyl carbonates may be readily formed by heating alcohols with carbonyl chloride in the presence of pyridine:

They may also be prepared by heating silver carbonate with alkyl iodide:

$$Ag_2CO_3 + 2RI \longrightarrow CO(OR)_2 + 2AgI$$

The alkyl carbonates are ethereal smelling liquids, readily soluble in water. In recent years they have found various uses in organic synthesis, *e.g.*,

(i) They may be used to introduce the carbalkoxy group, CO_2R, into ketones to produce β-ketoesters. The reaction is carried out by heating a ketone with sodium or sodium alkoxide in the presence of a large excess of alkyl carbonate:

$$R^2COCH_3 + (R^1O)_2CO + NaOR^1 \longrightarrow [R^2COCHOR^1]^-Na^+ + 2R^1OH$$

$$\xrightarrow{H^+} R_2COCH_2CO_2R^1$$

(ii) Alkyl carbonates react with primary alkyl cyanides in the presence of sodium alkoxide to form α-cyanoesters:

$$[R^2C(CO_2C_2H_5)_2]^- Na^+ + (R^1O)_2 CO \longrightarrow R^1R^2C(CO_2C_2H_5)_2 + R^1OCOONa$$

(iii) Alkyl carbonates may be used to alkylate sodiomalonic esters:

$$[R^2C(CO_2C_2H_5)_2]^- Na^+ + (R^1O)_2CO \longrightarrow R^1R^2C(CO_2C_2H_5)_2 + R^1OCOONa$$

The acid chloride of carbonic acid is *carbonyl chloride* (*phosgene*). It may be obtained by the action of chlorine on carbon monoxide under the influence of sunlight, or in the presence of heated charcoal (200°C) as catalyst; the latter is the commercial method.

Carbonyl chloride is a colourless liquid, b.p. 8°C, and has been used as a toxic gas in warfare. It is used in various organic syntheses, behaving as an acid chloride. It is very slowly decomposed by water:

$$COCl_2 + H_2O \longrightarrow CO_2 + 2HCl$$

When treated with slight excess of one equivalent of alcohol in the cold, carbonyl chloride forms chloroformic esters:

$$COCl_2 + ROH \longrightarrow ClCO_2R + HCl$$

When treated with an excess of alcohol in the presence of pyridine, alkyl carbonates are formed (see above). Carbonyl chloride reacts with ammonia to form urea:

$$COCl_2 + 2NH_3 \longrightarrow CO(NH_2)_2 + 2HCl$$

With primary or secondary amines, substituted ureas are formed, *e.g.*,

$$COCl_2 + 2RNH_2 \longrightarrow CO(NHR)_2 + 2HCl$$

Chloroformic acid (*chlorocarbonic acid*), $ClCO_2H$, is not known in the free state, but its esters have been prepared. These may be readily obtained by treating carbonyl chloride with slight excess of one equivalent of alcohol in the cold (see above). Chloroformic esters are acid-chloride esters, and the chlorine atom reacts with compounds containing active hydrogen; thus ethyl chloroformate, b.p. 94°C, reacts with water, alcohols, ammonia, primary and secondary amines; *e.g.*, are than is formed with ammonia:

$$ClCO_2C_2H_4 + NH_3 \longrightarrow NH_2CO_2C_2H_5 + HCl$$

Ethyl chloroformate is useful for introducing the carbethoxy group on a nitrogen atom; *e.g.*, ethyl *N*-tricarboxylate is formed when ethyl chloroformate is added to a mixture of urethan, ether and sodium, this mixture having been previously heated:

$$NH_2C\ O_2C_2H5 + 2ClCO_2H_2H_5 + 2Na \longrightarrow N(CO_2C_2H_5)_3 + 2NaCl + H_2 \quad (51–57\%)$$

Ethyl chloroformate may also be used to prepare ethyl esters by reaction with Grignard reagents.

Amides of carbonic acid. Since carbonic acid is dibasic, two amides are possible: the mono- and diamide. These are known, respectively, as *carbamic acid*, (1), and *urea*, (II):

$$O{=}C\begin{matrix} \diagup NH_2 \\ \diagdown OH \end{matrix} \qquad O{=}C\begin{matrix} \diagup NH_2 \\ \diagdown NH_2 \end{matrix}$$

(I) (II)

Carbamic acid is not known in the free state, but its salts and esters have been prepared. Ammonium carbamate is formed when dry ammonia reacts with dry carbon dioxide:

$$2NH_3 + CO_2 \longrightarrow NH_2COONH_4$$

It is a white crystalline solid, very soluble in water. When its aqueous solution is warmed to 60° C, ammonium carbamate is hydrolysed to ammonium carbonate:

$$NH_2COONH_4 + H_2O \longrightarrow CO(ONH_4)_2$$

Esters of carbamic acid are known as **urethans**. These may be prepared:

(i) By treating a chloroformic ester with ammonia, *e.g.*, ethyl chloroformate produces ethyl carbamate:

$$ClCO_2C_2H_5 + NH_3 \longrightarrow NH_2CO_2C_2H_5 + HCl$$

(ii) By refluxing urea in an alcohol, *e.g.*, n-butanol gives n-butyl carbamate when refluxed for 30 hours:

$$CO(NH_2)_2 \longrightarrow NH_3 + HNCO$$

$$HNCO + CH_3CH_2CH_2CHOH \longrightarrow NH_2CO_2CH_2CH_2CH_2CH_3$$

(75–76%)

(iii) N-substituted urethans may be prepared by the Curtius reaction. The acid azide is refluxed in benzene solution and then an alcohol is added:

$$R^1CON_3 \longrightarrow N_2 + R^1NCO \xrightarrow{R^2OH} R^1NHCO_2R^2$$

N-phenylurethans may be prepared by reaction between phenyl isocyanate and alcohol:

$$C_6H_5NCO + ROH \longrightarrow C_6H_5NHCO_2R$$

These are crystalline solids and may be used to characterise the alcohols.

Readily dehydrated alcohols, particularly t-alcohols, do not form urethans; alkenes and diphenyl-urea are produced:

$$R_2C(OH)CH_2CH_3 \longrightarrow R_2C = CHCH_3 + H_2O$$

$$PhNCO + H_2O \longrightarrow CO_2 + PhNH_2 \xrightarrow{PhNCO} (PhNH)_2CO$$

Ethyl carbamate, usually known as *urethan*, is a crystalline solid, m.p. 50°C. It reacts with ammonia to form urea:,

$$NH_2CO_2C_2H_5 + NH \longrightarrow CO(N\ H_2)_2 + C_2H_5OH$$

It is decomposed by aqueous sodium hydroxide on warming:

$$NH_2CO_2C_2H_5 + {}_2NaOH \longrightarrow Na_2CO_3 + NH_3 + C_2H_5OH$$

Ethyl carbamate has hypnotic properties; many urethans are valuable hypnotics, *e.g.*, aponal (t-amyl carbamate, t-amylurethan), $NH_2COOC(CH_3)_2(C_2H_5)$.

Urea (*carbamide*) is very important physiologically. It is the chief nitrogenous product of protein metabolism; adults excrete about 30 g per day in the urine, from which it can be extracted by evaporating the urine to small bulk and adding nitric acid, whereupon the slightly soluble urea nitrate, $CO(NH_2)_2{\cdot}HNO_3$, is precipitated.

Urea is historically very important because Wohler (1828) synthesised it by evaporating a solution containing potassium isocyanate and ammonium sulphate, ammonium isocyanate, which is formed first, undergoes molecular rearrangement:

$$NH_4NCO \rightleftharpoons CO(NH_2)_2$$

The reaction is reversible; the solution contains about 5 per cent ammonium isocyanate.

The mechanism of this rearrangement is uncertain. One possibility is as follows (via dissociation):

$$NH_4NCO \rightleftharpoons NH_3 + HNCO$$

$$H{-}N{=}C{=}O + \ddot{N}H_3 \rightleftharpoons H{-}\bar{N}{-}\underset{\underset{^{+}NH_3}{|}}{C}{=}O \rightleftharpoons H_2N{-}\underset{\underset{NH_2}{|}}{C}{=}O$$

Urea may be prepared in the laboratory by the action of ammonia on carbonyl chloride, alkyl carbonates, chloroformates or urethans, *e.g.*,

$$COCl_2 + 2NH_3 \longrightarrow CO(NH_2)_2 + {}_2HCl \quad (f)$$

$$(C_2H5O)_2CO + {}_2NH_3 \longrightarrow CO(NH_2)_2 + {}_2C_2H_5OH \quad (f)$$

Industrially, urea is prepared by allowing liquid carbon dioxide and liquid ammonia to interact, and heating the ammonium carbamate so formed to 130-150°C under about 35atm pressure:

$$2NH_3 + CO_2 \longrightarrow NH_2COONH_4 \longrightarrow CO(\ NH_2)_2 + H_2O$$

Structure of Urea

Although, urea appears to be a simple molecule, it is only recently that its structure has been ascertained with any degree of certainty. The diamide structure, $CO(NH_2)_2$, seemed to be indicated by its synthesis from carbonyl chloride and ammonia:

$$OC\begin{matrix} \diagup Cl \\ \diagdown Cl \end{matrix} + 2NH_3 \longrightarrow OC\begin{matrix} \diagup NH_2 \\ \diagdown NH_2 \end{matrix} + 2HCl$$

This structure, however, did not appear to explain all the reactions of urea and so gave rise to a great deal of controversy. Physical methods of structure determination have now shown that the diamide structure is correct, at least in the solid state. Crystal structure studies have shown that in solid urea both nitrogen atoms are identical. Bond-length measurements in urea give the C—N distance as 1-37Å. In aliphatic amines the C—N bond length is 1-47Å. This indicate; that the C—N bond in urea has some double bond character (about 28 per cent): this can be explained by resonance:

$$(H_2N)(NH_2)C{=}O \longleftrightarrow (H_2\overset{+}{N}{=})(NH_2)C{-}O^{-} \longleftrightarrow (H_2N)({=}\overset{+}{N}H_2)C{-}O^{-}$$

Both nitrogen atoms are identical in the hybrid molecule. Furthermore, the negatively charge oxygen atom is capable of co-ordinating with one proton (therefore urea will be a 'monoacidic' base), and thus the salt may be formulated as a resonance hybrid:

$$H_2\overset{+}{N}=C(NH_2)-OH \longleftrightarrow H_2N-C(=\overset{+}{N}H_2)-OH \longleftrightarrow H_2N-C(NH_2)=\overset{+}{O}H$$

NMR spectra of crystalline salts of urea (and thiourea) are consistent with protonation at the oxygen atom (or sulphur atom; Redpath *et al.*, 1962).

Properties and Reactions of Urea

Urea is a white crystalline solid, m.p. 132°C, soluble in water and ethanol, but insoluble in ether. It is used for preparing formaldehyde-urea plastics, barbiturates, as a fertiliser, etc. A recent use is for the manufacture of hydrazine; urea is treated with alkaline sodium hypochlorite (in effect, the Hofmann preparation of primary amines is applied to urea):

$$NH_2CONH_2 + NaOCl + 2NaOH \longrightarrow N_2H_4 + NaCl + Na_2CO_3 + H_2O$$

(i) Urea behaves as a 'monoacidic' base; the nitrate and oxalate are the most important, since neither is very soluble in water.

When urea nitrate is added to cold concentrated sulphuric acid, nitrourea is formed:

(ii) Urea is hydrolysed by boiling with acids or alkalis:

$$CO(NH_2)_2 + H_2O \longrightarrow CO_2 + 2NH_3$$

The enzyme urease (which occurs in soyabeans) produces the same change.

(iii) When gently heated, urea loses ammonia to form biuret:

$$(NH_2)_2\,CO \rightleftharpoons NH_3 + HNCO \xrightarrow{(NH_2)_2CO} NH_2CONHCONH_2$$

When an aqueous biuret solution is treated with sodium hydroxide solution and a drop of copper sulphate solution, a violet colour is produced. This is known as the *biuret reaction*, which is characteristic of all compounds containing the grouping —CONH—, *e.g.*, proteins.

When heated rapidly, urea evolves ammonia and forms cyanic acid which rapidly polymerises to cyanuric acid. When refluxed with alcohols, urea forms urethans (see above).

(iv) Nitrous acid reacts with urea with the liberation of nitrogen which, however, is not evolved quantitatively:

$$CO(NH_2)_2 + 2HNO_2 \longrightarrow CO_2 + 3H_2O + 2N_2$$

Nitrogen is also evolved, again not quantitatively, when urea is treated with excess of alkaline hypobromite:

(v) Acid chlorides and acid anhydrides react with urea to form weides, *e.g.*, acetyl chloride forms acetylurea:

$$CH_3COCl + NH_2CONH_2 \longrightarrow CH_3CONHCONH_2 + CHI$$

Many of these ureides are useful drugs, particularly when the acid group has a branched chain, *e.g.*, *bromural* (α-bromoisovalerylurea), $(CH_3)_2CHCHBrCONHCONH_2$.

Dicarboxylic acids react with urea in the presence of phosphoryl chloride to form cyclic ureides; *e.g.*, oxalic acid forms parabanic acid (oxalylurea):

$$\begin{matrix} CO_2H \\ | \\ CO_2H \end{matrix} + CO(NH_2)_2 \xrightarrow{POCl_3} + 2H_2O + \begin{matrix} CO{-}NH \\ | \\ CO{-}NH \end{matrix}\!\!\!\!>\!CO$$

Cyclic ureides may also be prepared by refluxing a di-ester with urea in ethanolic solution containing sodium ethoxide; *e.g.*, malonic ester forms barbituric acid (malonylurea):

$$CH_2\begin{matrix} \diagup CO_2C_2H_5 \\ \diagdown CO_2C_2H_5 \end{matrix} + \begin{matrix} H_2N \diagdown \\ H_2N \diagup \end{matrix} CO \xrightarrow{C_2H_5ONa} \text{barbituric acid (HN}^1\text{, C}^2\text{=O, N}^3\text{H, C}^4\text{=O, C}^5\text{, C}^6\text{=O)} + 2C_2H_5OH \quad (72\text{–}78\%)$$

Barbituric acid and its 5- or 5,5-derivatives are used in medicine as hypnotics and sedatives, *e.g.*, barbitone (5,5-diethylbarbituric acid), and *phenobarbitone* (5-phenyl-5-ethylbarbituric acid).

INCLUSION COMPLEXES

Several types of inclusion complexes are known, and in all of them molecules of one component, the 'guest' molecules, are physically imprisoned in the cavities of the crystalline structure of the second component, the 'host' molecules. Two important types of inclusion complexes are the *channel* (*canal*) *complexes* and the *clathrate* (*cage*) *complexes* (for a discussion of the latter).

Channel complexes are those cases where the host crystallises in a form with parallel, approximately cylindrical channels in which molecules of the guest are enclosed lengthwise.

Urea normally forms a crystal structure which is closely packed, but in the presence of various *straight-chain* molecules, *e.g.*, n-alkanes, n-alcohols, n-acids, n-esters, etc., urea crystallises in a more open structure which contains long channels enclosing the guest molecules. These channels contain a number of guest molecules and, in general, the number is inversely proportional to the length of the guest molecule. Branched-chain and cyclic structures cannot 'fit' into these channels, and so this property affords a means of separating straight-chain from branched-chain compounds (but see below). The complexes are decomposed by melting or by dissolving away the urea with water. The formula of the channel complexes is usually A_nB (A is urea) where n, usually not a whole number, has values of 4 or more and increases as the length of B increases. Furthermore, since channel complexes are characterised by the fact that guest molecules bear some structural resemblance to each other, it is not possible to separate homologues by channel complexes. For each homologous series there is a certain minimum chain-length for channel complex formation, *e.g.*, six carbons for alkanes, seven for alcohols, five for acids, etc.

Channel complex formation has been used to resolve racemic modifications, and also to separate geometrical isomers. It is also interesting to note that if a molecule is sufficiently long, then 'small' side-chains will not prevent such a molecule behaving as a guest, *e.g.*, Schiessler *et al.* (1957) have shown that 2-methyldecane forms a channel complex with urea.

X-ray analysis of these channel complexes has shown that the diameter of the channel is about 5 Å (Smith, 1952; see also thiourea).

Substituted ureas may be prepared by a reaction similar to Wohler's synthesis of urea; the hydrochloride or sulphate of a primary or secondary amine is heated with potassium isocyanate, *e.g.*, methylamine hydrochloride (I molecule) forms methylurea:

$$CH_3NH_2 \cdot HCl + KNCO \longrightarrow KCl + CH_3NH_2 \cdot HNCO \longrightarrow CH_3NHCONH_2$$

If excess of amine salt is used, N, N′-disubstituted ureas are obtained, *e.g.*, excess of aniline hydrochloride forms N, N′-diphenylurea::

$$C_6H_5NH_2{\cdot}HCl + KNCO \longrightarrow KCl + C_6H_5NH_2{\cdot}HNCO \longrightarrow$$

$$C_6H_5NHCONH_2 \xrightarrow{C_6H_5NH_2\cdot HCl} C_6H_5NHCONHC_6H_5 + NH_4Cl$$

Alternatively, N,N′-diphenylurea may be obtained by refluxing an aqueous solution of aniline hydrochloride with urea; the urea produces ammonia and isocyanic acid, and the reaction proceeds as shown above. Both phenylurea (38-40 per cent) and N,N′-diphenylurea (52-55 per cent) may be isolated.

N,N′-Substituted ureas may be obtained by the action of carbonyl chloride on a primary or secondary amine:

$$COCl_2 + 2RNH_2 \longrightarrow CO(NHR)_2 + 2HCl$$

The reaction between phenyl isocyanate and a primary or secondary amine also produces an 5-disubstituted urea, *e.g.*, with ethylamine, N-ethyl-N'-phenylurea is formed:

$$C_6H_5NCO + C_2H_5NH_2 \longrightarrow C_6H_5NHCONHC_2H_5$$

This reaction is used to characterise amines.

A very convenient method of preparing alkylureas is to evaporate an aqueous or ethanolic solution of an amine and nitrourea (see above); the latter decomposes into isocyanic acid and nitramide, NH_2NO_2, which readily decomposes into nitrous oxide:

$$NH_2CONHNO_2 \longrightarrow NHCO + NH_2NO_2$$

$$HNCO + R_2NH \longrightarrow R_2NCONH_2$$

$$NH_2NO_2 \longrightarrow N_2O + H_2O$$

When urea is treated with methyl sulphate in faintly alkaline solution, methylisourea is obtained:

$$NH_2CONH_2 \xrightarrow{Me_2SO_4} NH = C(OMe)NH_2$$

Succinic acid *(butanedioic acid)*, m.p. 185°C, was originally obtained by the distillation of amber (Latin: *succinum*, amber). It may be synthesised by the following methods:

(i) From ethylene dibromide:

$$\begin{array}{c} CH_2Br \\ | \\ CH_2Br \end{array} \xrightarrow{KCN} \begin{array}{c} CH_2CN \\ | \\ CH_2CN \end{array} \xrightarrow{H_3O^+} \begin{array}{c} CH_2CO_2H \\ | \\ CH_2CO_2H \end{array} \quad (80\%)$$

(ii) By the reaction between malonic ester (I molecule) and ethyl chloroacetate, or between malonic ester (2 molecules) and iodine.

Alkyl-substituted succinic acids may be prepared by using monoalkylmalonic ester and α-halogeno-acid esters or iodine.

Alternatively, they may be prepared by the addition of hydrogen cyanide to α, β-unsaturated esters and hydrolysing the Bcyanacomplex produced.

(iii) By heating malic acid, in a sealed tube, with constant boiling hydriodic acid and red phosphorus:

$$\begin{array}{l} CHOHCO_2H \\ | \\ CH_2CO_2H \end{array} + 2HI \longrightarrow \begin{array}{l} CH_2CO_2H \\ | \\ CH_2CO_2H \end{array} + I_2 + H_2O \quad (60\%)$$

Succinic acid is prepared industrially by the catalytic (or by the electrolytic) reduction of maleic acid:

$$\begin{array}{l} CHCO_2H \\ \| \\ CHCO_2H \end{array} + H_2 \xrightarrow{Ni} \begin{array}{l} CH_2CO_2H \\ | \\ CH_2CO_2H \end{array}$$

When heated, a large amount sublimes, the rest being converted into the *cyclic* anhydride, succinic anhydride:

$$\begin{array}{l} CH_2CO_2H \\ | \\ CH_2CO_2H \end{array} \longrightarrow \begin{array}{l} CH_2CO_2H \diagdown \\ | \qquad\qquad\quad O + H_2O \\ CH_2CO_2H \diagup \end{array}$$

When heated with excess of glycol, succinic acid forms high-polymer esters (*polyesters*, which belong to the group known as the *alkyl resins*). These esters are acidic, the end groups being succinic acid residues:

$$HO_2C(CH_2)_2CO_2H + HOCH_2CH_2OH + HO_2C(CH_2)_2CO_2H$$
$$+ HOCH_2CH_2OH + \ldots \longrightarrow HO_2C(CH_2)_2CO$$
$$—[—O(CH_2)_2OCO(CH_2)_2CO—]—OH$$

Dicarboxylic acids from malonic to adipic acid form these polyesters with ethylene glycol. These esters are linear polymers, but if glycerol is used instead of glycol, three dimensional polymeric esters are obtained.

Succinic anhydride (*butanedioic anhydride*) is obtained in excellent yield by distilling succinic acid with acetic anhydride, acetyl chloride, or phosphoryl chloride.

The anhydride, and not the acid chloride, is obtained when succinic acid is heated with thinly chloride.

Succinic anhydride, m.p. 119°C, when boiled with water or alkalis, is converted into succinic acid:

$$\begin{matrix} CH_2CO \\ | \\ CH_2CO \end{matrix}\!\!>O + H_2O \longrightarrow \begin{matrix} CH_2CO_2H \\ | \\ CH_2CO_2H \end{matrix}$$

When reduced with sodium and ethanol, succinic anhydride is converted first into γ-butyro- lactone, and finally into tetramethylene glycol:

$$\begin{matrix} CH_2CO \\ | \\ CH_2CO \end{matrix}\!\!>O \xrightarrow{e;\ H^+} \begin{matrix} CH_2CH_2 \\ | \\ CH_2CO \end{matrix} \xrightarrow{e;\ H^+} \begin{matrix} CH_2CH_2OH \\ | \\ CH_2CH_2OH \end{matrix}$$

If the reduction is carried out with sodium amalgam in *acid* solution, the lactone is obtained in good yield; some butyric acid is also formed.

Succinic anhydride is very useful for preparing the 'half-derivatives' of succinic acid, *e.g.*, with alcohols it forms the acid-ester:

$$\begin{matrix} CH_2CO \\ | \\ CH_2CO \end{matrix}\!\!>O + ROH \longrightarrow \begin{matrix} CH_2CO_2R \\ | \\ CH_2CO_2H \end{matrix}$$

With Grignard reagents, succinic anhydride forms γ-keto-acids:

$$\begin{matrix} CH_2CO \\ | \\ CH_2CO \end{matrix}\!\!>O + RMgX \longrightarrow RCOCH_2CH_2CO_2H$$

The carbonyl groups (C = O stretch) in *five*-membered ring anhydrides absorb in the regions 1870-1830 (s) and 1800-1760 cm^{-1}(s). Thus these anhydrides can be distinguished from acyclic anhydrides. The molecular formula should also help here for all cyclic anhydrides.

$$\begin{matrix} CH_2CO \\ | \\ CH_2CO \end{matrix}\!\!>O + NH_3 \longrightarrow \begin{matrix} CH_2CO \\ | \\ CH_2CO \end{matrix}\!\!>NH + H_2O \quad \text{(v.g.)}$$

Fig. 2.10 is the infrared spectrum of succinic anhydride (as a Nujol mull).

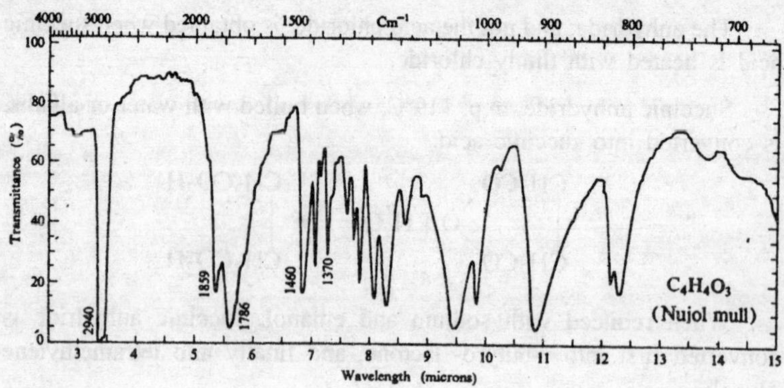

Fig. 2.10

2940, 1460 and 1370cm^{-1} : Bands for Nujol. 1859 and 1786cm^{-1}: C = O (str.) in 5-ring anhydrides

Succinimide (*butanimide*) is formed when succinic anhydride is heated in a current of dry ammonia:

Succinamic acid (m.p. 157°C), (I), and succinamide (m.p. 243°C), (II), are both readily converted into succinimide when heated:

$$\begin{array}{l} CH_2CONH_2 \\ | \\ CH_2CO_2H \end{array} \xrightarrow{-H_2O} \begin{array}{l} CH_2CO \diagdown \\ \quad\quad\quad\quad NH \\ CH_2CO \diagup \end{array} \xleftarrow{-NH_3} \begin{array}{l} CH_2CONH_2 \\ | \\ CH_2CONH_2 \end{array}$$

Succinimide, m.p. 125°C, when boiled with water or alkalis is converted into succinic acid:

$$\begin{array}{l} CH_2CO \diagdown \\ \quad\quad\quad\quad NH \\ CH_2CO \diagup \end{array} + 2H_2O \longrightarrow \begin{array}{l} CH_2CONH_2 \\ | \\ CH_2CO_2H \end{array} + NH_3$$

When succinimide is distilled with zinc dust, *pyrrole*, (III), is obtained. When reduced will sodium and ethanol, succinimide forms *pyrrolidine*, (IV), and when reduced electrolytical it forms *pyrrolidone*, (V):

$$\begin{array}{c} CH - CH \\ \| \quad\quad \| \\ CH \quad\quad CH \\ \diagdown NH \diagup \\ (III) \end{array} \qquad \begin{array}{c} CH_2 - CH_2 \\ | \quad\quad | \\ CH_2 \quad\quad CH_2 \\ \diagdown NH \diagup \\ (IV) \end{array} \qquad \begin{array}{c} CH_2 - CH_2 \\ | \quad\quad | \\ CH_2 \quad\quad CO \\ \diagdown NH \diagup \\ (V) \end{array}$$

Succinimide is acidic, *e.g.*, it reacts with potassium hydroxide to form potassium succinimide:

$$\begin{array}{l}CH_2CO\diagdown\\ \quad\quad\quad N^-\\ CH_2CO\diagup\end{array} \longleftrightarrow \begin{array}{l}CH_2C{-}O^-\\ \quad\quad\quad N^-\\ CH_2CO\end{array} \longleftrightarrow \begin{array}{l}CH_2CO\\ \quad\quad\quad N^-\\ CH_2CO{-}O^-\end{array}$$

Succinimide, however, is only very weakly acidic; the potassium salt is decomposed by carbon dioxide, the imide being regenerated.

A very important derivative of succinimide is N-*bromosuccinimide*, which may be prepared by the action of bromine on succinimide at 0°C in the presence of sodium hydroxide:

$$\begin{array}{l}CH_2CO\diagdown\\ \quad\quad\quad NH\\ CH_2CO\diagup\end{array} + Br_2 \xrightarrow{NaOH} \begin{array}{l}CH_2CO\diagdown\\ \quad\quad\quad NBr\\ CH_2CO\diagup\end{array} + HBr$$

N-bromosuccinimide (NBS) is a valuable reagent or brominating olefinic compounds in the allyl position (Ziegler. 1942).

$$-CH_2-CH{=}CH_2 + \begin{array}{l}CH_2CO\diagdown\\ \quad\quad\quad NBr\\ CH_2CO\diagup\end{array}$$

$$\longrightarrow -CHBrCH{=}CH_2 + \begin{array}{l}CH_2CO\diagdown\\ \quad\quad\quad NH\\ CH_2CO\diagup\end{array}$$

This reaction offers a means of splitting off three carbon atoms from a compound by oxidative degradation, provided the compound contains one double bond or can have one produced, *e.g.*,

$$RCH_2CH_2CH_2OH \xrightarrow[350^\circ C]{Al_2O_3} RCH_2CH_2CH = CH_2$$

$$\xrightarrow{\text{N-bromosuccinimide}} RCH_2CHBrCH = CH_2$$

$$\xrightarrow[\text{etanol}]{KOH} RCH = CHCH = CH_2 \xrightarrow{\text{ozonolysis}} RCHO$$

This method has found great use in steroid chemistry.

Normally, *N*-bromosuccinimide *substitutes* alkenes in the ally position, and this reaction is believed to take place by a free-radical mechanism; it is catalysed by peroxides and is promoted by light (both are free-radical producing agents). If there are two allylic positions then two monobromo-derivatives may be obtained. Since a secondary hydrogen is substitute more readily than a primary hydrogen atom, the former gives rise to the predominan product.

The details of the mechanism are uncertain. According to Goldfinger *et al.* (1953, 1956), the function of the *N*-bromosuccinimide is to supply a *low concentration of molecular bromine.*

$$>NBr + HBr \longrightarrow >NH + Br_2$$

$$Br_2 \rightleftharpoons Br\cdot$$

$$Br\cdot + >CHCH{=}CH_2 \longrightarrow >\dot{C}CH{=}CH_2 + HBr$$

$$Br_2 + >\dot{C}CH{=}CH_2 \longrightarrow \quad CBrCH{=}CH_2 + Br\cdot$$

N-bromosuccinimide also brominates unsaturated esters, *e.g.*,

$$CH_3CH{=}CHCO_2CH_3 \xrightarrow{>N{-}Br} BrCH_2CH{=}CH_2 + Br\cdot$$

In addition to substitution, Af-bromosuccinimide may also produce addition compounds, but these are usually formed only in small amount. Braude *et al.* (1952), however, have shown that the addition reaction is catalysed by tetraethylammonium salts, *e.g.*, *N*-bromosuccinimide and cyclohexene give 3-bromocyclohexene (Ziegler *et al.*, 1942), but in the presence of, *e.g.*, tetraethylammonium bromide, 1,2-dibromocvclohexane is the main product.

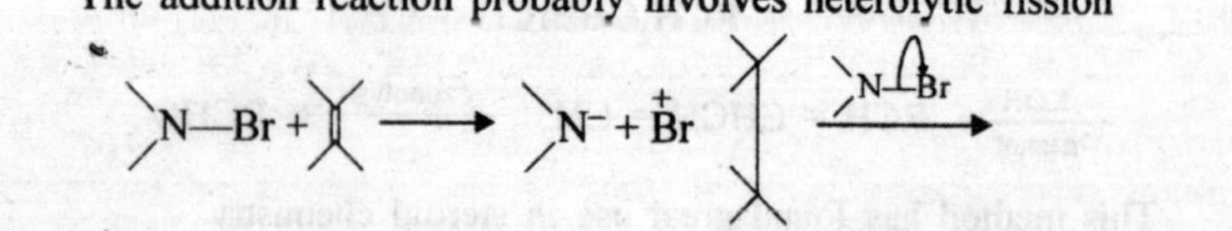

(80–90%) 3-bromocyclohexene $\xleftarrow{\text{substitution}}$ cyclohexene $\xrightarrow{\text{addition}}$ 1,2-dibromocyclohexane (74%)

Fig. 2.11

The addition reaction probably involves heterolytic fission

$$>N{-}Br + >C{=}C< \longrightarrow >N^- + \overset{+}{Br} \xrightarrow{>N{-}Br} $$

$$Br{-}C{-}C{-}Br + >N^- + >N^+ \longrightarrow 2\ >N^- \xrightarrow[\text{donor}]{\text{hydrogen}} 2\ >NH$$

Succinyl chloride (butanedioyl chloride). This is readily prepared by the action of phosphorus pentachloride on succinic acid. It was formerly believed that succinyl chloride existed in two forms, the chain isomer, (I), and the ring isomer, (II). However, all recent work indicates structure (1), *e.g.*, Schmid (1965, 1966), from NMR studies, showed that the molecule is symmetrical, and from inferred studies that a y-lactone oxo group (in II) is absent.

$$\underset{\text{(I)}}{(CH_2COCl)_2} \qquad \underset{\text{(II)}}{\begin{array}{l} CH_2CO \\ \quad\;\; \diagdown \\ \qquad\; O \\ \quad\;\; \diagup \\ CH_2CO \end{array}}$$

Isomeric with is succinic acid is *methylamine acid* or *isosuccinic acid*, m.p. 130°C. It may be readily prep ·red from sodiomalonic ester and methyl iodide, or as follows:

$$CH_3C_2CO_2H \xrightarrow{Br_2/P} CH_3CHBrCO_2H$$

$$\xrightarrow{KCN} CH_3CH(CN)CO_2H \xrightarrow{H^+} CH_3CH(CO_2H)_2$$

Oxalacetic ester is the diethyl ester of ketosuccinic acid, and may be prepared by the Claisen condensation between ethyl oxalate and ethyl acetate. It is a colourless liquid which can be distilled under reduced pressure (b.p. 132°C/24 mm), but at atmospheric pressure it eliminates a molecule of carbon monoxide to form malonic ester:

$$C_2H_5O_2CCOH_2CO_2C_2H_5 \longrightarrow CH_2(CO_2C_2H_5)_2 + CO$$

Hydrogen of the methylene group adjacent to the carbonyl group may be replaced by alkyl groups by reactions similar to those used for acctoacctic ester, *e.g.*,

$$C_2H_5O_2CCOCH_2CO_2C_2H_5 \xrightarrow{C_2H_5ONa}$$

$$[C_2H_5O_2CCOCHCO_2C_2H_5]^-Na^+ \xrightarrow{CH_3I}$$

$$C_2H_5O_2CCOCH(CH_3)CO_2C_2H_5 \xrightarrow[\text{(ii) } C_2H_5I]{\text{(i) } C_2H_5ONa}$$

$$C_2H_5O_2CCOC(CH_3)(C_2H_5)CO_2C_2H_5$$

Oxalacetic ester and its alkyl derivatives undergo 'acid hydrolysis' when boiled with alkalis, and 'ketonic hydrolysis' when boiled with dilute sulphuric acid.

'*Acid hydrolysis*'

$$C_2H5O_2CCOCHRCO_2C_2H_5 \xrightarrow[\text{(ii) } H^+]{\text{(i) NaOH}}$$

$$(CO_2H)_2 + RCH_2CO_2 + 2C_2H_5OH$$

'*Ketonic hydrolysis*'

$$C_2H_5O_2CCOCH_2CO_2C\,_2H_5 + 2H_2O \xrightarrow{HCl} [HO_2CCOCHRCO_2H]$$

$$\longrightarrow CO_2 + RCH_2COCO_2H$$

Hydrolysis of oxalacetic ester with concentrated hydrochloric acid in the cold gives oxalacetic acid (*ketosuccinic acid*):

$$C_2H_5O_2CCOCH_2CO_2H_5 + 2H_2O \xrightarrow{HCl} HO_2CCOCH_2CO_2H$$

$$+ 2C_2H_5OH$$

A simpler preparation of oxalacetic acid is to oxidise maleic acid with Fenton's reagent.

Oxalacetic acid is a fairly stable substance, soluble in water, the aqueous solution giving a red colour with ferric chloride, thus indicating the existence of an enolic form. Oxalacetic acid exists in two forms, one with m.p. 155°C and the other 184°C. These two forms are actually hydroxymaleic acid, (1), and *hydroxyfumaric acid*, (II), respectively:

```
HO   CO2H        HO2C   H
  \ /               \  /
   C                 C
   ||                ||
   C                 C
  / \               / \
HO   CO2H          H   CO2H
   (I)               (II)
```

(I) is convertible into (II) by 30 per cent sulphuric acid, and the salts of (II) give (I) on treatment with dilute acid.

According to Kumler *et al.* (1962), who have examined the NMR spectrum of oxalacetic acid in deuterium oxide, at equilibrium there is 50 per cent of keto form and 8 per cent of enol, the rest being present in some other form.

Glutaric acid (*pentanedioic acid*), m.p. 97°C (relationship to glutamic acid and tartaric acid gave rise to its name), may be prepared by a number of methods, *e.g.*:

(i) By refluxing trimethylene dicyanide with concentrated hydrochloric acid:

$$NC(CH_2)_3CN + 4H_2O \xrightarrow{HCl} HO_2C(CH_2)_3CO_2H + 2NH_3$$

(83–53%)

(ii) By the action of methylene di-iodide on sodiomalonic ester:

$$2[CH(CO_2C_2H_5)_2]^- Na^+ + CH_2I_2 \longrightarrow {}_2NaI + CH_2[CH(CO_2C_2H5)_2]_2$$

$$\xrightarrow[\text{(ii) HCl}]{\text{(i) KOH}} CH_2[CH(CO_2H_2)_2]_2$$

$$\xrightarrow{150-200°C} HO_2C(CH_2)3CO_2H + {}_2CO_2$$

(iii) By condensing formaldehyde with malonic ester in the presence of diethylamine:

$$2(C_2H_5O_2C)_2CH_2 + CH_2O$$

$$\xrightarrow[\text{reflux}]{(C_2H_5)_2NH} (C_2H5O_2C)_2CHCH_2CH(CO_2C_2H5)_2 \quad (61\%)$$

This tetracarboxylic ester may now be treated as in (ii), or may be converted into glutaric acid directly by refluxing with concentrated hydrochloric acid:

$$(C_2H_5O_2C)_2CHCH_2CH(CO_2H_5)_2$$

$$\xrightarrow{HCl} HO_2C(CH_2)_3CO_2H \qquad (76\text{–}80\ \%)$$

When heated with acetic anhydride or thionyl chloride, glutaric acid is converted into glutaric anhydride:

$$CH_2\begin{cases}CH_2CO_2H\\CH_2CO_2H\end{cases} \longrightarrow CH_2\begin{cases}CH_2\text{—}CO\\CH_2\text{—}CO\end{cases}O + H_2O$$

Succinic and glutaric acids differ from their higher homologues in that they readily form the cyclic anhydride when heated with acetic anhydride.

Adipic acid (*hexanedioic acid*), m.p. 150°C, received its name from .the fact that it was first obtained by the oxidation of fats (Latin: adeps, fat). It may be synthesised from sodiomalonic ester and ethylene dibromide, but is prepared industrially by the oxidation of cyclohexane (from the hydrogenation of benzene) by air in the presence of a catalyst, *e.g.*, cobalt acetate:

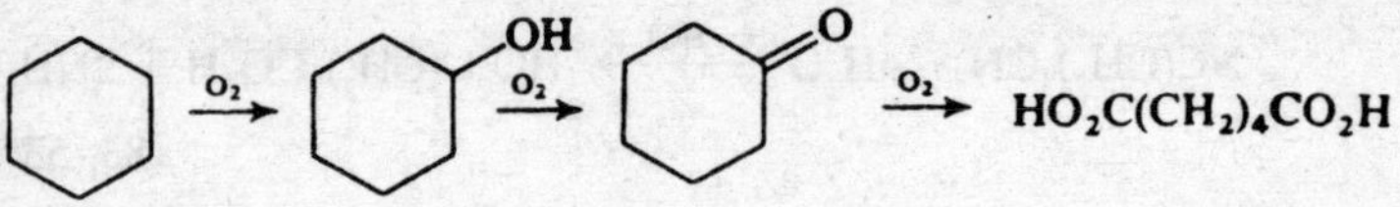

$$\text{cyclohexane} \xrightarrow{O_2} \text{cyclohexanol} \xrightarrow{O_2} \text{cyclohexanone} \xrightarrow{O_2} HO_2C(CH_2)_4CO_2H$$

Fig. 2.12

When adipic and pimelic acids are heated with acetic anhydride and the product distilled at 300°C, a cyclic ketone is obtained in each case:

Adipic acid is used for the preparation of resins (poly-esters), and is an intermediate in the manufacture of *nylon*. Nylon is used as the generic name for all synthetic fibre-forming polymeric amides having a protein-like structure. Nylon yarns and fabrics are practically non-inflammable. Nylon '66' is formed from adipic acid and hexamethylenediamine, and polymerisation is effected by melting the mixture in the presence of various catalysts.

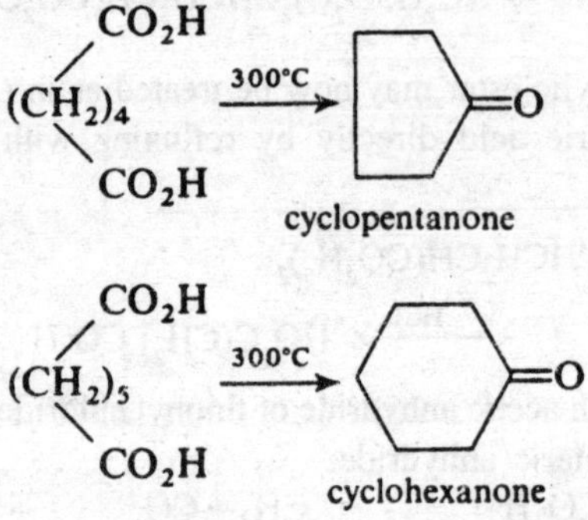

$$(CH_2)_4(CO_2H)_2 \xrightarrow{300°C} \text{cyclopentanone}$$

cyclopentanone

$$(CH_2)_5(CO_2H)_2 \xrightarrow{300°C} \text{cyclohexanone}$$

cyclohexanone

$$HO_2C(CH_2)_4CO_2H + H_2N(CH_2)_6NH_2 + HO_2C(CH_2)_4CO_2H + H_2N\text{—} \ldots \longrightarrow$$

$$+ HO_2C(CH_2)_4CO\text{—}[\text{—}NH(CH_2)_6NHCO(CH_2)_4CO\text{—}]_n\ NH(CH_2)_6NH_2$$

Nylon '6' is made from *caprolactam*, which is manufactured in several ways starting from cyclohexane:

The conversion of the oxime to caprolactam occurs via the Be Loann rearrangement, and the caprolactam is polymerised in the presence of a catalyst:

$$\text{caprolactam} \longrightarrow \text{—}[\text{—}CO(CH_2)_5NH\text{—}]_n\text{—}$$

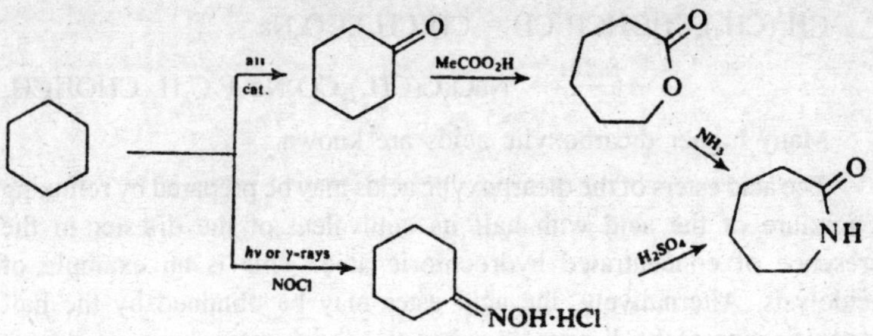

air
cat.
MeCOO$_2$H
O
O
O
NH$_3$
NH
O
hν or γ-rays
NOCl
H$_2$SO$_4$
NOH·HCl

Fig. 2.13

Pimelic acid (*heptanedioic acid*), m.p. 104°C, received its name from the fact that it was obtained originally from the oxidation of fats (Greek: *pimele*, fat).

It may be prepared by the hydrolysis of penta-methylene dicyanide (formed by the action of potassium cyanide on pentamethylene dibromide), or by the malonic ester synthesis, using trimethylene (*cf.* adipic acid). Pimelic acid may also be prepared by the reduction of salicylic acid with sodium and isopentanol, followed by the addition of water and then hydrochloric acid:

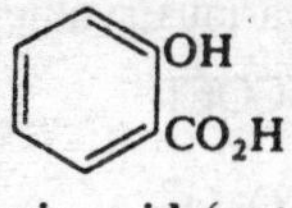

$$\text{C}_6\text{H}_4(\text{OH})(\text{CO}_2\text{H}) + \text{H}_2\text{O} \xrightarrow{e;\ H^+} \text{HO}_2\text{C}(\text{CH}_2)_5\text{CO}_2\text{H} \quad (43\text{–}50\%)$$

Subcric acid (*octanedioic acid*), $HO_2C(CH_2)_6CO_2H$, m.p. 144°C, may be synthesised by the electrolysis of potassium ethyl glutarate. It is prepared industrially by the oxidation of cork (Latin: *suber*, cork) with concentrated nitric acid.

Azelaic acid (*nonanedioic acid*), m.p. 107°C, may be synthesised from malonic ester and pentamethylene dibromide. It may be obtained .in the laboratory by refluxing castor oil with ethanolic potassium hydroxide, acidifying with sulphuric acid, and oxidising the crude ricinoleic acid so obtained with alkaline permanganate:

$$\text{CH}_3(\text{CH}_2)_5\text{CHOHCH}_2\text{CH} = \text{CH}(\text{CH}_2)_7\text{CO}_2\text{H}$$

$$\xrightarrow{[O]} \text{HO}_2\text{C}(\text{CH}_2)_7\text{CO}_2\text{H} \qquad (32\text{-}36\%)$$

Azelaic acid is prepared industrially by the ozonolysis folic acid.

Sebacic acid (*decanedioic acid*), m.p. 133°C, is prepared industrially by heating castor oil with sodium hydroxide.

$$CH_3(CH_2)_5CHOHCH_2CH = CH(CH_2)_7CO_2Na \xrightarrow{NaOH} NaO_2C(CH_2)_8CO_2Na + C_6H_{13}CHOHCH_3$$

Many higher dicarboxylic acids are known.

The acid esters of the dicarboxylic acids may be prepared by refluxing a mixture of the acid with half its equivalent of the di-ester in the presence of concentrated hydrochloric acid. This is an example of acidolysis. Alternatively, the acid ester may be obtained by the half saponification of the di-ester. In either case the di-ester is separated from the acid ester by fractional distillation.

Another method of preparing a half-ester is to heat equimolecular amounts of dicarboxylic acid and alcohol: the products are unchanged acid, half-ester and di-ester, which are then separated. If the dicarboxylic acid is substituted at one a-carbon atom, then the other carboxyl group is readily esterified by the Fischer-Speier method; this is probably a question of steric hindrance.

When a cyclic anhydride is available, this is a better way to prepare the half-ester. Here again, if one α-carbon atom is substituted, it is the half-ester of the other carboxyl group that is formed.

Half-esters are useful starting materials in various syntheses, *e.g.*,

(i)
$$\begin{array}{l} CH_2CH_2CO_2H \\ | \\ CH_2CH_2CO_2Et \end{array} \xrightarrow[\text{(ii) } Br_2]{\text{(i) } SOCl_2} \begin{array}{l} CH_2CHBrCOCl \\ | \\ CH_2CH_2CO_2Et \end{array}$$

$$\xrightarrow{ROH} \begin{array}{l} CH_2CHBrCO_2R \\ | \\ CH_2CH_2CO_2Et \end{array}$$

(ii) $HO_2C(CH_2)_8CO_2Et \xrightarrow[Br_2/CCl_4]{HgO} Br(CH_2)_nCO_2Et$

(iii) $HO_2C(CH_2)_nCO_2Et \xrightarrow{Na - EtOH} HO_2C\ (CH_2)_nCH_2OH$

When the calcium, barium, or better still, the thorium salts of the carboxylic acids from adipic to nonadecanedioic are distilled, varying yields of cyclic ketones are obtained; *e.g.*, adipic acid gives cyclopentanone:

$$(CH_2)_4 \begin{array}{l} \diagup CO_2 \diagdown \\ \diagdown CO_2 \diagup \end{array} Ca \longrightarrow \text{cyclopentanone}{=}O + CaCO_3$$

Based on the experimental results of the ease of formation of cyclic anhydrides and cyclic ketones, is **Blanc's rule** (1905). Blanc found that dicarboxylic acids, on heating with acetic anhydride, and then distilling at 300°C (or distilling directly at 300°C), gave cyclic anhydrides or cyclic ketones according to the relative positions of the two carboxyl groups. 1,4- and 1,5-dicarboxylic acids gave cyclic anhydrides. 1,6- and 1,7- gave cyclic ketones: this is Blanc's rule. By using Blanc's rule it is possible to determine the size of rings. A double bond is introduced into the ring, and the ring opened by oxidation to the corresponding dicarboxylic acid. This is then heated with acetic anhydride and distilled. If a cyclic anhydride is obtained, the acid is either 1,4- or 1,5-; if a cyclic ketone, either 1,6- or 1,7-; if there is no change, the acid is 1,8-or more. In naturally occurring compounds the rings are usually five- or six-membered. Hence the formation of the anhydride is taken to mean a five-membered ring, and of the ketone a six-membered ring. Blanc's rule, however, is only satisfactory for simple cyclic compounds. For substituted rings, the results may be misleading, since substituents in the chain bring about ring closure much more easily; *e.g.*, substituted adipic acids may give rise to the anhydride and not to the cyclic ketone.

COMPOUNDS RELATED TO UREA

Semicarbazide (*aminourea*), m.p. 96°C, may be prepared by treating hydrazine sulphate with potassium cyanate:

$$H_2NNH_2{\cdot}HNCO \longrightarrow NH_2CONHNH_2$$

It is manufactured by heating urea with hydrazine hydrate.

$$NH_2CONH_2 + N_2H_4{\cdot}H_2O \longrightarrow NH_2CONHNH_2 + NH_3 + H_2O$$

A more recent method is the electrolytic reduction of nitrourea in sulphuric acid solution using a lead anode:

$$NH_2CONHNO_2 \xrightarrow{e;H^+} NH_2CONHNH_2 + 2H_2O \ (61\text{–}69\%)$$

Semicarbazide is an important reagent for the identification of carbonyl compounds, with which it forms *semicarbazones*; it is also used in the Wolff-Kishner reduction.

Guanidine (*ammomethanamidine*), m.p. 50° C, is found in beet juice, and is one of the degradation products of the purines. It may be prepared by heating cyanamide with ammonium chloride:

$$NH_2CN + NH_4Cl \longrightarrow (NH_2)_2C = NH{\cdot}HCl$$

Guanidine is a *strong 'monoacid'* base, even forming a carbonate. Its strength as a base may be explained by resonance. There are unlike charges in the neutral molecule, whereas the ion has only one positive charge which, is spread, and at the same time all the resonating structures are equivalent (cf. urea).

Neutral molecule

$$(H_2N)_2C{=}NH \longleftrightarrow (H_2N^+)(H_2N)C{=}NH^- \longleftrightarrow (H_2N)(H_2N^+)C{=}NH^-$$

Ion

$$(H_2N)_2C{=}NH_2^+ \longleftrightarrow (H_2N^+)(H_2N)C{=}NH_2 \longleftrightarrow (H_2N)(H_2N^+)C{=}NH_2$$

X-ray analysis of guanidinium iodide shows that the three nitrogen atoms are symmetrically placed round the carbon atom. Furthermore, the C—N distance has been found to be 1.38Å (Theilacker, 1935; cf. urea). These facts are in keeping with the assumption that the guanidinium ion is a resonance hybrid.

Careful hydrolysis with barium hydroxide solution converts guaridine into urea:

$$(NH_2)_2C = NH + H_2O \longrightarrow (NH_2)_2C = O + NH_3$$

Guanidine nitrate may be prepared by heating dicyandiamide with ammonium nitrate:

$$(NH_2)_2C = NCN + 2NH_4NO_3 \longrightarrow 2(NH_2)_2C = NH{\cdot}HNO_3 \ (85\%)$$

When treated with concentrated sulphuric acid, guanidine nitrate is converted into nitroguanidine, $(NH_2)_2C = NNO_2$ (cf. urea nitrate). Nitroguanidine is used for making flashless powders.

Thiourea (*thiocarbamide*), m.p. 180°C, may be prepared by heating ammonium thiocyanate at 170°C for some time:

The structure of thiourea is similar to that of urea; it behaves as a 'monoacidic' base and, like urea, forms channel complexes. However, in this case, the diameter of the channel is about 7Å, and so can accommodate larger guest molecules (than can urea).

Thus, thiourea readily encloses branched-chain Carbohydrates and Proteins and even cycloalkanes. Apparently the cross-section of straight-chain alkanes is too small to support the channels in thiourea and consequently these compounds do not form channel complexes. When heated with alkalis, thiourea is hydrolysed:

$$S = C(NH_2)_2 + 2H_2O \xrightarrow{NaOH} CO_2 + H_2S + 2NH_3$$

Oxidation with *alkaline* permanganate or alkaline hydrogen peroxide converts thiourea into urea:

$$S = C(NH_2)_2 + [O] \longrightarrow O = C\ (NH_2)_2 + S$$

Oxides of lead, silver or mercury remove a molecule of hydrogen sulphide from thiourea into room temperature to form cyanamide:

$$(NH_2)_2CS + HGO \longrightarrow NH_2CN + HgS + H_2O$$

When treated with alkyl halide, thiourea forms S-*alkyl*-ψ-thiouronium salts (S-*alkyliso- thiouronium salts*):

$$CH_3I + S{=}C(NH_2)_2 \longrightarrow \left[CH_3SC\begin{matrix}{/\!\!/}NH_2\\ \diagdown NH_2\end{matrix}\right]I^-$$

These compounds are used to characterise sulphonic acids, with which they form insoluble salts. They may also be used to prepare thioalcohols, and also to prepare sulphonyl chlorides by oxidation with chlorine-water.

$$RSC\begin{matrix}{/\!\!/}\overset{+}{N}H_2Cl^-\\ \diagdown NH_2\end{matrix} \quad \xrightarrow[H_2O]{Cl_2} \quad RSO_2Cl$$

s-*Diphenyllhiourea* or *diphenylthiocarbanilide*, $(C_6H_5NH)_2CS$, which is used as a rubber accelerator, may be prepared by heating aniline with carbon disulphide in an ethanolic solution containing potassium hydroxide.

Dithiocarbamic acid, $S = C(NH_2)SH$, although unstable in the free state, gives rise to stable salts. These may be prepared by heating a primary or secondary mine with carbon disulphide. This reaction is complicated and the mechanism is uncertain.

A possibility is via the formation of the unstable N-alkyldithioca basic acid which forms the stable dithiocarbamic acid salt with another molecule of amine:

$$RNH_2 \; C(=S)S \longrightarrow \left[R{-}\overset{H}{\underset{H}{N^+}}{-}C(=S)(S^-) \right]$$

$$\xrightarrow{RNH_2} \left[RNH{-}C(=S)S^- \quad RNH_3^+ \right]$$

The Hofmam mustard oil reaction is believed to take place via the formation of a dithio-carbamic acid salt, which is then decomposed by the mercuric chloride, possibly as follows:

$$\left[RNH{-}C(=S)S^- \quad RNH_3^+ \right]$$

$$\xrightarrow{HgCl_2} RNH_2^+Cl^- + R{-}\overset{H}{N}{-}C(=S){-}S{-}Hg{-}Cl$$

$$\longrightarrow RN{=}C{=}S + HgS + HCl$$

EXERCISES

1. Convert n-$C_5H_{11}CO_2H$ into n-$C_4H_9CO_2H$ by three methods.
2. Which of the following would be most and least readily hydrolysed with NaOH, and why?

   (i) $MeCO_2Me$; (ii) Me_2CHCO_2Me; (iii) $MeCO_2Bu^t$.
3. Using MeI, $\overset{*}{M}eI$ NaCN, Na$\overset{*}{C}$N ($\overset{*}{C} = {}^{14}C$) and any other chemicals, suggest a synthesis for each of the following:

   (i) $\overset{*}{C}H_3CO_2H$; (ii) $CH_3\overset{*}{C}O_2H$

   (iii) $\overset{*}{C}H_3\overset{*}{C}O_2H$; (iv) $\overset{*}{C}H_3\overset{*}{C}O\overset{*}{C}H_3$;

   (v) $\overset{*}{C}H_3\overset{*}{C}HO$;
4. Complete the equations and comment:

(i) $CHCl_2COCl \xrightarrow{LAH} ?$

(ii) $HOCH_2CH_2CN \xrightarrow{HBr-H_2O} ?$

(iii) $B_{AC}2$ mech. $Me_3CCH_2OAc + OH^- \longrightarrow ?$

(iv) $B_{AL}1$ mech. $Me_3CCH_2OAc + OH^- \longrightarrow ?$

(v) $ClCH_2CO_2H + NaI \xrightarrow{acetone} ?$

(vi) $Me_2CBrCONH_2 \xrightarrow[NaOH]{Br_2} ?$

5. Calculate $\Delta G^{\ominus}$ for the equilibrium:

$$AcOH + EtOH \quad AcOEt + H_2O$$

moles	1/3	1/3	2/3	2/3

What information does the value of $\Delta G^{\varnothing}$ give on the time required for this equilibrium to be reached?

6. Draw the energy profile for the $A_{AC}2$ esterification mechanism and comment.

7. A solution of Ph_3CCO_2H in conc. H_2SO_4 gives $MeOCPh_3$ when poured into methanol. Discuss.

8. Convert:

   (i) n-BuOH into n-$PrNH_2$;

   (ii) Et_2CHCO_2H into Et_2CHCHO.

9. Can $MeCH_2COCl$, $MeCHClCO_2H$ and $ClCH_2CH_2CO_2H$ be differentiated by NMR spectroscopy? Give reasons for your answer.

10. Draw the diagrammatic NMR spectra of:

    (i) $MeCO_2Et$; (ii) $EtCO_2Me$.

11. Give the fragmentation paths that could produce the following

3

ETHERS AND POLYCARBONYL COMPOUNDS

INTRODUCTION TO ETHERS

The general formula of the ethers is $C_nH_{2n+2}O$ (which is the same as that for the monohydric alcohols), and since their general structure is R—O—R, they may be regarded as alkyl oxides or the anhydrides of the alcohols.

When the two alkyl groups in an ether are the same, the ether is said to be symmetrical or simple, *e.g.*, diethyl ether, C_2H—O—C_2H_5. When the two alkyl groups are different, the ether is said to be unsymmetrical of mixed, *e.g.*, ethyl methyl ether, CH_3—O—C_2H_5.

Nomenclature

1. In this system of nomenclature all the members are known as ethers, and the individuals are named according to the alkyl groups attached to the oxygen atom, *e.g.*, CH_3—O—CH_3, dimethyl ether, C_2H_5—O—$CH(CH_3)_2$, ethyl isopropyl ether.
2. According to the I.U.P.A.C. system of nomenclature, the ethers are regarded as hydrocarbons in which a hydrogen atom is replaced by an alkoxyl group, —OR, the larger group being chosen as the alkane. For symmetrical ethers, method 1 is to be used, *e.g.*, $C_2H_5OC_2H_5$, diethyl ether; $CH_3OC_2H_5$, methoxyethane.

General Methods of Preparation

1. Ethers may be prepared from alcohols in acid media, and the generally accepted mechanisms are that straight-chain primary alcohols react by the bimolecular mechanism (S_N2), t-alcohols by the unimolecular mechanism (S_N1), and s-alcohols by either of these mechanisms (the alcohol is the nucleophilic reagent):

$$S_N2 \quad ROH + H^+ \underset{}{\overset{\text{fast}}{\rightleftharpoons}} ROH_2^+$$

$$R{-}\ddot{O}(H) \;\; R{-}OH_2^+ \overset{\text{slow}}{\rightleftharpoons} R{-}\overset{+}{O}(H){-}R + H_2O \xrightarrow{\text{fast}} ROR + H_3O^+$$

$$S_N1 \quad ROH + H^+ \overset{\text{fast}}{\rightleftharpoons} ROH_2^+ \overset{\text{slow}}{\rightleftharpoons} R^+ + H_2O$$

$$R^+ \;\; \ddot{O}(H){-}R \xrightarrow{\text{fast}} R{-}\overset{+}{O}(H){-}R \xrightarrow[\text{fast}]{H_2O} ROR + H_3O^+$$

On the other hand, it still appears to be uncertain whether the preparation of ethers from primary alcohols and *concentrated* sulphuric acid proceeds by the S_N2 mechanism or via an intermediate alkyl hydrogen sulphate.

Dehydration of alcohols to ethers may also be carried out by passing the alcohol vapour over a heated catalyst such as alumina, aluminium phosphate, etc.

2. Ethers may be prepared by the addition of alcohols to alkenes in the presence of acid, *e.g.*, ethanol and isobutene give t-butyl ethyl ether:

$$Me_2C{=}CH_2 + H^+ \overset{H_2SO_4}{\rightleftharpoons} Me_3C^+ \overset{EtOH}{\rightleftharpoons} Me_3C{-}\overset{+}{O}(H)Et \xrightarrow{-H^+} Me_3COEt$$

This reaction gives very good yields with alkenes which can produce t-carbonium ions, and is very useful for preparing mixed ethers.

3. **By Williamson's synthesis**, in which sodium or potassium alkoxide is heated with an alkyl halide (S_N2):

$$R^1O^- \;\; R^2{-}X \rightleftharpoons \overset{\delta}{R^1O}\cdots R^2 \cdots \overset{\delta-}{X} \longrightarrow R^1OR^2 + X^- \quad (g.)$$

This method is particularly useful for preparing mixed ethers, and it is best to use the alkoxide of the secondary or tertiary alcohol, and primary alkyl halide because secondary and tertiary alkyl halides readily undergo E2 elimination in the presence of a strong base to form alkenes.

Williamson's synthesis proves the structure of the ethers.

When R^2 is a methyl or ethyl group, methyl or ethyl sulphate, respectively, can be used instead of the corresponding alkyl halide *e.g.*,

$$Na^+\{C_2H_5O^- \;\; CH_3{-}O{-}SO_3CH_3 \longrightarrow C_2H_5OCH_3 + CH_3OSO_3^-Na^+$$

This has been modified to a *general* method for preparing ethers by refluxing an alcohol with an ester of toluene-*p*-sulphonic acid in the presence of sodium:

$$R^1O^- \; R^2—O—SO_2—C_6H_4—Me \xrightarrow{Na} R^1OR^2 + Me—C_6H_4—SO$$

4. From halogeno-ethers and Grignard reagents.

5. Methyl ethers are readily prepared by the action of diazomethane on alcohols.

General Properties of the Ethers

The lower members are gases or volatile liquids, and their vapours are highly inflammable. There boiling points are much lower than those of the alcohols containing the same number of carbon atoms, and this is due to the fact that ethers cannot associate through hydrogen bonding. All the ethers are less dense than water in which they are not very soluble, but their solubility is very much increased in the presence of small amounts of alcohol.

The C—O group (stretch) in acyclic ethers containing two primary alkyl groups, *i.e.*, those containing the group $—CH_2—O—CH_2—$, absorbs in the region 1150–1060 cm^{-1}(s).

Fig: 7.1 is the infra-red spectrum of di-n-butyl ether (as a film).

The NMR spectra of a number of methyl ethers have been examined, and it appears that the methoxyl group in alkyl ethers shows one band with a τ-value of 6.60–6.80 p.p.m. Hence the presence of this band is very strong evidence that there is a methoxyl group in the molecule. In aromatic ethers, the τ-value is around 6.30 p.p.m.

Mass spectra. The molecular ion of ethers is weak, and the principal modes of fission occur through α- and β-cleavage:

$$R^1_3C^+ + R^2O\cdot \xleftarrow{\alpha} [R^1 \overset{\beta}{—} CR^1_2 \overset{\alpha}{—} OR^2]^{+} \xrightarrow{\beta} R^{1\cdot} + CR^1_2 \overset{+}{O}R^2$$

Ethers containing α-substituted alkyl groups also undergo *double* cleavage and rearrangement, and the resulting oxygen-containing fragment ions have high intensity and may even be the base peaks:

$$R^1_2C(R^1)—\overset{+}{O}—CH_2—CHR^2(H) \longrightarrow R^{1\cdot} + CR^1_2{=}\overset{+}{O}H + R^2CH{=}CH_2$$

This is the mechanism for the formation of ions *m/e* 45, 59, etc.

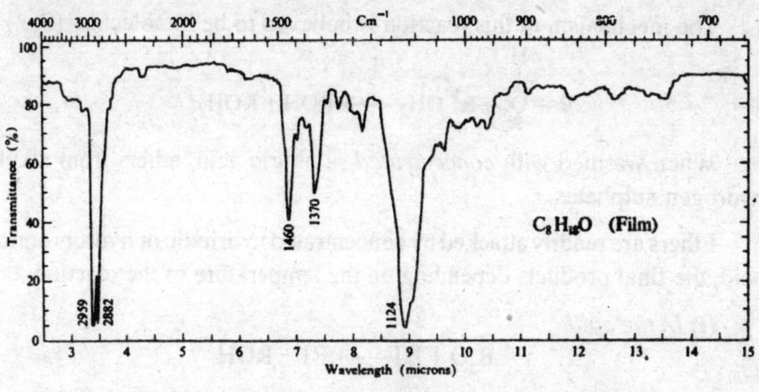

Fig. 3.1

1124 cm^{-1} : C—O (str.); this is characteristic of the group —CH_2—O—CH_2—. 2959 and 2882 cm^{-1}: C—H (str). in Me and/or CH_2. 1460 cm^{-1} : C—H (def.) in Me.

GENERAL REACTIONS OF ETHERS

1. Ethers dissolve in concentrated solutions of strong inorganic acids to form oxonium salts, *i.e.*, ethers behave as Bronsted-Lowry bases *e.g.*,

$$R_2O + H_2SO_4 \rightleftharpoons [R_2OH]^+HSO_4^-$$

It is the presence of the lone-pair electrons on the oxygen atom which predominantly characterises the reactions of the ethers. Thus, ethers also readily form co-ordination complexes (*etherates*) with Lewis acids, and treatment of an etherate with an alkyl fluoride produces a tertiary oxonium salt, *e.g.*,

$$R_2^1O + BF_3 \longrightarrow R_2^1\overset{+}{O}—\overset{-}{B}F_3 \xrightarrow{R^2F} [R_2^1OR^2]^+ BF_4^-$$

These trialkyloxonium fluoroborates are powerful alkylating reagents for nucleophilic substrates such as carboxylic acids, phenols, thioethers, etc., *e.g.*, thioethers are readily converted into trialkylsuphonium fluoroborates:

$$R_2^2\ddot{S}\ R^1—\overset{+}{O}R_2^1\}\ BF_4^- \longrightarrow [R_2^2S—R^1]^+BF_4^- + R_2^1O$$

2. When heated with dilute sulphuric acid under pressure, ethers form the corresponding alcohols:

$$R_2O + H_2O \xrightarrow{H_2SO_4} 2ROH$$

The mechanism of this reaction is believed to be bimolecular (S_N2):

$$R{-}\overset{\overset{\displaystyle H}{|}}{O^+}{-}R \;\; \ddot{O}H_2 \longrightarrow ROH + ROH_2^+$$

When warmed with *concentrated* sulphuric acid, ethers from alkyl hydrogen sulphates.

Ethers are readily attacked by concentrated hydriodic or hydrobromic acid, the final products depending on the temperature of the reaction.

(*i*) *In the cold:*

$$R_2O + HI \longrightarrow RI + ROH$$

The mechanism of this reaction presumably could be S_N1 or S_N2.

$$\mathbf{S_N1} \quad R_2OH^+ \underset{}{\overset{\text{slow}}{\rightleftharpoons}} ROH + R^+ \xrightarrow[\text{fast}]{I^-} RI$$

$$\mathbf{S_N2} \quad R(OH)^+R + I^- \rightleftharpoons \overset{\delta-}{I}\cdots R\cdots \overset{\delta+}{(OH)R} \longrightarrow I{-}R + ROH$$

When mixed ethers are used, the alkyl iodide produced depends on the nature of the alkyl groups. If one group is Me and the other a prim.- or s-alkyl group, it is methyl iodide which is produced. This can be explained on the assumption that the mechanism is S_N2, and because of the steric effect of the larger group, I^- attacks the smaller (methyl) group. This finds support from the following reaction. When optically active s-butyl methyl ether is decomposed with hydrogen iodide, methyl iodide and optically active s-butanol are obtained. Thus the s-Bu-O bond is *not* broken in this reaction.

When the substrate is a methyl-t-alkyl ether, the products are t-RI and MeOH. This can be explained by an S_N1 mechanism, the carbonium ion produced being the t-R since this type is more stable than a prim.- or s-R carbonium ion.

(*ii*) *When heated:*

$$R_2O + 2HI \longrightarrow 2RI + H_2O$$

In this case, (*i*) is followed by:

$$ROH + HI \longrightarrow RI + H_2O$$

Reactions (*i*) and (*ii*) are very useful for identifying the groups present in an ether. Furthermore, since (*i*) occurs very easily, it is possible to 'protect' a hydroxyl group in a polyfunctional compound by converting the hydroxyl group into an ether, which is later treated with concentrated hydriodic acid to regenerate the hydroxyl group.

Decomposition of ethers by concentrated hydriodic acid is the basis of the *Zeisel* method for estimating methoxyl and ethoxyl groups (p. 374).

Ethers are readily split by Lewis acids,

$$R_2O + BCl_3 \xrightarrow{\text{heat}} RCl + ROBCl_2$$
$$\xrightarrow{R_2O} RCl + (RO)_2BCl$$
$$\xrightarrow{R_2O} RCl + (RO)_3B$$

3. When treated with chlorine or bromine, ethers undergo substitution, the extent of which depends on the conditions. Hydrogen joined to the carbon directly attached to the oxygen atom is most readily replaced, *e.g.*, diethyl ether reacts with chlorine in the dark to form 1,1'-dichlorodiethyl ether:

$$CH_3CH_2OCH_2CH_3 \xrightarrow{Cl_2} CH_3CHClOCH_2CH_3$$
$$\xrightarrow{Cl_2} CH_3ClOCHClCH_3$$

In the presence of light, perchlorodiethyl ether is obtained:

$$(C_2H_5)_2O \xrightarrow{Cl_2} (C_2Cl_5)_2O$$

The reaction proceeds by a free-radical mechanism, and α-substitution occurs readily because of resonance stabilisation of the intermediate radical.

$$R^1\ddot{O}—CH_2R^2 \xrightarrow{Cl\cdot} R^1\ddot{O}—\dot{C}HR^2 \longleftrightarrow R^1\dot{O}=CHR^2$$
$$\xrightarrow{Cl_2} R^1O—CHClR^2 + Cl\cdot;\ \text{etc.}$$

4. Acid chlorides react with ethers when heated in the presence of anhydrous zinc chloride, aluminium chloride, etc.; *e.g.*,

$$MeCOCl + AlCl_3 \rightleftharpoons MeCO^+ + AlCl_4^-$$

$$R_2O + MeCO^+ \longrightarrow R—\overset{+}{O}(R)—COMe \longrightarrow$$
$$MeCO_2R + R^+ \xrightarrow{AlCl_4^-} RCl + AlCl_3$$

Acid anhydrides also split ethers to form esters:

$$R_2O + (CH_3CO)_2O \xrightarrow{ZnCl_2} 2CH_3CO_2R$$

5. Ethers react with carbon monoxide at 125–180°C and at a pressure of 500 atmospheres, in the presence of boron trifluoride plus a little water:

$$R_2O + CO \longrightarrow RCO_2R$$

6. Many ethers are attacked by alkyl derivatives of the alkali metals (and sodamide), *e.g.*,

$$Na^+C_2H_5^- \; H{-}CH_2{-}CH_2{-}O{-}C_2H_5 \longrightarrow$$

$$C_2H_6 + C_2H_4 + C_2H_5O^-Na^+$$

Only very strong bases are capable of abstracting hydrogen from ethers. The alkanes are extremely weak acids, but their conjugate bases are extremely strong bases.

Dimethyl ether (*methyl ether*) is prepared industrially by passing methanol vapour at 350–400°C, and a pressure of 15 atmospheres, over aluminium phosphate as catalyst. It is a gas, b.p. – 23.6°C, and is used as a refrigerating agent.

Diethyl ether (*ether, 'sulphuric ether'*) is prepared in the laboratory and industrially by the '*continuous etherification process*', *i.e.*, heating excess of ethanol with concentrated sulphuric acid. It is also obtained industrially: (*i*) as a by-product in the preparation of ethanol from ethylene and sulphuric acid; (*ii*) by passing ethanol vapour, under pressure, over heated alumina or aluminium phosphate (*cf.* dimethyl ether)

Diethyl ether is a colourless liquid, b.p. 34.5°C. It is fairly soluble in water, and is miscible with ethanol in all proportions. It is highly inflammable and forms explosive mixtures with air; this is a great disadvantage in its use as an industrial solvent for oils, fats, gums, resins, etc., and as an extracting solvent. It is also used in surgery as an anaesthetic, and is the usual solvent for carrying out Grignard reactions. In the presence of air and light, ether forms ether peroxide, $CH_3CH(OOH)OC_2H_5$, which is a heavy, pungent, oily liquid, and is explosive. Since its boiling point is higher than that of ether, it is left in the residue after ether distillations, and may cause explosions. Addition of a small amount of a cuprous compound, *e.g.*, cuprous oxide, has been recommended for avoiding the formation of ether peroxide. The chief impurity in ether is ethanol, and this has the property of preventing the formation of ether peroxide.

An important derivative of ether is 2,2'-dichlorodiethyl ether, which is prepared by heating ethylene chlorohydrin with concentrated sulphuric acid at 100°C:

$$2ClCH_2CH_2OH \xrightarrow[-H_2O]{H_2SO_4} ClCH_2CH_2OCH_2CH_2Cl$$

It may also be prepared by passing a mixture of ethylene and chlorine into ethylene chlorohydrin at 80°C:

$$HOCH_2CH_2Cl + C_2H_4 + Cl_2 \longrightarrow (ClCH_2CH_2)_2O + HCl$$

2,2'-Dichlorodiethyl ether is used as a solvent, as a soil fumigant and also as the starting point of many chemicals.

2,2'-Dichlorodiethyl ether is also named as bis-(2-chloroethyl) ether. The prefix bis (Latin, twice) indicates that there are *two identical groups* attached to a given atom; it is generally used for complex groups.

Di-isopropyl ether is obtained industrially as a by-product in the preparation of isopropanol from propene and sulphuric acid. It is also prepared by passing propene into 75 per cent sulphuric acid at 75–125°C under a pressure of 3–7 atmospheres; very little isopropanol is formed under these conditions.

Di-isopropyl ether is a colourless liquid, b.p. 69°C. It is used as an industrial solvent for extraction operations, and for decreasing the knocking properties of petrol which, mixed with di-isopropyl ether, acquires a high octane number.

Di-n-butyl ether, b.p. 142°C, is manufactured by heating n-butanol with concentrated sulphuric acid. It is used as an industrial solvent and as a solvent in Grignard reactions for which higher temperatures are required than can be obtained by using diethyl ether.

Di-isoamyl ether, $[(CH_3)_2CHCH_2CH_2—)_2O$, is prepared by the action of concentrated sulphuric acid on isopentanol. It is a colourless liquid, b.p. 172°C, and has a pear-like odour. It is used as an industrial solvent, and as a solvent in Grignard reactions.

Mixed ethers of the following types have been prepared: primary-primary; primary-secondary; primary-tertiary; secondary-secondary; secondary-tertiary; tertiary-tertiary.

UNSATURATED ETHERS

The simplest unsaturated ether is methyl vinyl ether, which is prepared industrially by passing acetylene into methanoi at 160-200°C in the presence of 1-2 per cent potassium methoxide, and under pressure sufficient to prevent boiling:

$$C_2H_2 + CH_3OH \longrightarrow CH_3OCH = CH_2$$

The acetylene is diluted with nitrogen to prevent explosions.

Methyl vinyl ether is a very reactive gas, b.p. 5-6°C. It is hydrolysed rapidly by dilute acid at room temperature to give methanoi and acetaldehyde (c/. the ethers), but is stable in alkaline solution. It undergoes many addition reactions at the double bond, *e.g.*,

It readily polymerises, and is used for making polyvinyl ether plastics.

Divinyl ether (*vinyl ether*), b.p. 39°C, may be prepared by heating ethylene chlorohydrin with sulphuric acid, and then passing the product, 2,2′-dichlorodiethyl ether, over potassium hydroxide at 200-240°C:

It is used as an anaesthetic, being better for this purpose than diethyl ether.

INTRODUCTION OF TAUTOMERISM

Acetoacetic ester or **ethyl acetoacetate** (E.A.A.) is the ethyl ester of acetoacetic acid, CH_3COCH_2H, a β-ketonic acid. Acetoacetic ester was fist discovered by Geuther (1863), who prepared it by the action of sodium on ethyl acetate, and suggested the formula $CH_3C(OH){=}CHCO_2C_2H_5$ (β-hydroxycrotonic ester). In 1865, Fankland and Duppa, who, independently of Geuther, also prepared acetoacetic ester by the action of sodium on ethyl acetate, proposed the formula CH_3COCH_2-$CO_2C_2H_5$ (β-ketobutyric ester). These two formulae immediately gave rise to two schools of thought, one upholding the Geuther formula, and the other the Frankland-Duppa formula, each school bringing forward evidence to prove its own claim, *e.g.*,

1. **Evidence in favour of the Geuther Formula** (*reactions of an unsaturated alcohol*). (*i*) When acetoacetic ester is treated with sodium, hydrogen is evolved and the sodium derivative is formed. This indicates the presence of a hydroxyl group.

(*ii*) When acetoacetic ester is treated with an ethanolic solution of bromine, the colour of the latter is immediately discharged. This indicates the presence of an olefinic double bond.

(*iii*) When acetoacetic ester is treated with ferric chloride, a reddish-violet colour is produced.

This is characteristic or compounds containing the group

$-C(OH)=C\langle$ (*cf.* phenols).

2. **Evidence in favour of the Frankland-Duppa formula** (*reactions of a ketone*).

(i) Acetoacetic ester forms a bisulphite compound with sodium hydrogen sulphite.

(*ii*) Acetoacetic ester forms a cyanohydrin with hydrogen cyanide.

Thus, the remarkable position arose where it was possible to show that a given compound had two different formulae, each of which was based on a number of *particular* reactions. The controversy continued until about 1910, when chemists were coming to the conclusion that both formulae were correct, and that the two compounds existed together in equilibrium in solution (or in the liquid state):

$$CH_3COCH_2CO_2C_2H_5 \rightleftharpoons CH_3\overset{\overset{\Large OH}{|}}{C}=CHCO_2C_2H_5$$

When a reagent which reacts with ketones is added to acetoacetic ester, the ketone form is removed. This upsets the equilibrium, and in order to restore the equilibrium mixture, the hydroxy-form of acetoacetic ester changes into the ketone form. Thus, provided sufficient reagent is added, acetoacetic ester reacts completely as the ketone from. Similarly, when a reagent which reacts with alkenes or with hydroxy-compounds is added in sufficient quantity, acetoacetic ester reacts completely as the hydroxy-form.

The problem was finally settled by Knorr (1911), who succeeded in isolating both forms. He cooled a solution of acetoacetic ester in light petrol to – 78°C, and obtained crystals which melted at –39°C. This substance gave no coloration with ferric chloride and did not combine with bromine, and was therefore the pure ketone form corresponding to the Frankland-Duppa formula.

Knorr then suspended the sodium derivative of acetoacetic ester in light petrol cooled to – 78°C, and treated his suspension with just enough hydrogen chloride to decompose the sodium salt. He now obtained a product which did not crystallise, but set to a glassy solid when cooled. This substance gave an intense coloration with ferric chloride, and was therefore the pure hydroxy-form corresponding to the Geuther formula.

Thus acetoacetic ester is a substance that does the duty of two structural isomers, each isomer being capable of changing rapidly into the other when the equilibrium is disturbed, *e.g.*, by the addition of certain reagents. This is a case of *dynamic isomerism*, and the name *tautomerism* (Greek: *same parts*) was given to this phenomenon by Laar (1885). The two forms are known as *tautomers* or *tautomerides*, the ketone isomer being called the *keto* form, and the hydroxy isomer the *enol* form. Hence this type of tautomerism is known as *keto-enol* tautomerism. (In the I.U.P.A.C. system of nomenclature the suffix *-en* indicates the presence of a double bond, and the suffix *-ol*, a hydroxyl group. The word *enol* is a combination of these suffixes and indicates the structure of this form.)

When one tautomer is more stable than the other under ordinary conditions, the former is known as the *stable* form, and the latter as the *labile* form. In practice, it is generally difficult to say which is the labile form, since very often a slight change in the conditions, *e.g.*, temperature or solvent, shifts the equilibrium from keto to enol or *vice-versa* . Tautomerism in the solid state is rare, and hence, in the solid state, one or other tautomer is normally stable, but in the liquid or gaseous state, or in solution, the two forms usually exist as an equilibrium mixture.

It has been found that the enol form is more volatile than the keto, ad that he change from enol to keto is extremely sensitive to catalysts. Meyer *et al.* (1920, 1921) found that traces of basic compounds were very effective catalysts. Thus the found that soft glass vessels were unsuitable for the separation of the keto and enol forms, since, when fractionated, the more volatile enol form rapidly changed into the original keto-enol mixture under he catalytic influence of the walls of the containing vessel. Meyer, however, succeeded in separating the enol form from the keto by carrying out the fractional distillation under reduced pressure in silica apparatus which had been thoroughly cleaned (free from dust, moisture, etc.). Distillation under these conditions is known as *aseptic distillation.*

In tautomerism in general thee may be an equilibrium between two or more forms. One may predominate, or all may be present to about the same extent, the concentration of each form depending on the temperature and the solvent (if in solution). Although it may not be possible to separate tautomers owing to the ease and rapidity of their interconversion or, as in many cases, due to one form being almost completely absent, the presence of more than one compound may be shown by special properties of each isomer. Experiments using deuterium exchange reactions have also the presence of keto-enol mixtures, *e.g.*, Klar (1934) showed that hydrogen is exchanged shown the presence of keto-enol mixtures, *e.g.*, Klar (1934) showed that hydrogen is exchanged slowly by deuterium when acetaldehyde is dissolved in D_2O. Since the C—H bond in alkanes is stable under these conditions, the inference is that a hydroxyl group is present in acetaldehyde, *e.g.*, some enol form is present. Acetone was found to undergo this exchanged more rapidly.

The methods used for the quantitative estimation of each form in any tautomeric equilibrium mixture, of which the keto-enol system is only one example, fall into two distinct groups, physical and chemical.

THEORIES OF TAUTOMERISM

Ingold (1927) suggested the name *cationotropy* for all those cases of tautomerism which involve the separation of a *cation*; and the name *anionotropy* for those cases which involve the separation of an *anion*. Lowry (1923) suggested the name *prototropy* for those cases in which a proton separates, and called such systems *prototropic* systems. Using Ingold's generalised classification of tuatomeric systems, it can be seen that prototropy is a special case of cationotropy. Braude and Jones (1944) have proposed the term *oxotropy* for anionotropic rearrangements involving only the migration of a *hydroxyl* group (see allylic rearrangement).

$$B + R_2CHCOR \longrightarrow BH^+ + R_2\bar{C}-\overset{O}{\overset{\|}{C}}R \longleftrightarrow R_2C{=}\overset{O^-}{\overset{|}{C}}R$$

(I)

Acid-catalysed enolisation involves the removal of a proton from the conjugate acid of the ketone, (II), and this process is dependent mainly on the hyperconjugation by the alkyl groups is the transition state for the formation of the carbon-carbon double bond.

$$H^+ + R_2\overset{H}{\overset{|}{C}}-\overset{O}{\overset{\|}{C}}-R \longrightarrow R_2\overset{H}{\overset{|}{C}}-\overset{OH}{\overset{|+}{C}}-R \longrightarrow H^+ + R_2C{=}\overset{OH}{\overset{|}{C}}-R$$

Evidence for these mechanisms is that alkyl groups depress base-catalysed reaction rates.

The mechanisms described above are step-wise mechanisms, and they appear to operate in certain cases. On the other hand, there is evidence that in other cases, acid-and base-catalysis of enolisation take place by a *concerted* mechanism, *i.e.*, the molecule undergoing change is attacked simultaneously at two places. Thus the enolisation of, *e.g.*, acetone, proceeds by, the simultaneous removal of a proton from a n α-carbon and the addition of a proton to the oxygen of the carbonyl group. This may be represented as follows (B is a base and HA is an acid):

$$B\cdots\!\!\rightarrow H\!-\!CH_2\!-\!C(CH_3)\!=\!O\leftarrow\!\cdots HA \longrightarrow \underset{\text{transition state}}{B\cdots H\cdots CH_2\!\overset{\cdot\cdot}{-}C(CH_3)\!\overset{\cdot\cdot}{=}O\cdots HA} \longrightarrow B\overset{+}{H} + CH_2\!=\!C(CH_3)\!-\!OH + A^-$$

In this type of mechanism, the solvent water molecules are believed to be involved, one acting as a proton acceptor and other a proton donor. Furthermore, the above mechanism is termolecular, and Swain (1950) has presented evidence to show that both the acid- and base-catalysed enolisation of acetone are termoleular reactions. Emmons *et al.* (1956) have obtained evidence supporting Swain's termolecular mechanism for ketone enolisation, but propose that the transition state for both acid- and base-catalysed enolisation may be represented by (III). Under these conditions in either acid-or base-catalysed reactions, the transition state will be very close to the enol and hence will be stabilised by hyper-conjugation. In the acid-actalysed reaction, bond *b* is relatively tight and bond *a* is relatively loose. Hence in acid-catalysed enolisation, steric hindrance around bond *a* is not observed in most cases, and hyper-conjugation is therefore a dominating factor in the enolisation of a ketone. In the base-catalysed reaction, bond *a* is relatively tight and bond *b* is relatively loose. Under these conditions a steric factor becomes important and is probably the reason why alkyl groups depress base-catalysed reaction rates.

The tremendous increase in speed by acid and base catalysts is well illustrated by the fact that the pure keto and enol forms of ethyl acetoacetate change very slowly into the equilibrium mixture (several weeks), whereas the latter is obtained rapidly (several minutes) by the addition of acid or base catalysts (lowry *et al.*, 1924).

$$\begin{array}{c} \overbrace{O\cdots HA}^{b} \\ RCH_2\diagdown\ \ \vdots \\ C\overset{\cdot\cdot}{=}CR \\ \diagup\ \ \vdots \\ RCH_2\ \ \underbrace{B\cdots H}_{a} \end{array}$$

(III)

Position of Equilibrium in Keto-enol Tautomerism

Since, $\Delta G^{\ominus} = \Delta H^{\ominus} - T\Delta S^{\ominus} = -RT \ln K$

a knowledge of the values of the enthalpy and entropy changes would enable us to calculate $\Delta G^{\ominus}$, and from this, *K*, the equilibrium constant of the reaction:

$$\text{keto enol}; \rightleftharpoons K = [\text{enol}]/[\text{keto}]$$

The sum of the bond energies of the — CH_2CO— group is 1870 and that of the — CH=C(OH)— groups is 1820 KJ (from Table 2.2):

$$2\text{C—H}(828.4) + \text{C — C}(347.3) + \text{C=O}(694.5) = 1870 \text{ KJ}$$

$$\text{C—H}(414.2) + \text{C= C}(606.7) + \text{C — O}(334.7) + \text{O — H}(464.4) = 1820 \text{ KJ}$$

This means that the *enthalpy of formation* of the CH_2CO group is – 1870 KJ and that of the CH=C(OH) group is – 1820 KJ. Thus the keto from is more stable than the enol by 50 KJ mol^{-1}, and hence there must be some driving force to bring about enolisation.

If the equilibrium mixture contains 1 per cent enol, then.

$$\Delta G^{\ominus} = -5.7 \log 1/99 = 11.42 \text{ kJ}$$

If the equilibrium mixture contains 99 per cent enol, then

$$\Delta G^{\ominus} = -5.7 \log 99 = -11.42 \text{ kJ}$$

Thus a difference of 22.84 kJ mol^{-1} in ΔG will have large effects on the position of the equilibrium. Hence any attempt to calculate *K* from *estimated* values of ΔH and ΔS is almost certainly doomed to failure. However, a semi-quantitative approach is instructive. Any factors that affect $\Delta H^{\ominus}$ and $\Delta S^{\ominus}$ will affect the value of $\Delta G^{\ominus}$. Consider the following equilibria:

The enol form, in each case, is stabilised by intramolecular hydrogen bonding (say, 29.3 kJ mol^{-1}).

Thus, in the absence of any other factor affecting ΔH, we have:

$$\Delta H^{\ominus} = \Sigma H^{\ominus}(\text{products}) - \Sigma H^{\ominus}(\text{reactants})$$

$$= -(1820 + 29.3) - (-1870) = +20.7 \text{ kJ}$$

$$\Delta G^{\ominus} = 20.7 - T\Delta S^{\ominus} = -RT \ln K$$

Now let us consider the resonance energies of crotonaldehyde (I), ethyl acetate (II) and ethyl *acetylacetone* $\mathrm{Me{-}\overset{\overset{\displaystyle O}{\|}}{C}{-}CH_2{-}\overset{\overset{\displaystyle O}{\|}}{C}{-}Me \rightleftharpoons MeC(OH){=}CH{-}C(=O)Me}$ (chelated enol, O–H···O)

ethyl acetoacetate $\mathrm{Me{-}\overset{\overset{\displaystyle O}{\|}}{C}{-}CH_2{-}\overset{\overset{\displaystyle O}{\|}}{C}{-}OEt \rightleftharpoons MeC(OH){=}CH{-}C(=O)OEt}$ (chelated enol, O–H···O)

ethyl malonate $\mathrm{EtO\overset{\overset{\displaystyle O}{\|}}{C}{-}CH_2{-}\overset{\overset{\displaystyle O}{\|}}{C}{-}OEt \rightleftharpoons EtOC(OH){=}CH{-}C(=O)OEt}$ (chelated enol, O–H···O)

	MeCH=CHCHO	$\mathrm{Me\overset{\overset{\displaystyle O}{\|}}{C}{-}OEt}$	$\mathrm{CH_2{=}C(Me){-}C({=}O){-}OEt}$
R.E. (kJ)	(I) 10.04	(II) 75.31	(III) 75.31

methacrylate (III). The important point here is that in (III) the ethylenic bond does not enter into resonance with the CO_2Et group. (III) contains crossed conjugation , and the contribution to resonance is only the path that leads to the greater R.E., which, in this case, is the carbethoxy group.

Inspection of the enol formulae shows that the ring contains a conjugated system:

$$\mathrm{H{-}O{-}\overset{|}{C}{=}CH{-}\overset{|}{C}{=}O \longleftrightarrow H{-}\overset{+}{O}{=}\overset{|}{C}{-}CH{=}\overset{|}{C}{-}O^-}$$

The R.E. value of this is difficult to assess, but it can be expected to be considerably less than 73.31 kJ. Thus, in E.A.A., only the remaining group, HO—C=CH— , will make any contribution to the 'ring' resonance. It therefore follows that the enol form of E.A.A. is less stabilised by resonance than that of acetylacetone, and so will be present in smaller amount. It might also be noted here that if enolisation in E.A.A. occurred at the CO group of CO_2Et, then the R.E. of the system produced is far less. Thus there is no tendency to enolise in this direction. From this it can be seen that D.E.M. will have far less enol form than

E.A.A. In the case of simple carbonyl compounds, *e.g.*, acetone, there is no intramolecular hydrogen bonding to stabilise the enol. Thus there is very little tendency for these carbonyl compounds to enolise.

Just as it is difficult to estimate $\Delta H^{\ominus}$, so it is more difficult to estimate $\Delta S^{\ominus}$. In the diketo form, the molecule is acyclic and consequently can assume various conformations. In the enol form, however, the presence of a double bond prevents change in conformation at this part of the molecule and, at the same time, conformational change is further restricted by hydrogen bonding, since this tends to maintaining structure. Thus both of these factors will tend to cause a *decrease* in entropy when the diketo compound enolises. Some support for this argument is that Geo (1961) has found that cyclic monoketones contain more enol than the corresponding acyclic 2-one. In the cyclic ketones, change from keto to enol involves a relatively small change in strain in the ring due to the introduction of a double bond. For acyclic ketones, the introduction of the double bond has a much greater effect on the freedom to take up different conformations.

Another factor that determines the enol content is the steric factor. Let us consider the enol content of acetylacetone and α-methylacetylacetone; the values given are for the gas phase (Conant *et al.*, 1932). In the latter compound, thee is much greater steric repulsion due to the presence of the α-methyl group. Thus, the α-methyl enol form has greater internal strain than the enol of acetylacetone and so the former is less favourable energetically than the latter. Hence less enol will be present in the former equilibrium mixture than in the latter. A striking example for strain affecting the position of equilibrium is the case of triacylmethanes, (RCO)3CH. Nonhebel (1967) has shown, by means of i.r., u.v., and NMR studies, that as the size of R increases, the amount of enol content decreases, *e.g.*, $(Bu^tCO)_3CH$ is *completely* in the keto form.

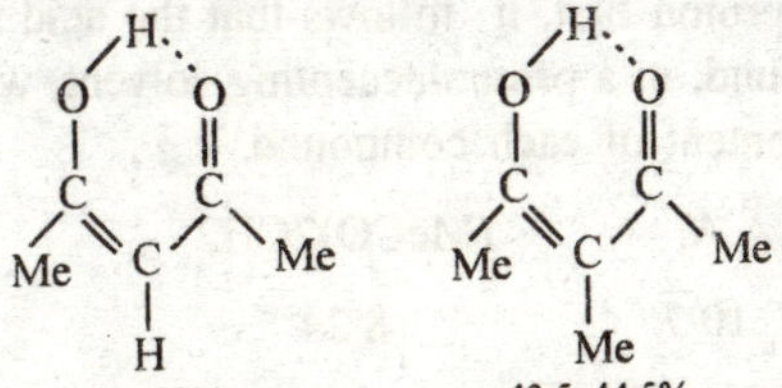

91–93% 43.5–44.5%

Finally, let us consider the effect of the nature of the solvent on the enol content. If one form is more solvated than the other, then there may be considerable entropy effects connected with the nature of the solvent (*cf.* acids). Any solvent that can form hydrogen bonds with the carbonyl group of the keto form will cause a decrease in entropy. The enol form, however, since it forms an intramolecular bond, will be largely prevented from forming hydrogen bonds with the solvent, and consequently there will be a smaller decrease in entropy. Thus, the term ΔS becomes important in deciding the position of keto–enol equilibrium, *e.g.*, solvents such as water, methanol, acetic acid, etc., tend to reduce the enol content. On the other hand, in solvents such as hexane, benzene, etc., the enol content will be larger, *e.g.*, the enol content of acetylacetone in hexane is 92 per cent.

It has been shown that enol content of E.A.A. (pure liquid) decreases as the pressure increases (le Noble, 1960). From this it follows that the volume of the enol form is greater than that of the keto.

THE ACIDITY OF KETO COMPOUNDS

The proton transfer for these compounds may be written:

$$-CH_2-\overset{\overset{\displaystyle O}{\|}}{C}- + B \rightleftharpoons BH^+ + -\ddot{C}H-\overset{\overset{\displaystyle O}{\|}}{C}- \longleftrightarrow -CH=\overset{\overset{\displaystyle O^-}{|}}{C}-$$

The carbanion produced is a resonance hybrid, and the transference of charge to the oxygen atom which has a higher electronegativity than the carbon atom will make this resonating structure a greater contributor to the resonance hybrid; the result is an ambident anion. The presence of the carbonyl group facilitates the release of a proton from the α-carbon atom, and the resonance hybrid nature of the conjugate base stabilises this with respect to its acid. Both factors are acting in the 'same' direction, and consequently such compounds will be acidic. However, the pK_a of the compound will depend on how strong these two driving forces are. As we have seen above, since enolisation involves a removal of a proton first, it follows that the acid strengths of these carbonyl compound, in a proton-accepting solvent, will run parallelled with the enol content of each compound, *e.g.*,

	E.A.A.	(MeCO)2CH2
pK_a	10·7	8·24

Infra-red Spectra of β-diketones

If β-diketones existed as such, one would expect to find absorption in the region of a single carbonyl group (1725–1700 cm^{-1}), but with greater intensity because two such groups are present. In practice, those β-diketones which exist partly (or wholly) in the enol from (as indicated by the coloration with ferric chloride) show the carbonyl absorption in the region 1640–1540 cm^{-1} (s); the enolic form of β-keto-esters absorbs in the region 1655-1635 cm^{-1} (s), α,β-Unsaturated ketones, *i.e.*, those containing the grouping C=C—C=O absorb in the region 1685–1665 cm^{-1} (s). Enols contain this grouping, but the shift that occurs in these compounds is attributed to hydrogen bonding as follows: —O···O=C— (see chelated structures of the enols described above). At the same time, the true alcoholic hydroxyl band near 3700 cm^{-1} is absent in enols, but there is a band near 2700 cm^{-1} (s) which is attributed to the *chelated* hydroxyl group. Thus, infra-red spectral studies of β-dicarbonyl compounds may be used to ascertain their structures.

NMR spectra studies have been very useful in the examination of keto -enol tautomeric mixtures, *e.g.*, Fig 3.2 is the NMR spectrum of ethyl acetoacetate (in CCl_4 at 60 MHz). and it can be seen that the CH_2 and =CH protons give separate signals. Furthermore, very careful integration will give the proton ratio of these two groups and so it is possible to estimate the amount of enol content.

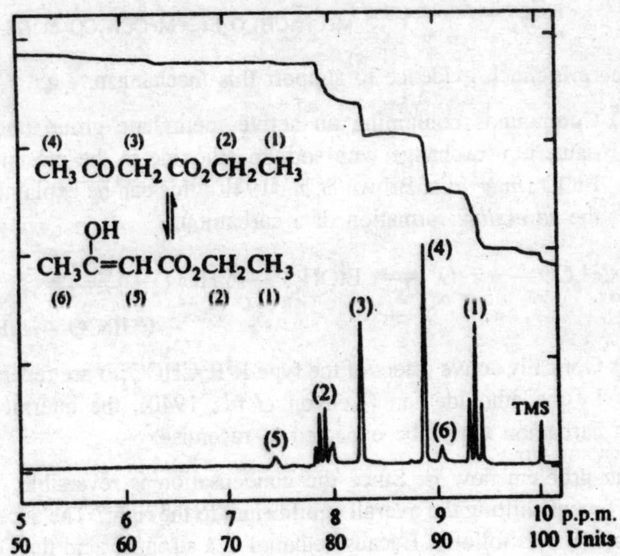

Fig. 3.2

THE PREPARATION OF ETHYL ACETOACETATE

In the preparation of ethyl acetoacetate, condensation is effected between two molecules of the same ester. This is one example of the **Claisen condensation** (1887), in which a keto-ester is formed by the reaction between two molecules of an ester containing α-hydrogen atoms. The Claisen condensation may also take place between an ester and a ketone to form a 1,3-diketone. The condensation is brought about by sodium ethoxide, sodamide, triphenylmethylsodium, etc.

Many mechanisms have been proposed for the formation of ethyl acetoacetate. The one most widely accepted is (*cf.* aldol condensation):

$$EtO^- + H{-}CH_2CO_2Et \rightleftharpoons EtOH + \bar{C}H_2CO_2Et$$

$$MeC(=O)(OEt) + \bar{C}H_2CO_2Et \rightleftharpoons MeC(O^-)(OEt){-}CH_2CO_2Et \rightleftharpoons$$

$$MeC(=O){-}CH(H)CO_2Et + EtO^- \rightleftharpoons EtOH + MeC(\bar{O}){=}CHCO_2Et \xrightarrow{MeCO_2H}$$

$$MeC(OH){=}CHCO_2Et \rightleftharpoons MeC(=O)CH_2CO_2Et \quad (28\text{–}29\%)$$

There is much evidence to support this mechanism, *e.g.*;

(*i*) Compounds containing an active methylene group undergo deuterium exchange with sodium ethoxide in the presence of EtOD (*inter alia*, Brown *et al.*, 1940); this can be explained by the *reversible* formation of a carbanion,

$$—CH_2CO— + EtO^- \rightleftharpoons EtOH + —\bar{C}HCO— \xrightleftharpoons{EtOD} —CHDCO— + EtOH$$

(*ii*) Optically active esters of the type $R^1R^2CHCO_2Et$ are racemised by the ethoxide ion (Kenyon *et al.*, 1940); the intermediate carbanion would be expected to racemise.

The problem now is; Since the condensation is reversible, what factors favour shifting the overall equilibrium to the right? The accepted explanation is as follows. Because ethanol is a stronger acid than ethyl acetate, the equilibrium of the first stage lies on the left. However, E.A.A. is a stronger acid pK_a 10.68) than ethanol (pK_a 15.5). Thus, the former is present mainly as its anion, *i.e.*, conjugate base, in the last stage of the condensation, and so the equilibrium lies largely on the right. In this way, the E.A.A. anion is 'removed' from the reaction, and so the overall equilibrium is displaced to the right. This argument is supported by the fact that ethyl isobutyrate does *not* undergo the Claisen condensation. The β-keto ester that would be produced, $Me_2CHCOCMe_2CO_2Et$, does not contain an active α-hydrogen atom, and consequently the last reversible stage of the reaction given for ethyl acetate as substrate is now absent. Hence, for ethyl isobutyrate, the overall equilibrium lies almost completely on the left. If a very strong base such as the triphenylmethyl anion (as sodium salt), $Ph_3C^-Na^+$ (pK_a of Ph_3CH is about 33) is used, then the equilibrium of the first stage with ethyl isobutyrate lies on the right, and the Claisen condensation will proceed, since the position of equilibrium of the last stage will no longer matter.

Ethyl acetoacetate is also prepared industrially by polymerising ketone in acetone solution to diketone, which is then treated with ethanol. A possible mechanism is :

$$\begin{matrix} CH_2{=}C{=}O \\ CH_2{=}C{=}O \end{matrix} \longrightarrow \begin{matrix} CH_2{=}C{-}O \\ | \quad\;\; | \\ CH_2{-}C{=}O \end{matrix} \xrightarrow{EtOH} \begin{matrix} CH_2{=}C{-}O \quad H \\ | \quad\;\; | \;\; / \\ CH_2{-}C{-}\overset{+}{O} \\ \quad | \quad\; \backslash \\ \quad O^- \quad OEt \end{matrix} \longrightarrow$$

$$\left[\begin{matrix} O^-\cdots H{-}\overset{+}{O}{-}Et \\ | \qquad\qquad | \\ CH_2{=}C{-}CH_2{-}C{=}O \end{matrix}\right] \longrightarrow \left[\begin{matrix} O{-}H\cdots O \\ | \qquad\quad \| \\ CH_2{=}C{-}CH_2{-}C{-}OEt \end{matrix}\right] \longrightarrow CH_3\overset{O}{\overset{\|}{C}}CH_2\overset{O}{\overset{\|}{C}}{-}OEt$$

In 'mixed ethyl acetoacetate condensations' between two different esters, a mixture of all four possible products is usually obtained; *e.g.*, with ethyl acetate and ethyl propionate:

$$CH_3CO\boxed{OC_2H_5 + H}{-}CH_2CO_2C_2H_5 \xrightarrow{C_2H_5ONa} CH_3COCH_2CO_2C_2H_5 + C_2H_5OH$$

$$CH_3CH_2CO\boxed{OC_2H_5 + H}{-}\underset{}{\overset{CH_3}{\overset{|}{C}H}}{-}CO_2C_2H_5 \xrightarrow{C_2H_5ONa} CH_3CH_2CO\overset{CH_3}{\overset{|}{C}H}CO_2C_2H_5 + C_2H_5OH$$

$$CH_3CO\boxed{OC_2H_5 + H}{-}\overset{CH_3}{\overset{|}{C}H}CO_2C_2H_5 \xrightarrow{C_2H_5ONa} CH_3CO\overset{CH_3}{\overset{|}{C}H}CO_2C_2H_5 + C_2H_5OH$$

$$CH_3CH_2CO\boxed{OC_2H_5 + H}{-}CH_2CO_2C_2H_5 \xrightarrow{C_2H_5ONa} CH_3CH_2COCH_2CO_2C_2H_5 + C_2H_5OH$$

Esters with no α-hydrogen atoms will condense only 'one way' with compounds containing an active methylene group. Three important esters of this type are formic, benzoic and oxalic esters, and these give, respectively, the formyl, benzoyl and oxalyl derivatives Ethyl carbonate also does not contain any α-hydrogen atoms, and is very useful for introducing a carbethyoxyl group into compounds with an active methylene group.

β-Keto esters, other than E.A.A., may be prepared by the Claisen condensation between ketones and ethyl carbonate or ethyl oxalate (*cf.* decarbonylation of ethyl pyruvate):

$$RCOMe + EtO_2CCO_2Et \xrightarrow{EtONa} RCOCH_2COCO_2Et$$

$$\xrightarrow{heat} RCOCH_2CO_2Et + CO$$

Properties of Ethyl Acetoacetate

Ethyl acetoacetate is a colourless, pleasant-smelling liquid, p.b. 181°C (with slight decomposition), sparingly soluble in water, but miscible with most organic solvents. It is readily reduced by sodium amalgam, or catalytically to β-hydroxybutyric ester:

$$CH_3COCH_2COC_2H_5 \longrightarrow CH_3CH(OH)CH_2CO_2C_2H_5$$

This is also produced by using lithium aluminium hydride in pyridine. In the absence of pyridine, the product is butane-1,3-diol (Buchta *et al.*, 1951).

$$CH_3COCH_2COC_2H_5 \xrightarrow{LiAlH_4} CH_3CHOHCH_2CH_2OH \text{ (30\%)}$$

Kent *et al.* (1963) have shown that sodium borohydride in boiling ethanol also produces the diol.

Ethyl acetoacetate is neutral to litmus, but is soluble in dilute sodium hydroxide solution: it is the enol form which dissolves to form the sodium salt. It does *not* give the haloform reaction. When ethyl acetoacetate is hydrolysed in the cold with dilute sodium hydroxide, the solution then acidified, extracted with ether and the ether removed in the cold under reduced pressure, free acetoacetic acid is obtained. Krueger (1952) has prepared acetoacetic acid as a crystalline solid, m.p. 36–37°C. As prepared in the usual way it is unstable, readily decomposing into acetone ad carbon dioxide:

$$CH_3COCH_2CO_2H \longrightarrow CH_3COCH_3 + CO_2$$

All β-ketonic acids readily decompose in this manner, but it is interesting to note that trifluoroacetoacetic acid, $CF_3COCH_2CO_2H$, is quite stable and can be distilled without much decomposition. Ethyl acetoacetate forms α-oximinoacetoacetate when treated with nitrous acid:

$$CH_3COCH_2CO_2C_2H_5 \xrightarrow{HNO_2} CH_3COC(=NOH)CO_2C_2H_5$$

On the other hand, monosubstituted derivatives of ethyl acetoacetate are split by nitrous acid; the nitroso-compound is formed first and then rearranges to the oximino form with fission of the molecule.

$$CH_3COCHRCO_2C_2H_5 \xrightarrow{HNO_2} CH_3CO\underset{\displaystyle NO}{\underset{|}{C}}RCO_2C_2H_5 \longrightarrow R\underset{\displaystyle NO}{\underset{||}{C}}CO_2C_2H_5$$

Ethyl acetoacetate forms a green copper compound which is soluble in organic solvents. This indicates that the copper compound is not an ionic but a chelate compound (the alternative is a π-delocalised structure). Bromination of E.A.A. is catalysed by the presence of cupric ions; this is due to the stabilisation of the enol form by chelate formation.

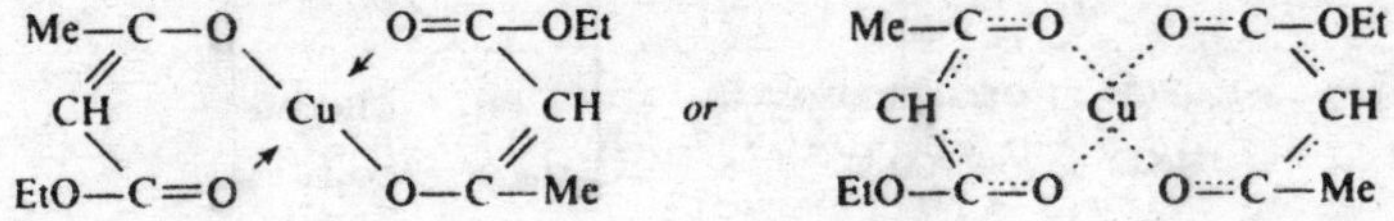

```
Me—C—O         O=C—OEt              Me—C⋯O     O⋯C—OEt
  //    \    ↙      \                  ⁄⁄⋮  ⋱  ⋰   ⋮\\
 CH      Cu          CH      or      CH      Cu       CH
  \    ↗    \       //                 \\⋮  ⋰  ⋱   ⋮⁄⁄
EtO—C=O        O—C—Me              EtO—C⋯O     O⋯C—Me
```

E.A.A. reacts with ammonia (and primary amines) to form an enamine:

$$Me-\overset{\displaystyle O}{\overset{||}{C}}-CH_2CO_2Et + NH_3 \longrightarrow Me-\overset{\displaystyle OH}{\overset{|}{\underset{\displaystyle NH_2}{\underset{|}{C}}}}-CH_2CO_2Et \xrightarrow{-H_2O}$$

$$Me-\underset{\displaystyle NH_2}{\underset{|}{C}}=CHCO_2Et$$

Monoalkyl derivatives of E.A.A. also behave in a similar manner, but dialkyl derivatives form amides, $MeCOCR_2CONH_2$. In the latter, there is no α-hydrogen atom.

Ethyl acetoacetate reacts with hydoxylamine to form an isoxazolone.

$$CH_2COCH_2CO_2C_2H_5 + NH_2OH \longrightarrow CH_3\underset{\displaystyle NOH}{\underset{||}{C}}CH_2CO_2C_2H_5 \longrightarrow$$

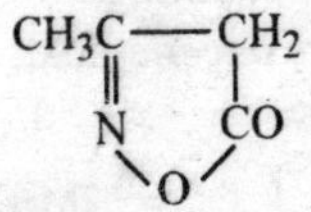

```
CH3C——CH2
  ||    |      + C2H5OH
  N     CO
   \   /
     O
```

With phenylhydrazine, ethyl acetoacetate forms a pyrazolone.

Acetoacetic ester undergoes the Knoevengel reaction due to the presence of an 'active' methylene group, and also couples with diazonium salts. It reacts with Grignard reagents to form the hydrocarbon, which indicates that it reacts in the enol form, *e.g.*,

$$CH_3\overset{\displaystyle OH}{\overset{|}{C}}{=}CHCO_2C_2H_5 + CH_3MgI \longrightarrow CH_3\overset{\displaystyle OH}{\overset{|}{C}}{=}CHCO_2C_2H_5 + CH_4$$

The *O*-methyl derivative is also produced by the action of diazomethane on ethyl acetoacetate. Thus the latter compound reacts in the enol form.

When acetoacetic ester is heated under reflux with a trace of sodium hydrogen carbonate, *dehydroacetic acid*, m.p. 109°C, is obtained.

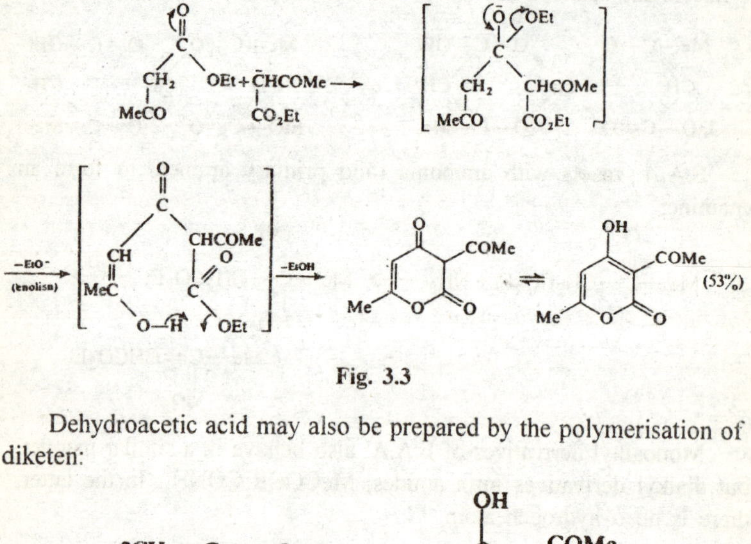

$$MeCOCH_2C(=O)OEt + \bar{C}HCOMe(CO_2Et) \longrightarrow [MeCOCH_2C(\bar{O})(OEt)CH(COMe)CO_2Et]$$

$$\xrightarrow[\text{(enolisn)}]{-EtO^-} [MeC(OH){=}CH{-}C(=O){-}CH(COMe){-}C(=O)OEt] \xrightarrow{-EtOH} \text{Me-pyranone with COMe (keto form)} \rightleftharpoons \text{OH, COMe, Me pyranone}\ (53\%)$$

Fig. 3.3

Dehydroacetic acid may also be prepared by the polymerisation of diketen:

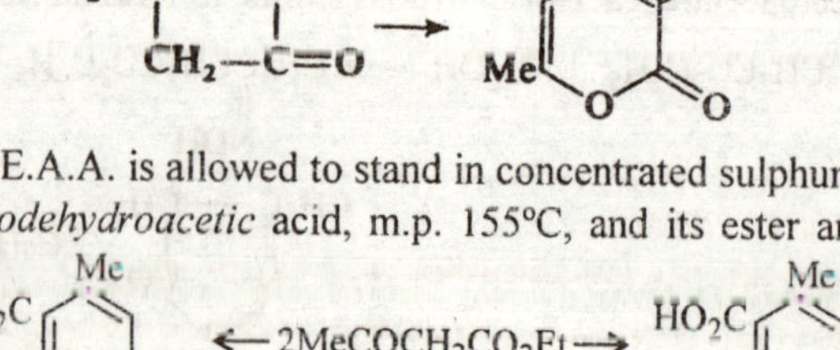

$$2\,CH_2{=}C{-}O{-}C(=O){-}CH_2 \longrightarrow \text{4-hydroxy-3-acetyl-6-methyl-2-pyrone (OH, COMe, Me)}$$

When E.A.A. is allowed to stand in concentrated sulphuric acid (5–6 days), *isodehydroacetic* acid, m.p. 155°C, and its ester are formed:

$$\underset{(27\text{–}36\%)}{\text{Me, } EtO_2C,\ \text{Me-2-pyrone}} \longleftarrow 2MeCOCH_2CO_2Et \longrightarrow \underset{(22\text{–}27\%)}{\text{Me, } HO_2C,\ \text{Me-2-pyrone}}$$

Both acids are α-pyrone deivatives.

Physical Methods

Physical methods of not disturb the equilibrium, for they do not depend on the removal of one form, and they should therefore be used wherever possible.

(*i*) The refractive index of the equilibrium mixture is determined experimentally. The refractive indices of both the keto and the enol forms are calculated (from a table of atomic refractions), and from these figures it is then possible to calculate the amount of each form present in the equilibrium mixture. In some cases the refractive index of each form may be obtained directly by isolating it, *e.g.*, acetoacetic ester.

(*ii*) If *one* from is an electrolyte, the electrical conductivity of the mixture is determined experimentally, and the amount of this form present may be calculated from the results, *e.g.*, nitromethane.

(*iii*) The composition of the equilibrium mixture may be determined by means of optical rotation measurements (*cf.* mutarotation).

(*iv*) The position of equilibrium in b-diketones and β-keto-esters has been determined by proton magnetic resonance.

Chemical Methods

Since chemical methods cause the removal of one form, it is necessary to use a reagent that reacts with this form faster than the rate of interconversion of the tautomers. Meyer (1911, 1912) found that in the case of keto enol tautomerism, bromine reacts instantaneously with the enol form, and so slowly with the keto form in comparison with the enol, that the keto reaction may be ignored. Meyer introduced two procedures, the *direct* and the *indirect* method. In the *direct* method a weighed sample of the keto-enol mixture dissolved in ethanol is *rapidly* titrated with a dilute ethanolic solution of bromine at 0°C (to slow down the interconversion of the tautomers). The first appearance of excess of bromine indicates the end point.

The titration must be carried out rapidly; otherwise the keto form changes into the enol during the time taken for the titration. In any case, it has been found impossible to carry out the titration sufficiently rapidly to avoid the conversion of some keto into enol, and so this method always results in too high a value for the conversion of some keto into enol, and so this method always results in too high a value for the enol form.

More reliable results may be obtained by the *indirect* method. An *excess* of dilute ethanolic solution of bromine is added rapidly to the weighed sample dissolved in ethanol, and then an excess of 2-nanphthol dissolved in ethanol is added *immediately*. By this means, the excess of bromine is removed almost instantaneously, and so the keto-enol equilibrium is not given time to be disturbed. Potassium iodide solution and hydrochloric acid are now added, and the liberated iodine is titrated with standard thiosulphate. The overall equation is:

$$CH_3COCHBrCO_2Et + 2I^- + H^+ \longrightarrow CH_3COCH_2CO_2Ft + I_2 + Br^-$$

The mechanism of bromination of ethyl acetoacetate is probably (*cf.* acetone):

$$Br{-}Br \;\; CII{=}C(OH)CH_3(CO_2Et) \longrightarrow Br^- + CHBr(CO_2Et)C(={}^{+}OH)CH_3 \xrightarrow{-H^+} CH_3COCHBrCO_2Et$$

Gero (1960, 1961) has determined the enol content of monoketones by adding an excess of iodine monochloride in methanol to a mixture of the dry ketone and sodium hydrogen carbonate, then adding an excess of aqueous sodium iodide and titrating with standard thiosulphate.

Schwarzenbach and Wittwer (1947) have introduced the *flow-method for* the estimation of the enol mixture and an acidified bromide-bromate solution are simultaneously mixed and diluted in a mixing chamber, and the mixture made to flow past a platinum electrode. The relative amounts of he two solutions are adjusted so that the potential measured at the platinum electrode shows a sharp rise–this corresponds to the end-point in the titration of the enol form by the bromine. The method gives good results of cnol contents as low as 10^{-5} per cent. *e.g.*, the enol content of acetone in aqueous solution was found to be 2.5×10^{-4} per cent.

The enol content in keto-enol mixtures (in dipropyl ether) has been determined by means of lithium aluminium hydride; the hydrogen liberated is estimated (Hofling *et al.*, 1952). The values obtained are about 10 per cent higher than those obtained by the bromine titration or by physical methods.

Compound	*Per cent enol (in ethanol)*
$CH_3COCH_2CO_2CH_3$	4.8
$CH_3COCH_2CO_2C_2H_5$	7.5
$CH_3COCH_2COCH_3$	76
$CH_3COCH(CH_3)COCH_3$	31
$C_6H_5COCH_2COC_6H_5$	96
$CH_2(CO_2C_2H_5)_2$	trace
Aldehydes of type RCH_2CHO	trace
Ketones of type RCH_2COCH_2R	race

ENOLISATION

The phenomenon of enolisation is exhibited by compounds containing either a methylene group, —CH_2— , or a methyne group, $>$CH— adjacent to a carbonyl group. The actual amount of enol from present depends on a number of factors; these are discussed later.

When the methylene or methine group is attached to two or three carbonyl groups, the hydrogen atom might migrate equally well to one or other carbonyl group.

This is not found to be so in practice for unsymmetrical compounds, one enol form being present exclusively, or largely predominating; *e.g.*, in acetoacetic ester the hydrogen atom migrates exclusively, or largely predominating *e.g.*, in acetoacetic ester the hydrogen atom migrates exclusively to the acetyl carbonyl group (see also below).

When two or more enol forms are theoretically possible, ozonolysis may be used to ascertain the structure of the form present; *e.g.*, in hexane-2,4-dione, $CH_3COCH_2COCH_2CH_3$, the two possible enols are:

$$\underset{\text{(I)}}{CH_3\overset{\overset{OH}{|}}{C}=CHCOCH_2CH_3} \qquad \underset{\text{(II)}}{CH_3COCH=\overset{\overset{OH}{|}}{C}CH_2CH_3}$$

Ozonolysis of (I) will give CH_3CO_2H and CH_3CH_2COCHO; (II) will give CH_3COCHO and $CH_3CH_2CO_2H$. Identification of these compounds will decide whether the enol is (I) or (II), or both.

The type of tautomerism discussed above is known as known as the **keto-enol trial system**. (In this system a hydrogen atom migrates from atom 1 (*oxygen*) to atom 3 (*carbon*):

$$\overset{3}{>CH}-\overset{2}{C}=\overset{1}{O} \rightleftharpoons \overset{3}{>C}-\overset{2}{C}=\overset{1}{OH}$$

Enols resemble phenols in a number of ways; *e.g.*, both from soluble sodium salts; both give characteristic colorations with ferric chloride; and both couple with diazonium salts.

The keto-enol type of tautomerism is only one example of a triad system. The trial system is the most important class of tautomeric systems, and the following, which are exemplified in the text, are the commonest:

(*i*) *Three-carbon systems:*

$$CH-C=C \rightleftharpoons C=C-CH$$

(*ii*) *Nitroso-oximino system:*

$$CH-NO_2 \rightleftharpoons C=NOOH$$

(*iii*) *Nitroso-oximino system:*

$$CH-N=O \rightleftharpoons C=NOH$$

(*iv*) *Amidine system:*

$$NH-C=N \rightleftharpoons N=C-NH$$

(*v*) *Amido-imidol system:*

$$NH-C=O \rightleftharpoons N=C-OH$$

(*vi*) *Azo-hydozone system:*

$$N=N-CH \rightleftharpoons HN-N=C$$

(*vii*) *Diazo-amino (triazen) system:*

$$N=N-NH \rightleftharpoons NH-N=N$$

(*viii*) *Diazzo-nitrosamine system:*

$$Ar-N=NOH \rightleftharpoons Ar-NH-N=O$$

In addition to the 'open' systems of tautomerism, there is also, *e.g.*, *ring-chain tautomerism*.

The Use of Acetoacetic Ester in the Synthesis of Ketones and Monocarboxylic Acids

The synthetic use of acetoacetic ester depends on two chemical properties:

1. (*a*) When treated with sodium ethoxide, acetoacetic ester forms sodioacetoacetic ester:

$$CH_3\overset{\overset{\displaystyle OH}{|}}{C}{=}CHCO_2C_2H_5 + C_2H_5ONa \longrightarrow \left[CH_3\overset{\overset{\displaystyle O^-}{|}}{C}{=}CHCO_2C_2H_5\right]Na + C_2H_5OH \; (ex.)$$

(*b*) So acetoacetic ester readily reacts with primary and secondary alkyl halides (vinyl and aryl halides do *not* react) to produce alkyl derivatives of acetoacetic ester in which the alkyl group is attached to carbon. This fact has given rise to considerable speculation regarding the mechanism of the alkylation process. The problem is still not settled, but a highly favoured theory is that the negative ionis a resonance hybrid.

$$\underset{(I)}{CH_3\overset{\overset{\displaystyle O^-}{|}}{C}{=}CHCO_2C_2H_5} \longleftrightarrow \underset{(II)}{CH_3\overset{\overset{\displaystyle O^-}{\|}}{C}{-}\ddot{C}HCO_2C_2H_5}$$

i.e., this ion is ambident. Kornblum *et al.* (1955) have pointed out that the tendency for alkylation at the more electronegative atom of an ambident anionusually increases with the S_N1 character of the reaction. As far as the alkylation of E.A.A. is concerned, the evidence is in favour of an S_N2 mechanism and consequently attack occurs at the carbon atom.

$$\begin{matrix} CH_3{-}C{=}O \\ | \\ HC{:}^- \; R{-}X \\ | \\ CO_2C_2H_5 \end{matrix} \longrightarrow \begin{matrix} CH_3{-}C{=}O \\ | \\ HC{-}R + X^- \\ | \\ CO_2C_2H_5 \end{matrix}$$

For the sake of simplicity the alkylation of sodioacetoacetic ester will, in future, be represented as:

$$[CH_3COCHCO_2C2H_5]^-Na^+ + RX \longrightarrow CH_3COCHRCO_2C_2H_5 + NaX (v.g.)$$

After one alkyl group has been introduced, the whole process may then be repeated to give the dialkyl derivative of acetoacetic ester:

$$CH_3COCHR^1CO_2C_2H5 \xrightarrow{C_2H_5ONa} [CH_3COCR^1CO_2C_2H_5]^-Na^+ \xrightarrow{R^2X} CH_3COCR^1R_2CO_2C_2H_5 \; (g.–v.g.)$$

Formerly, it was not considered possible to prepare *disubsituted* derivatives of acetoacetic ester in *one* step. Sandberg (1957), however, has now carried out this one-step reaction in the preparation of ethyl β-acetotricarballylate from acetoacetic ester (0.5 mole), ethyl bromoacetate (1.2 mole) and sodium hydride (1.2 mole) in benzene solution.

$$CH_3COCH_2CO_2C_2H_5 + 2BrCH_2CO_2C_2H_5 \xrightarrow{2NaH} CH_3COC(CH_2CO_2C_2H_5)_2(CO_2C_2H_5) \quad (77\%)$$

2. Acetoacetic ester and its alkyl derivatives can undergo two types of hydrolysis with potassium hydroxide:

(*a*) **Ketonic hydrolysis.** Ketonic hydrolysis, so called because a ketone is the chief product, is carried out by boiling with *dilute* aqueous or ethanolic potassium hydroxide solution, and is usually followed by acidification (dilute sulphuric acid) and warming:

$$CH_3COCHRCO_2C_2H_5 \xrightarrow[\text{(ii) } H_2SO_4]{\text{(i) KOH}} [CH_3COCHRCO_2H] \xrightarrow{-CO_2} CH_3COCH_2R\ (g.)$$

The ketone obtained is acetone or its derivatives, and the latter *alway contain the goup* CH_3CO—.

Dehn and Jackson (1933) found that 85 per cent phosphoric acid was a very good catalyst for the ketonic hydrolysis of acetoacetic ester and its alkyl derivatives, the yield of ketone reaching 95 per cent.

The mechanism of the decarboxylation of E.A.A. has been the subject of much discussion. Evidence obtained by several workers has led to the proposal that decarboxylation occurs through a cyclic transition state, and that the product is originally produced in the enol from.

$$Me(C{=}O)CMe_2CO\text{–}O\text{–}H \longrightarrow [\text{cyclic transition state}] \xrightarrow{-CO_2} Me\text{–}C(OH){=}CMe_2 \rightleftharpoons Me\text{–}C(=O)\text{–}CHMe_2$$

(*b*) **Acid hydrolysis.** Acid hydrolysis, so called because an acid is the chief product, is carried out by boiling with *concentrated* ethanolic potassium hydroxide solution, *e.g.*,

$$CH_3COCH_2CO_2C_2H_5 + 2KOH \longrightarrow CH_3CO_2K + CH_3CO_2K + C_2H_5OH\ (f.g.)$$

$$CH_3COCH_2CO_2C_2H_5 + 2KOH \longrightarrow CH_3CO_2K + R_2CHCO_2K + C_2H_5OH \ (f.g.)$$

The acid obtained is acetic acid or its derivatives as the potassium salt. Form this the free acid is readily obtained by treatment with inorganic acid.

The mechansim of this cleavage is possibly:

$$Me-\overset{O}{\overset{\|}{C}}-CHR-CO_2Et + EtO^- \rightleftharpoons Me-\underset{OEt}{\underset{|}{\overset{O^-}{\overset{|}{C}}}}-CHR-CO_2Et \rightleftharpoons$$

$$MeCO_2Et + CHRCO_2Et \xrightleftharpoons{EtOH} CH_2RCO_2Et + EtO$$

This is the reverse of the Claisen condensation. However, the esters formed are hydrolysed by the potassium hydroxide to their salts, and since do not undergo the condensation, the final equilibrium is driven completely to the right.

Examples of the Synthesis of Ketones

The formula of the ketone is written down, and *provided it contains the group* CH_3CO—, the ketone can be synthesised via acetoacetic ester as follows. The acetone nucleus is picked out, and the alkyl groups attached to it are then introduced into acetoacetic ester one at a time; this is followed by ketonic hydrolysis. It is usually better to introduce the larger group before the smaller (steric effect):

(*i*) $\boxed{CH_3COCH_2}CH_3$

$$CH_3COCH_2CO_2C_2H_5 \xrightarrow{C_2H_5ONa} [CH_3COCHCO_2C_2H_5]^-Na^+$$

$$\xrightarrow{CH_3I} CH_3COCH(CH_3)CO_2C_2H_5 \xrightarrow[\text{(ii) } H^+]{\text{(i) dil. KOH}} CH_3COCH_2CH_3$$

(*ii*) 3-*Methylentan-2-one.* $\boxed{CH_3CO\overset{CH_3}{\overset{|}{C}H}}CH_2CH_3$

$$CH_3COCH_2CO_2C_2H_5 \xrightarrow{C_2H_5ONa} [CH_3COCHCO_2C_2H_5]^-Na^+$$

$$\xrightarrow{C_2H_5I} CH_3COCH(C_2H_5)CO_2C_2H_5 \xrightarrow{C_2H_5ONa}$$

$$[CH_3COCH(C_2H_5)CO_2C_2H_5]^- Na^+ \xrightarrow{CH_3I} CH_3COC(CH_3)$$

$$(C_2H_5)CO_2C_2H_5 \xrightarrow[\text{(ii) } H^+]{\text{(i) dil. KOH}} CH_3COCH(CH_3)CH_2CH_3$$

Examples of the Synthesis of Monocarboxylic Acids

The approach is similar to that for ketones except that the acetic acid nucleus is picked out, and the acetoacetic ester derivative is subjected to 'acid' hydrolysis. The acetoacetic ester acid synthesis is usually confined to the preparation of straight-chain acids or brached-chain acids where the branching occurs on the α-carbon atom:

(*i*) n-*Butyric acid.* $CH_3CH_2\boxed{CH_2CO_2H}$

$$CH_3COCH_2CO_2C_2H_5 \xrightarrow{C_2H_5ONa} [CH_3COCHCO_2C_2H_5]^-Na^+$$

$$\xrightarrow{C_2H_5I} CH_3COCH(C_2H_5)CO_2C_2H_5$$

$$\xrightarrow[\text{(ii) } H^+]{\text{(i) c. KOH}} CH_3CH_2CH_2CO_2H$$

(*ii*) a-*Methyl-n-valeric acid.* $CH_3CH_2CH_2\boxed{CH(CH_3)CO_2H}$

$$CH_3COCH_2CO_2C_2H_5 \xrightarrow{C_2H_5ONa} [CH_3COCHCO_2C_2H_5]^-Na^+$$

$$\xrightarrow{C_3H_7Br} CH_3COCH(C_3H_7)CO_2C_2H_5 \xrightarrow{C_2H_5ONa}$$

$$[CH_3COC(C_3H_7)CO_2C_2H_5]^- Na^+ \xrightarrow{CH_3I} CH_3COC(CH_3)$$

$$(C_3H_7)CO_2C_2H_5 \xrightarrow[\text{(ii) } H^+]{\text{(i) c. KOH}} CH_3CH_2CH_2CH(CH_3)CO_2H$$

It has been found that 'ketonic' hydrolysis and 'acid' hydrolysis of acetoacetic ester *always take place simultaneously*, but one or the other can be made to predominate by adjusting the concentration of the potassium hydroxide. Thus, in the preparation of acids, there will always be some ketone formed as a by-product; and vice versa. For this reason it is better to use ethyl malonate to synthesise acids since the yields are greater. However, Ritter *et al.* (1962) have improved the preparation of acids via E.A.A. by heating the mono-alkyl derivatives with catalytic amounts of alkoxide in excess of absolute ethanol:

$$MeCOCHRCO_2Et+EtOH \xrightarrow{OEt^-} MeCO_2Et+RCH_2CO2Et(70\text{-}90\%)$$

In addition to the acetoacetic ester ketone synthesis, ketones may be prepared:

(*i*) by the alkylation of simpler ketones (sodamide followed by RI). The difficulty with this method is that over-alkylation usually occurs. This synthesis has been modified to prepare t-carboxylic acids from acetophenone.

$$CH_3COC_6H_5 \xrightarrow[\text{(ii) } R^1I]{\text{(i) } NaNH_2} R^1CH_2COC_6H_5 \xrightarrow[\text{(ii) } R^2I]{\text{(i) } NaNH_2} R^1R^2CHCOC_6H_5 \xrightarrow[\text{(ii) } R^3I]{\text{(i) } NaNH_2}$$

$$R^1R^2R^3CCOC_6H_5 \xrightarrow{NaNH_2} R^1R^2R^3CCONH_2 \xrightarrow{HNO_2} R^1R^2R^3CCO_2H$$

(*ii*) Stork *et al.* (1963) have introduced a new method of alkylating aldehydes and ketones, and so this affords a means of preparing more complicated aldehydes and ketones from simpler ones. An aliphatic primary amine is condensed with an aldehyde or ketone containing at least one active a-hydrogen atom, and the imine produced is treated with ethylmagnesium bromide in boiling tetrahydrofuran. The magneisum salt so produced readily reacts with alkyl halides, *e.g.*, a-benzyliso-butyraldehyde:

$$Me_2CHCHO \xrightarrow{Me_3CNH_2} Me_2CHCH{=}NCMe_3 \xrightarrow{3EtMgBr} Me_2C{=}CHN(MgBr)(CMe_3)$$

$$\xrightarrow{PhCH_2Cl} Me_2C(CH_2Ph)CH{=}NCMe_3 \xrightarrow{HCl} PhCH_2CMe_2CHO \quad (80\%)$$

(*iii*) β-Keto-acids, $R^1COCHR^2CO_2H$, can be prepared by α-carboxylation of ketones, $R^1COCH_2R^2$, with magensium methyl carbonate. The intermediate magnesium salt can be alkylated *in situ* and then decarboxylated to give ketones, $R^1COCHR^2R^2$ (Stiles *et al.*, 1959).

Sodioacetoacetic ester reacts with many other halogen compounds besides alkyl halides, and so may be used to synthesise a variety of compounds.

(*i*) 1,3–*Diketones*. In the synthesis of 1,3–diketones the halogen compound used is an acid chloride. Since acid chlorides react with ethanol, the reaction cannot be carried out in this solvent in the usual way. The reaction, however, may conveniently be carried out by treating acetoacetic ester in benzene solution with magnesium and the acid chloride, *e.g.*, pentane-2,4-dione may be obtained by the ketonic hydrolysis of the intermediate product ethyl diacetylacetate:

$$CH_3COCH_2CO_2C_2H_5 + CH_3COCl \xrightarrow{Mg} (CH_3CO_2)CHCO_2C_2H_5 \xrightarrow[(ii)\ H^+]{(i)\ dil.\ KOH} CH_3COCH_2CH_3$$

If sodioacetoacetic ester or acetoacetic ester itself is treated with acetyl chloride in pyridine as solvent, the *O-acetyl* derivative of acetoacetic ester, acetoxycrotonic ester, is obtained, and not the carbon-linked compound (as above):

$$[CH_3COCHCO2C_2H_5]^- Na^+ + CH_3COCl \xrightarrow{pyridine} CH_3C(OCOCH_3){=}CHCO_2C_2H_5 + NaCl$$

If we use Kornblum's generalisation with ambident ions, then the first reaction presumably occurs by the S_N2 mechanism and the latter by the S_N1.

(*ii*) *Dicarboxylic acids.* Dicarboxylic acids may be prepared by interaction of sodioacetoacetic ester and a halogen derivative of an ester, *e.g.*, succinic acid from ethyl chloroacetate:

$$[CH_3COCHCO_2C_2H_5]^-Na^+ + ClCH_2CO_2C_2H_5 \longrightarrow CH_3COCH(CH_2CO_2C_2H_5)CO_2C_2H_5 \xrightarrow[(ii)\ H^+]{(i)\ c.KOH} \begin{array}{c} CH_2CO_2H \\ | \\ CH_2CO_2H \end{array}$$

Ketonic hydrolysis of this acetoacetic ester derivative gives the γ-keto-acid ester (ethyl ester of laevulic acid):

$$CH_3COCH(CH_2CO_2C_2H_5)CO_2C_2H_5 \xrightarrow[(ii)\ H^+]{(i)\ dil.KOH} CH_3COCH_2CH_2CO_2C_2H_5$$

(*iii*) *Long-chain monocarboxlic acids.* This method involves a combination of methods (*i*) and (*ii*) described above (Mrs. Robinson, 1930):

$$[CH_3COCHCO_2C_2H_5]^- Na^+ \xrightarrow{Br(CH_2)_xCO_2C_2H_5} CH_3COCH((CH_2)_xCO_2C_2H_5)CO_2C_2H_5$$

$$\xrightarrow[\text{(ii) } CH_3(CH_2)_yCOCl]{\text{(i) } C_2H_5ONa} CH_3CO\underset{\substack{|\\ CO(CH_2)_yCH_3}}{\overset{\substack{(CH_2)_xCO_2C_2H_5\\ |}}{C}}HCO_2C_2H_5$$

$$\xrightarrow[\text{hydrolysis}]{\text{graded}} CH_3(CH_2)_yCOCH_2(CH_2)_xCO_2H$$

These keto-acids are readily reduced by means of the Clemmensen reduction. Improved modifications of this method have now been developed.

(*iv*) The acetyl group in acetoacetic ester may be replaced by another acyl group under suitable conditions, *e.g.*, ethyl benzoylacetate:

$$MeCOCH_2CO_2Et + PhCOCl \xrightarrow{NaOH} PhCOCH(COMe)CO_2Et$$

$$\xrightarrow{\text{aq. } NH_4Cl} PhCOCH_2CO_2Et + MeCO_2NH_4$$

(v) Alicyclic compounds and heterocyclic compounds may be prepared by the use of E.A.A.

Malonic Ester Syntheses

Malonic ester, $CH_2(CO_2C_2H_5)_2$, which is the diethyl ester of malonic acid, $CH_2(CO_2H)_2$, is prepared by dissolving potassium cyanoacetate in ethanol, adding concentrated hydrochloric acid, and warming the mixture on the water-bath:

$$NCCH_2—CO_2K + 2C_2H_5OH + 2HCl \longrightarrow CH_2(CO_2C_2H_5)_2 + KCl + NH_4Cl \; (v.g.)$$

It is a pleasant-smelling liquid, b.p. 199°C. Its use as a synthetic reagent depends on two chemical properties.

(1) With sodium ethoxide it forms a sodium derivative, sodiomalonic ester, which reacts with compounds containing a reactive halogen atom, *e.g.*, alkyl halides, acid chlorides, halogensubstituted esters, etc. (*cf.* acetoacetic ester). In all cases the yields are *g.–v.g.*

$$C_2H_5O—\overset{\overset{O}{\|}}{C}—CH_2—\overset{\overset{O}{\|}}{C}—OC_2H_5 \xrightarrow{C_2H_5ONa}$$

$$\left[C_2H_5O-\overset{\bar{O}}{\overset{|}{C}}=CH-\overset{O}{\overset{\|}{C}}-OC_2H_5\right]Na^+ \longleftrightarrow$$

$$\left[C_2H_5O-\overset{O}{\overset{\|}{C}}-\bar{C}H-\overset{O}{\overset{\|}{C}}-OC_2H_5\right]Na^+ \xrightarrow{RX}$$

$$C_2H_5O-\overset{O}{\overset{\|}{C}}-\bar{C}HR-\overset{O}{\overset{\|}{C}}-OC_2H_5 + NaX$$

The process may then be repeated to produce the disubstituted derivative of malonic ester:

$$R^1CH(CO_2C_2H_5)_2 \xrightarrow{C_2H_5ONa} [R^1C(CO_2C_2H_5)_2]^-Na^+$$
$$\xrightarrow{R^2X} R^1R^2C(CO_2C_2H_5)_2$$

These disubstituted derivatives of malonic ester can readily be prepared in *one* step by treating the ester with *two* equivalents of sodium ethoxide and then with *two* equivalents of alkyl halide.

This procedure is only used if it is required to introduce two identical alkyl groups.

$$CH_2(CO_2C_2H_5)_2 \xrightarrow[\text{(ii) } 2RX]{\text{(i) } 2EtONa} R_2C(CO_2C_2H_5)_2 + NaX$$

(2) Malonic acid and its derivatives eliminate a molecule of carbon dioxide when heated just above the melting point of the acid (usually between 150° and 200°C) to form acetic acid or its derivatives (the yields are *v.g.-ex.*):

$$HO_2CCH_2CO_2H \longrightarrow CH_3CO_2H + CO_2$$

$$HO_2CCHRCO_2H \longrightarrow RCH_3CO_2H + CO_2$$

Decarboxylation may also be effected by refluxing malonic acid or its derivatives in sulphuric acid solution.

The mechanism of decarboxylation is probably similar to that of acetoacetic acid:

$$R_2C\begin{matrix}\overset{OH}{\overset{|}{C}}=O \\ \\ \underset{\underset{O}{\|}}{C}-O\cdots H\end{matrix} \xrightarrow{-CO_2} \left[R_2C=C\begin{matrix}OH\\OH\end{matrix}\right] \longrightarrow R_2CHCO_2H$$

In a number of cases ethyl cyanoacetate (b.p. 207°C) may be used instead of malonic ester in many syntheses. The cyano-group is readily converted into the carboxyl group on hydrolysis.

Synthesis of Monocarboxylic Acids

Malonic ester is preferable to acetaocetic ester in synthesising acids, and should be used wherever possible. The structural formula of the acid required is written down, the acetic acid nucleus picked out, and the required alkyl groups introduced into sodiomalonic ester. The substituted ester is then refluxed with potassium hydroxide solution, acidified with hydrochloric acid, and the precipitated salt may be refluxed with sulphuric acid:

(*i*) n-*Valeric acid.* $\boxed{CH_3CH_2CH_2}CH_2CO_2H$

$$CH_2(CO_2C_2H_5)_2 \xrightarrow{C_2H_5ONa} [CH(CO_2C_2H_5)_2]^-Na^+ \xrightarrow{C_3H_7Br} C_3H_7CH(CO_2C_2H_5)_2 \xrightarrow{KOH}$$

$$C_3H_7CH(CO_2K_5)_2 \xrightarrow{HCl} C_3H_7CH(CO_2H)_2 \xrightarrow{150-200°C} CH_3CH_2CH_2CH_2CO_2H$$

(*ii*) *Dinethylacetic acid.* $CH_3\boxed{\overset{\overset{\displaystyle CH_3}{|}}{C}HCO_2H}$

$$CH_2(CO_2C_2H_5)_2 \xrightarrow{C_2H_5ONa} [CH(CO_2C_2H_5)_2]^-Na^+ \xrightarrow{CH_3I}$$

$$CH_3CH(CO_2C_2H_5)_2 \xrightarrow{C_2H_5ONa} [CH_3C(CO_2C_2H_5)_2]^-Na^+ \xrightarrow{CH_3I}$$

$$(CH_3)_2C(CO_2C_2H_5)_2 \xrightarrow{KOH} (CH_3)_2C(CO_2K)_2 \xrightarrow{HCl} (CH_3)_2C(CO_2H)_2 \xrightarrow{150-200°C} (CH_3)_2CHCO_2H$$

Since two methyl groups are required for this synthesis, their introduction an be carried out in one step (see above):

$$CH_2(CO_2C_2H_5)_2 \xrightarrow[2CH_3I]{2C_2H_5ONa} (CH_3)_2C(CO_2C_2H_5)_2 \xrightarrow{KOH,\ etc.} (CH_3)_2CHCO_2H$$

Since aryl groups cannot be introduced directly into malonic ester, aryl-substitued derivatives must be prepared indirectly. A very good method uses the Claisen condensation followed by decarboxylation *e.g.*, ethyl phenylmalonate:

$$PhCH_2CO_2Et + (CO_2Et)_2 \xrightarrow{EtONa} EtOH +$$

$$PhCH(CO_2Et)COCO_2Et \xrightarrow{heat} PhCH(CO_2Et)_2 + CO \ (80\text{-}85\%)$$

This method may also be used of the preparation of long-chain alkylmalonic esters; the ester of a longchain monocarboxylic acid with two α-hydrogen atoms is used instead of the ester of phenylacetic acid given above.

Synthesis of Dicarboxylic Acids

Dicaboxylic acids of the type $R^1R^2C(CO_2H)_2$ are readily prepared from malonic ester as shown above.

Dicarboxylic acids of the type $HO_2C(CH_2)N_nCO_2H$ are very important, and the malonic ester synthesis is particularly useful for their preparation. The actual procedure depends mainly on the value of *n*, and the following examples illustrate this point:

(*i*) *Adipic acid.* $\boxed{HO_2CCH_2}$ CH_2CH_2 CH_2CO_2H

The acetic acid is blocked off at *one* end, and the remaining fragment is considered from the point of view of accessibility. In this example the fragment required is g-bromobutyric ester.

This is not readily accessible, and so the procedure now is to block off the acetic acid nucleus at the *other* end of the adipic acid molecule, and then to use *two* molecules of malonic ester and the a, w-*dihalide* of the polymethylene fragment that joins together the two acetic acid nuclei.

In this case it is ethylene dibromide, and since this is readily accessible, the synthesis of adipic acid may be carried out as follows:

$$[(C_2H_5O_2C)_2CH]^-Na^+ + BrCH_2CH_2Br + Na^+ [CH(CO_2C_2H_5)_2]^- \longrightarrow (C_2H_5O_2C)_2CHCH_2CH_2CH(CO_2C_2H_5)_2 \xrightarrow[\text{(ii) HCl}]{\text{(i) KOH}}$$

$$(HO_2C)_2CHCH_2CH_2CH(CO_2H)_2 \xrightarrow{150\text{–}200°C} HO_2CCH_2CH_2CH_2CH_2CO_2H$$

Should the fragments required by either route be inaccessible, then a method not involving the use of malonic ester may be more satisfactory.

(*ii*) *Succinic acid.* $\boxed{HO_2CCH_2}$ CH_2CO_2H

If *one* acetic acid nucleus is blocked off, the fragment required is ethyl chloroacetate. This is readily accessible and hence succinic acid may be synthesised from *one* molecule of malonic ester as follows:

$$[CH(CO_2C_2H_5)_2]^-Na^+ + ClCH_2CO_2C_2H_5 \longrightarrow$$

$$(C_2H_5O_2C)_2CHCH_2CO_2C_2H_5 \xrightarrow[\text{(ii) HCl}]{\text{(i) KOH}} (HO_2C)_2CHCH_2CO_2H$$

$$\xrightarrow{150-200°C} HO_2CCH_2CH_2CO_2H$$

If the *other* acetic acid nucleus in succinic acid is blocked off, there is no intervening fragment left. However, the union of two malonic ester molecules may be effected by means of iodine.

$$[(C_2H_5O_2C)_2CH]^-Na^+ + I_2 + Na^+ [CH(CO_2C_2H_5)_2]^- \longrightarrow$$

$$(C_2H_5O_2C)_2CHCH(CO_2C_2H_5)_2 \xrightarrow[\text{(ii) HCl}]{\text{(i) KOH}} (HO_2C)_2CHCH(CO_2H)_2$$

$$\xrightarrow{150-200°C} HO_2CCH_2CH_2CO_2H$$

HYDROXYALDEHYDES AND HYDROXYKETONES

When naming a compound which contains more than one functional group, it is necessary to choose one as the principal function. The compound is then named by using the suffix of the principal function and the prefixes of the other functions. The carboxylic and the sulphonic acid group are *always* chosen as principal functions, and the usual order for choosing the principal function is:

Carboxylic, sulphonic, acid halide, amide, imdie, aldehyde, cyanide, isocyanide, ketone, alcohol, phenol, thioalcohol, amine, imine.

Examples. Acetoacetic acid contains a ketonic group and a carboxyl group, and since the latter is the principal function, acetoacetic acid is named as a ketone acid, *viz.*, β-ketobutyric acid, 3-ketobutanoic acid, or 3-oxobutanoic acid. Acetoacetic acid may also be named propan-2-one-1-carboxylic acid—each functional group is indicated by its appropriate suffix.

CH_3OHCH_2CHO. This is both an alcohol and an aldehyde, so that, bearing in mind the order of preference, the name of the compound will be β-hydroxypropionaldehyde or 3-hydroxypropanl.

The Table indicates the prefixes and suffixes used for designating the functions (in alphabetical order).

Function	*Prefix*	*Suffix*
Acid amide	carbamoyl	amide or carbonamide
Acid	carboxy	carboxylic or -oic
Acid chloride	chloroformyl	oyl or carbonyl
Alcohol	hydroxy	ol
Aldehyde	oxo, aldo (for aldehydic O) or formyl (for CHO)	al
Amine	amino	amine
Azo-derivative	azo	—
Azoxy-derivative	azoxy	—
Carbonitrile (nitrile)	cynao	carbonitrile
Double bond	—	ene
Ester	alk(yl)oxycarbonyl	carboxylate
Ether	alkoxy	—
Ethylene oxide, etc.	epoxy	—
Halogenide (halide)	halogeno (halo)	—
Hydrazine	hydrazino	hydrazine
Ketone	oxo or keto	one
Nitro-derivative	nitro	—
Nitroso-derivative	nitroso	—
'Quinquevalent' nitrogen	—	onium, inium
Sulphide	sulphonyl	—
Sulphonic derivative	sulpho	sulphonic
Sulphoxide	sulphinyl	—
Sulphonic derivative	sulpho	sulphonic
Sulphoxide	sulphinyl	—
Thioalcohol (Mercaptan)	mercapto	thiol
Triple bond	—	yne
Urea	ureido	urea

The simplest hydroxyladehyde is **glycoladehyde** (*hydroxyethanal*). It may be prepared by the oxidation of glycol with *Fenton's reagent* (hydrogon peroxide and ferrous sulphate).

$$HOCH_2CH_2OH + H_2O_2 \xrightarrow{FeSO_4} HOCH_2CHO + 2H_2O$$

It may also be prepared by ozonolysis of allyl alcohol:

$$CH_2{=}CHCH_2OH \xrightarrow{O_3} \begin{array}{c} CH_2-O-CHCH_2OH \\ |\qquad\qquad | \\ O\text{———}O \end{array} \xrightarrow[Zn]{H_2O} OCHCH_2OH$$

The most convenient means of preparing glycoladehyde, however, is by heating dihydroxymaleic acid, which is obtained from the oxidation of tartaric acid (*q.v.*):

$$\begin{array}{c} HO_2C-C-OH \\ \| \\ HO_2C-C-OH \end{array} \longrightarrow 2CO_2 + \left[\begin{array}{c} H-C-OH \\ \| \\ H-C-OH \end{array}\right] \longrightarrow \begin{array}{c} CHO \\ | \\ CH_2OH \end{array}$$

Glycolaldehyde exists in the solid form as the dimer, m.p. 96°C; but in aqueous solution it exists as the monomer which forms the stable hydrate, $CH(OH)_2CH_2OH$ (*cf.* chloral). Careful oxidation of glycolaldehyde with bromine water produces glycollic acid:

$$HOCH_2CHO + [O] \longrightarrow HOCH_2CO_2H$$

Glycolaldehyde is a powerful reducing agent, reducing ammoniacal silver nitrate and Fehling's solution at room temperature. With phenhylhydrazine it forms the *osazone*. This osazone is identical with that formed from glyoxal.

Glycolaldehyde undergoes the aldol condensation in the presence of alkali; with sodium hydroxide solution a tetrose sugar is formed, and with sodium carbonate solution, a hexose sugar:

$$2HOCH_2CHO \xrightarrow{NaOH} HOCH_2CHOHCHOHCHO$$

$$3HOCH_2CHO \xrightarrow{Na_2OH_3}$$
$$HOCH_2CHOHCHOHCHOHCHOHCHO$$

Glycolaldehyde is formed in small amounts with formaldehyde is allowed to stand in the presence of calcium carbonate:

$$2HCHO \xrightarrow{CaCO_3} HOCH_2CHO$$

Aldol (*acetaldol, β--hydroxybutyraldehyde, 3-hydroxybutanal*), p.b. 83°C/20 mm., may be prepared by the aldol condensation of acetaldehyde (p. 217):

$$2CH_3CHO \xrightarrow{NaOH} CH_3CHOHCH_2CHO$$

When heated, aldol is dehydrated to crotonaldehyde:

$$CH_3CHOHCH_2CHO \longrightarrow CH_3CH = CHCHO + H_2O$$

There is some doubt about the structure of aldol. Recent work suggests it is an equilibrium mixture of β-hydroxybutyraldehyde and the cyclic hemiacetal:

$$CH_3CHOHCH_2CHO \rightleftharpoons CH_3\overset{\ulcorner\ O\ \urcorner}{CHCH_2CHOH}$$

This is an example of ring-chain tautomerism.

The simplest hydroxyketone is **hydroxyacetone** (*acetol, pyruvic alcohol*), b.p. 145°C. The best method of preparation is by heating bromoacetone with potassium hydroxide in methanolic solution, and adding ethyl formate:

$$HCO_2C_2H_5 + KOH \longrightarrow HCO_2K + C_2H_5OH$$

$$CH_3COCH_2Br + HCO_2K \longrightarrow CH_3COCH_2OOCH + KBr$$

$$CH_3COCH_2OOCH + CH_3COCH_2OH \longrightarrow CH_3COCH2OH + HCO_2CH_3 \text{ (54-88\%)}$$

This is an example of alcoholysis, hydroxyacetone being replaced by methanol.

Hydroxyacetone reduces ammoniacal silver nitrate, thereby being oxidised to DL-lactic acid, and reduces Fehling's solution, thereby being oxidised to a mixture of formic and acetic acids. Although ketones do not normally reduce ammoniacal silver nitrate and Fehling's solution, α-hydroxyketones are exceptions. Similarly, ketones form phenylhydrazones with phenylhydrazine, but α-hydroxyketones form osazones; thus, hydroxyacetone gives the same osazone as methylglyoxal (*cf.* glycolaldehyde):

$$\begin{array}{c} CH_3 \\ | \\ CO \\ | \\ CH_2OH \end{array} \xrightarrow{C_6H_5NHNH_2} \begin{array}{l} CH_3 \\ | \\ C{=}NNHC_6H_5 \\ | \\ CH{=}NNHC_6H_5 \end{array}$$

Hydroxyacetone appears to exist as an equilibrium mixture of hydroxyketone and cyclic hemiacetal forms (ring-chain tautomerism).

$$CH_3COCH_2OH \rightleftharpoons \begin{array}{c} CH_3C\text{——}CH_2 \\ |\ \diagdown\ \diagup \\ |\quad O \\ OH \end{array}$$

Acyloins are α-hydroxyketones, and may be readily prepared by reduction of esters. Acetoin, p.b. 148°C, the simplest member, may be prepared from ethyl acetate.

Diacetone alcohol (*4-hydroxy-4-methylpentan-2-one*), p.b. 164°C, may be prepared by the condensation of acetone in the presence of barium hydroxide (p. 218):

$$2CH_3COCH_3 \xrightarrow{Ba(OH)_2} (CH_3)_2C(OH)CH_2COCH_3$$

When heated with a trace of acid or iodine, diacetone alcohol eliminates a molecule of water to form mesityl oxide:

$$(CH_3)_2C(OH)CH_2COCH_3 \xrightarrow{I_2} (CH_3)_2C = CHCOCH_3 + H_2O$$

Diacetone alcohol is oxidised by sodium hypobromite to β-hydroxyisovaleric acid (*cf.* haloform reaction):

$$(CH_3)_2C(OH)CH_2COCH_3 \xrightarrow{[O]} (CH_3)_2C(OH)CH_2CO_2H$$

DIALDEHYDES, KETOALDEHYDES AND DIKETONES

Dialdehydes

The simplest dialdehyde is **glyoxal** (*oxaldehyde, ethanedial*). It may be obtained by the oxidation of ethanol, acetaldehyde or glycol with nitric acid, but the yields are poor. It is most conveniently prepared, as the bisulphite compound, by refluxing a mixture of paraldehyde, 50 per cent aqueous acetic acid, dioxan and selenous acid (yield 72-74 per cent on the selenous acid):

$$(CH_3CHO)_3 + 3H_2SEeO_3 \longrightarrow 3OCHCHO + 3Se + 6H_2O$$

$$\xrightarrow{NaHSO_3} OCHCHO{\cdot}2NaHSO_3{\cdot}H_2O$$

Glyoxal is manufactured by the vapour-phase oxidation of glycol with air 250-300°C in the presence of copper as actalyst.

Glyoxal exists as a colourless polymer giving, on distillation, the monomer. This is a green vapour, which condenses to yellow crystals (m.p. 15°C) which polymerise on standing to a colourless solid of unknown molecular weight. Glyoxal exists in aqueous solution as the dihydrate, $CH(OH)_2CH(OH)_2$, which has been isolated (*cf.* chloral hydrate).

Glyoxal undergoes many of the reactions of a dialdehyde. *e.g.*, it reduces ammoniacal silver nitrate, and forms addition compounds with two molecules of hydrogen cyanide and sodium hydrogen sulphite. It does *not* reduce Fehling's solution. Since it has no α-hydrogen atom, glyoxal undergoes an intramolecuelar Cannizzaro reaction in the presence of alkali to form glycollic acid.

$$H-\overset{O}{\overset{\|}{C}}-\overset{O}{\overset{\|}{C}}-H + OH^- \rightleftharpoons H-\overset{O}{\overset{\|}{C}}-\underset{H}{\underset{|}{\overset{OH}{\overset{|}{C}}}}-O^- \xrightarrow[\text{Shift}]{H-} H-\underset{H}{\underset{|}{\overset{O^-}{\overset{|}{C}}}}-\overset{OH}{\overset{|}{C}}=O \longrightarrow H-\underset{H}{\underset{|}{\overset{OH}{\overset{|}{C}}}}-CO_2^-$$

With phenylhydrazine, glyoxal forms the osazone (this is identical with that from glycoladehyde):

$$\begin{array}{c}CHO\\|\\CHO\end{array} + 2C_6H_5NHNH_2 \longrightarrow \begin{array}{c}CH{=}NNHC_6H_5\\|\\CH{=}NNHC_6H_5\end{array} + 2H_2O$$

It also combines with *o*-phenylenediamines of form *quinoxaline* (heterocyclic compounds); *e.g.*, with *o*-phenylenediamine it forms quinoxaline itself:

$$C_6H_4(NH_2)_2 + \begin{array}{c}O{=}CH\\|\\O{=}CH\end{array} \longrightarrow \text{quinoxaline} + 2H_2O$$

Methylglyoxal (pyruvaldehyde) may be prepared by the oxidation of hydroxyacetone with hydrogen peroxide, or of acetone with selenium dixoide, or by the hydrolysis of oximinoacetone:

$$\left.\begin{array}{l}CH_3COCH_2OH \xrightarrow{H_2O_2}\\ CH_3COCH_3 \xrightarrow{SeO_2}\\ CH_3COCH{=}NOH \xrightarrow{H_3O^+}\end{array}\right\} \longrightarrow CH_3COCHO$$

It is manufactured by the vapour-phase oxidation of propene glycol with air at 250-300°C in the presence of copper as catalyst:

$$CH_3CHOHCH_2OH \xrightarrow[Cu]{O_2} CH_3COCHO + 2H_2O$$

It is a yellow oil with a pungent odour, and beings to boil at 72°C to give a light green vapour.

The liquid form of methylglyoxal is the dimer, and dimer, and this slowly polymerises at room temperature to form a glassy mass of unknown molecular weight (*cf.* glyoxal).

Malondialdehyde is prepared commercially as its tetra-acetal, 1,1,3,3-tetaethoxypropane, $(EtO)_2CHCH_2CH(OEt)_2$, p.b. 100°C, is formed when ethyl formate undergoes the Claisen condensation with acetone.

$$CH_3COCH_3 + HCO_2Et \xrightarrow{EtONa} CH_3COCH_2CH$$

Fomylacetone is strongly acidic, *e.g.*, it dissolves in sodium hydroxide solution to form the sodium salt, and gives a violet colour with ferric chloride. Thus it is enolised. Moreover, infra-red spectra studies by Boileau (1954) have shown that it is the aldehyde group that has enolised almost 100 per cent *i.e.*, formylacetone is hydroxymethyleneacetone, MeCOCH = CHOH. It appears that all β-ketoaldehydes behave in this way. The interesting point about their direction of enolisation is that they were the first exceptions to the *Erlenmeyer rule* (1880, 1892) that a terminal = CHOH group (in acyclic compounds) is so unstable that it always rearranges to the aldehyde (*cf.* 'vinyl alochol'). More recently, Garbisch, Jr. (1963) has shown, from his NMR spectra studies of cyclic β-keto-aldehydes, that the direction of enolisation depends on their structure.

Succinaldehyde (succindialdehyde, butanedial) may be prepared by the ozonolysis of hexa-1,5-diene:

$$CH_2{=}CHCH_2CH_2CH{=}CH{=}CH_2 \xrightarrow{ozonolysis} OCHCH_2CH_2CHO$$

Another method of preparation is to allow pyrrole and hydroxylamine to interact, and to treat the product, succinaldoxime, with aqueous nitrous acid:

$$\begin{matrix} CH{=}CH \diagdown \\ | \quad\quad NH \\ CH{=}CH \diagup \end{matrix} \xrightarrow{NH_2OH} \begin{matrix} CH_2CH{=}NOH \\ | \\ CH_2CH{=}NOH \end{matrix} \xrightarrow{HNO_2} \begin{matrix} CH_2CHO \\ | \\ CH_2CHO \end{matrix}$$

Barton *et al.* (1964) have obtained succinaldehye (yield: 80 per cent) from butane-1,4-diol via the alkyl chloroformate.

Succinaldehyde is a colourless oil, p.b. 170°C, which readily polymerises. It undergoes the usual reactions of a (di)aldehyde; *e.g.*, it reduces ammoniacal silver nitrate and Fehling's solution, forms addition compounds with two equivalents of hydrogen cyanide ad sodium hydrogen sulphite; etc. When treated with phosphorus pentoxide, ammonia and phosphorus pentasulphide, it gives respectively furan, pyrrole and thiophen:

$$\begin{matrix} CH_2CHO \\ | \\ CH_2CHO \end{matrix} \begin{cases} \xrightarrow{P_2O_5} \text{furan (O)} \\ \xrightarrow{NH_3} \text{pyrrole (NH)} \\ \xrightarrow{P_2S_5} \text{thiophen (S)} \end{cases}$$

Diketones

These are classified as α, β, γ . . . diketones according as the two carbonyl groups are in the 1,2, 1,3, 1,4, . . . positions, respectively.

The infra-red absorption region of α-diketones is 1730-1710cm^{-1} (s), and that of γ-diketones is 1725-1705 cm^{-1}(s). Both of these are in the absorption region of acyclic monoketones (1725-1700 cm^{-1}) but of greater intensity, and since they are completely different from the region of enolisable β-diketones, it therefore means that α-and γ-diketones are like acetone, *i.e.*, are in the keto form.

Butane-2,3-dione, dimethylglyoxal (*diacetyl*), is the simplest α-diketone. It may be prepared: (*i*) By the hydrolysis of oximinobutan-2-one formed by the action of nitrous acid on the ketone:

$$CH_3COCH_2CH_3 \xrightarrow{HNO_2} CH_3COC(=NOH)CH_3$$

$$\xrightarrow{HCl} CH_3COCOCH_3$$

(ii) By the oxidation of butan-2-one with selenium dioxide:

$$CH_3COCH_2CH_3 + SeO_2 \longrightarrow CH_3COCOCH_3 + Se + H_2O$$

(iii) Acetoin may be readily oxidised by bismuth trioxide in acetic acid; the yield of dimethylglyoxal is almost quantitative (Rigby, 1951). This reagent is specific for this reaction, *i.e.*, for the oxidation of acyloins to α-diketones.

Butane-2,3-dione is a yellow oil, b.p. 88°C. It gives the usual reactions of a diketone; *e.g.*, it forms the monoxime and di-oxime with hydroxylamine, the bisulphite compound with sodium hydrogen sulphite, etc. It is oxidised by hydrogen peroxide to acetic acid:

$$CH_3COCOCH_3 \xrightarrow{H_2O_2} 2CH_3CO_2H$$

Reduction with lithium aluminium hydride produces the corresponding diol.

The dioxime of butane-2,3-dione, *i.e.*, *dimethylglyoxime*, $CH_3C(=NOH)C(=NOH)CH_3$, forms chelate compounds with many metals, and is used to estimate nickel, whose salts produce a red precipitate when treated with dimethylglyoxime. According to Brady *et al.* (1930), the structure of bisdimethylglyoxime nickel is (I). The planer configuration round the nickel ion has been demonstrated (Sugden *et al.*, 1935), and confirmed by X-ray analysis (Rundle *et al.*, 1953).

$$\begin{array}{c} O\cdots H-O \\ CH_3C=N \nearrow \quad \searrow N=CCH_3 \\ \searrow Ni \swarrow \\ \nearrow \quad \nwarrow \\ CH_3C=N \quad N=CCH_3 \\ \searrow O\cdots H-O \nearrow \end{array} \quad \text{(I)}$$

Pentane-2,4-dione, acetylacetone, is the simplest β-diketone. It may be prepared:

(*i*) By means of the Claisen condensation between ethyl acetate and acetone (yield 38-45 per cent on the acetone):

$$CH_3CO_2C_2H_5 + CH_3COCH_3 \xrightarrow{C_2H_5ONa} CH_3COCH_2COCH_3$$

Bloomfield (1962) has shown that β-diketones may be obtained in very good yields by this method by using dimethyl sulphoxide as solvent and sodium hydride as the base.

$$CH_3COCH_2CO_2C_2H_5 \xrightarrow[Mg]{CH_3COCl} CH_3CO\underset{}{\overset{COCH_3}{|}}{CH}_2C_2H_5 \xrightarrow[\text{(ii) } H^+]{\text{(i) dil.KOH}} CH_3COCH_2COCH_3$$

(*iii*) By the condensation of acetic anhydride with acetone in the presence of boron trifluoride as catalyst (yield 80-85 per cent on the acetone):

$$CH_3COCH_3 + (CH_3CO)_2O \xrightarrow{BF_3} CH_3COCH_2COCH_3 + CH_3CO_2H$$

Pentane-2,4-dione is a colourless liquid, p.b. 139°C/746 mm. It exhibits tautomerism, *e.g.*, it gives a red colour with ferric chloride.

$$CH_3COCH_2COCH_3 \rightleftharpoons CH_3COCH{=}C(OH)CH_3$$

It is readily oxidised to acetic acid, and is converted into a mixture of acetone and acetic acid when heated with potassium hydroxide solution (*cf.* 'acid hydrolysis' of acetoacetic ester). It forms pyrazoles when treated with hydrazine or its derivatives; *e.g.*, with phenylhydrazine it forms 3,5-dimethyl-1-penylpyrazole:

$$CH_3C(=O)-CH_2-C(=O)CH_3 + PhNHNH_2 \longrightarrow \text{3,5-dimethyl-1-phenylpyrazole } (CH_3,\ CH_3,\ N\text{-}Ph) + 2H_2O$$

Pentane-2,4-dione forms chelated compounds with various metals, *e.g.*, iron, aluminium, copper, etc. (*cf.* E.A.A.):

$$CH_3-C=O \rightarrow Cu \leftarrow O-C-CH_3 \quad \text{(chelate: } HC, CH_3-C-O, O=C-CH_3, CH\text{)} \quad or \quad CH_3-C\cdots O \cdots Cu \cdots O\cdots C-CH_3 \text{ (with } CH \text{ and } CH_3-C\cdots O, O\cdots C-CH_3)$$

All β-diketones are split by aqueous alkali, and when they are unsymmetrical, fission can occur in two different ways:

$$R^1COCH_2-C(O^-)(OH)R^2 \xleftarrow{OH^-} R^1COCH_2COR^2 \xrightarrow{OH^-} R^1C(O^-)(OH)-CH_2COR^2$$

$$\downarrow \qquad\qquad\qquad\qquad \downarrow$$

$$R^1COCH_3 + R^2CO_2^- \qquad\qquad R^1CO_2^- + CH_3COR^2$$

In general, fission occurs in the direction that produces the stronger acid in greater yield.

Pentane-2,4-done is oxidised by selenium dioxide to pentane-2,3,4-trione:

$$MeCOCH_2COMe \xrightarrow{SeO_2} MeCOCOCOMe \ (29\%)$$

Hexane-2,5-dione, acetonylacetone, p.b. 188°C, is the simplest γ-diketone. It may be prepared by menas of the acetoacetic ester synthesis:

$$[CH_3COCHCO_2C_2H_5]^-Na^+ + BrCH_2COCH_3 \longrightarrow$$

$$CH_3COCH(CH_2COCH_3)CO_2C_2H_5 \xrightarrow[(ii)\ H^+]{(i)\ dil.KOH} CH_3COCH_2CH_2COCH_3$$

An alternative and more convenient acetoacetic ester synthesis is as follows:

$$2[CH_3COCHCO_2C_2H_5]^-Na^+ \xrightarrow{I_2} \begin{array}{c} CH_3COCHCO_2C_2H_5 \\ | \\ CH_3COCHCO_2C_2H_5 \end{array} \xrightarrow[(ii)\ HCl]{(i)\ KOH}$$

$$\begin{array}{c} CH_3COCHCO_2H \\ | \\ CH_3COCHCO_2H \end{array} \xrightarrow{heat} \begin{array}{c} CH_3COCH_2 \\ | \\ CH_3COCH_2 \end{array} + 2CO_2$$

It readily forms five-membered rings: with phosphorus pentoxide, 2,5-dimethylfuran; with ammonia, 2,5-dimethylpyrrole; and with phosphorus pentasulphide, 2,5-dimethylfuran; with ammonia, 2,5-

dimethylpyrrole; and with phosphorus pentasulphide, 2,5-dimethylthophen (*cf.* succinaldehyde).

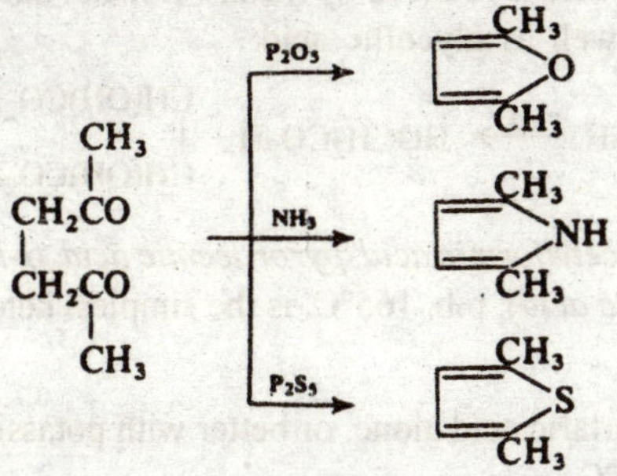

$$\begin{array}{c} CH_3 \\ | \\ CH_2CO \\ | \\ CH_2CO \\ | \\ CH_3 \end{array} \begin{array}{l} \xrightarrow{P_2O_5} \text{2,5-dimethylfuran} \\ \xrightarrow{NH_3} \text{2,5-dimethylpyrrole} \\ \xrightarrow{P_2S_5} \text{2,5-dimethylthiophen} \end{array}$$

Fig. 3.4

Hexane-2,5-dione is oxidised by selenium dioxide to hex-3-ene-2,5-dione (*cf.* pentane-2,4-dione):

$$MeCOCH_2CH_2COMe \xrightarrow{SeO_2} MeCOCH{=}CHCOMe$$

Aldehydic and Ketonic Acids

Glyoxylic acid (*glyoxalic acid, oxoethanoic acid*) is the simplest aldehydic acid. It occurs in unripe fruits, *e.g.*, gooseberries, and disappears during ripening; it also occurs in animal tissues and fluids. It may be prepared by the oxidation of ethanol, glycol or glycollic acid with nitric aid (yields are poor). It may also be prepared by the reduction of oxalic acid either electrolytically or by means of magnesium and sulphuric acid, but it is most conveniently prepared by the hydrolysis of dichloroacetic acid with water:

$$CHCl_2CO_2H + H_2O \longrightarrow OCHCO_2H + 2HCl$$

Glyoxylic acid crystallises from water with one molecule of water which is combined as water of constitution, $CH(OH)_2CO_2H$, *i.e.*, dihydroxyacetic acid (*cf.* chloral hydrate). The anhydrous acid may be obtained as a thick syrup by evaporating the aqueous solution over phosphorus pentoxide *in vacuo*.

Glyoxylic acid gives all the reactions of aldehyde and an acid; *e.g.*, it reduces maniacal silver nitrate, forms a bisulphite compound, etc. Since it has no α-hydrogen atoms, glyoxylic acid undergoes the Cannizzaro reaction to form glycollic and oxalic acids:

$$2OCHCO_2H \xrightarrow{NaOH} HOCH_2CO_2Na + (CO_2Na)_2$$

When glyoxylic acid is reduced by means of metal and acid, tartaric acid is obtained as well as glycollic acid:

$$OCHCO_2H \xrightarrow{e;\ H^+} HOCH_2CO_2H + \begin{array}{l} CH(OH)CO_2H \\ | \\ CH(OH)CO_2H \end{array}$$

Pyruvic acid (*acetylformic acid, pyroracemic acid, α-ketopropionic acid, 2-oxopropanoic acid*), p.b. 165°C, is the simplest keto-acid. It may be prepared:

(*i*) By heating tartaric acid alone, or better with potassium hydrogen sulphate at 210-220°C.

The reaction is believed to take place via the formation of hydroxyamleic acid (I), which rearranges to oxalacetic acid, (II):

$$\begin{array}{l} CH(OH)CO_2H \\ | \\ CH(OH)CO_2H \end{array} \xrightarrow{-H_2O} \begin{array}{l} CHCO_2H \\ \| \\ C(OH)CO_2H \end{array} \longrightarrow \begin{array}{l} CH_2CO_2H \\ | \\ COCO_2H \end{array} \longrightarrow$$

$$CH_3COCO_2H + CO_2 \quad (50 - 55\%)$$

This is the best method for preparing pyruvic acid, and it was this method which gave rise to the name pyroracemic acid:

(*ii*) Other methods of preparation are the oxidation of lactic (α-hydroxypropionic) acid, the hydrolysis of α, α-dibromopropionic acid or acetyl cyanide, *e.g.*,

$$CH_3COCl \xrightarrow{KCN} CH_3COCN \xrightarrow{H^+} CH_3COCO_2H$$

A general method of preparing α-keto-acids uses the Claisen condensation between ethyl oxalate and monocarboxylic esters:

$$EtO_2CCO_2Et + RCH_2CO_2Et \xrightarrow{EtONa} EtO_2CCOCHRCO_2Et$$

$$\xrightarrow{10\%\ H_2SO_4} [HO_2CCOCHRCO_2H] \xrightarrow{-CO_2} RCH_2COCO_2H (94\%)$$

Pyruvic acid behaves as a ketone and as an acid; it forms an oxime, hydrazone, etc. It is reduced by sodium amalgam to alctic acid and dimethyltartaric acid (*cf.* glyoxylic acid, above):

$$3CH_3COCO_2H \xrightarrow{e;\ H^+} CH_3CH(OH)CO_2H + \begin{array}{l} CH_3C(OH)CO_2H \\ | \\ CH_3C(OH)CO_2H \end{array}$$

Pyruvic acid reduces ammoniacal silver nitrate, itself being oxidised to acetic acid:

$$CH_3COCO_2H + [O] \longrightarrow CH_3CO_2H + CO_2$$

It is oxidised by warm nitric acid to oxalic acid:

$$CH_3COCO_2H + [O] \longrightarrow (CO_2H)_2 + CO_2 + H_2O$$

Pyruvic acid is easily decarboxylated with warm dilute sulphuric acid to give acetaldehyde:

$$CH_3COCO_2H \longrightarrow CH_3CHO + CO_2$$

The mechanism of this reaction is uncertain, but since it occurs only with α-keto-acids, it suggests that the -I effect of the α-acrbonyl group plays an important part (*cf.* trichloroacetic acid). On this basis, the following mechanism could be postulated, involving a carbene intermediate.

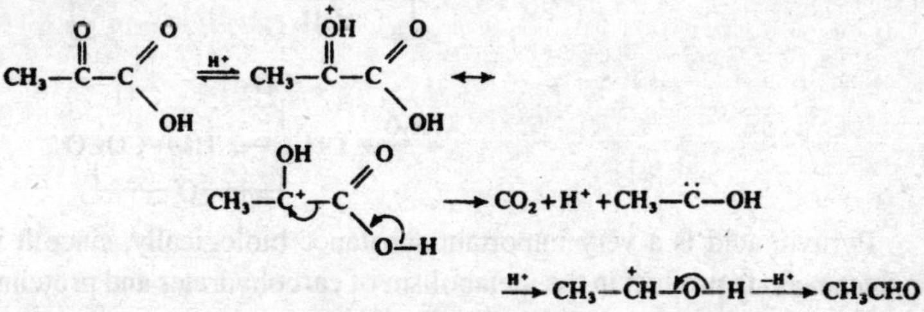

$$CH_3-\overset{\overset{O}{\|}}{C}-C\begin{matrix}\nearrow O \\ \searrow OH\end{matrix} \overset{H^+}{\rightleftharpoons} CH_3-\overset{\overset{\overset{+}{O}H}{\|}}{C}-C\begin{matrix}\nearrow O \\ \searrow OH\end{matrix} \longleftrightarrow$$

$$CH_3-\overset{\overset{OH}{|}}{C^+}-C\begin{matrix}\nearrow O \\ \searrow O-H\end{matrix} \longrightarrow CO_2 + H^+ + CH_3-\ddot{C}-OH$$

$$\xrightarrow{H^+} CH_3-\overset{+}{C}H-\ddot{O}-H \xrightarrow{-H^+} CH_3CHO$$

Fig. 3.5

On the other hand, when warmed with *concentrated* sulphuric acid, pyruvic acid undergoes *decarboylation, i.e.*, eliminates a molecule of carbon monoxide, to form acetic acid:

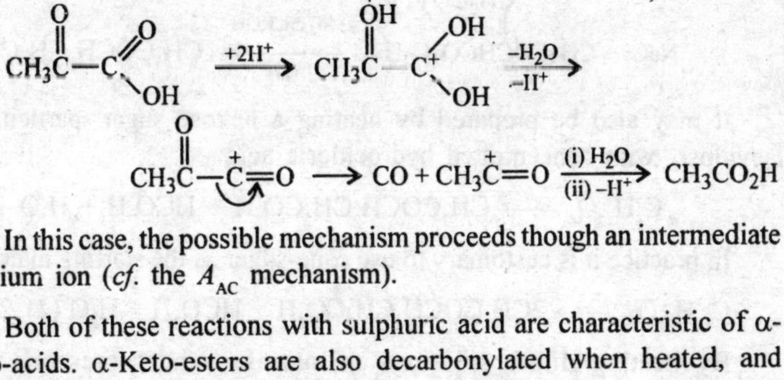

$$CH_3\overset{\overset{O}{\|}}{C}-C\begin{matrix}\nearrow O \\ \searrow OH\end{matrix} \xrightarrow{+2H^+} CH_3\overset{\overset{\overset{+}{O}H}{\|}}{C}-\overset{+}{C}\begin{matrix}\nearrow OH \\ \searrow OH\end{matrix} \xrightarrow[-H^+]{-H_2O}$$

$$CH_3\overset{\overset{O}{\|}}{C}-\overset{+}{C}\equiv O \longrightarrow CO + CH_3\overset{+}{C}=O \xrightarrow[\text{(ii) }-H^+]{\text{(i) }H_2O} CH_3CO_2H$$

In this case, the possible mechanism proceeds though an intermediate acylium ion (*cf.* the A_{AC} mechanism).

Both of these reactions with sulphuric acid are characteristic of α-keto-acids. α-Keto-esters are also decarbonylated when heated, and Calvin *et al.* (1940), using ethyl pyruvate labelled with ^{14}C, showed that the carbon monoxide came from the carbethoxyl group. In this case, the results may be explained by he formation of an epoxide intermediate:

$$CH_3-\overset{\overset{O}{\|}}{C}-\underset{\underset{:OEt}{|}}{\overset{\overset{O}{\|}}{C}} \xrightarrow{110°C} CH_3-\overset{\overset{^-O}{|}}{C}-\overset{\overset{O}{\|}}{C}-\overset{+}{O}(Et) \longrightarrow CH_3-\underset{\underset{OEt}{|}}{\overset{\overset{O}{\|}}{C}} + CO$$

The methyl group is made reactive the adjacent carbonyl group, and so pyruvic acid undergoes many condensation reactions characteristic of a compound containing an active methylene group; *e.g.*, in the presence of dry hydrogen chloride, pyruvic acid forms α-keto-γ-valerolactone-γ-carboxylic acid:

$$CH_3\overset{\overset{CO_2H}{|}}{C}=O + CH_3COCO_2H \xrightarrow{HCl} \left[CH_3\underset{\underset{OH}{|}}{\overset{\overset{CO_2H}{|}}{C}}-CH_2-COCO_2H \right]$$

$$\xrightarrow{-H_2O} CH_3\overset{\overset{CO_2H}{|}}{C}-CH_2-COCO \quad (\text{ring closed through } -O-)$$

Pyruvic acid is a very important substance biologically, since it is an intermediate product in the metabolism of carbohydrates and proteins.

Laevulic acid (β-*acetylpropionic acid, γ-keovaleric acid, butan*-3-*one-1-carboxylic acid*), m.p. 34°C, is the simplest γ-keto-acid. It may be prepared via acetoacetic ester as follows:

$$[CH_3COCHCO_2C_2H_5]^- Na^+ + ClCH_2CO_2C_2H_5 \longrightarrow$$

$$NaCl + CH_3CO\overset{\overset{CH_2CO_2C_2H_5}{|}}{C}HCO_2C_2H_5 \xrightarrow[\text{(ii) } H^+]{\text{(ii) dil. KOH}} CH_3COCH_2CH_2CO_2H$$

It may also be prepared by heating a hexose sugar, particularly laevulose, with concentrated hydrochloric acid:

$$C_6H_{12}O_6 \longrightarrow CH_3COCH_2CH_2CO_2H + HCO_2H + H_2O$$

In practice it is customary to use cane-sugar as the starting material:

$$C_{12}H_{22}O_{11} \longrightarrow 2CH_3COCH_2CH_2CO_2H + HCO_2H + H_2O \text{ (21-22\%)}$$

By heating dilute solutions of sucrose sunder pressure with hydrochloric acid, the yield is increased to 50 per cent.

Laevulic acid behaves as a ketone (forms oxime, etc.) and as an acid (forms esters, etc.). On the other hand, many of the reactions of laevulic acid indicate that it exists in the *lactol* form, *i.e.*, hydroxylactone (this is an example of ring-chain tautomerism), *e.g.*,

$$\begin{array}{l} CH_2COCH_3 \\ | \\ CH_2CO_2H \end{array} \rightleftharpoons \begin{array}{l} CH_2-C(OH)CH_3 \\ \quad\quad\quad \diagdown \\ \quad\quad\quad\quad O \\ \quad\quad\quad \diagup \\ CH_2-CO \end{array} \xrightarrow{Ac_2O} \begin{array}{l} CH_2C(CH_3)OAc \\ \quad\quad\quad \diagdown \\ \quad\quad\quad\quad O \\ \quad\quad\quad \diagup \\ CH_2CO \end{array}$$

When heated for some time, laevulic acid is converted into α-and β-*angelica* lactones (derived from the lactol form):

$$\underset{\alpha-}{\begin{array}{l} CH=C-CH_3 \\ \quad\quad \diagdown \\ \quad\quad\quad O \\ \quad\quad \diagup \\ CH_2-CO \end{array}} \qquad \underset{\beta-}{\begin{array}{l} CH-CH-CH_3 \\ \quad\quad \diagdown \\ \quad\quad\quad O \\ \quad\quad \diagup \\ CH_2-CO \end{array}}$$

A general method for preparing γ-keto-acids is to treat a ketone with lithium amide and lithium bromoacetate; the ketone undergoes α-carboxymehylation (Puterbaugh *et al.*, 1959):

$$R^1COCH_2R^2 \xrightarrow[CH_2BrCO_2Li]{LiNH_2} R^1COCHR^2CH_2CO_2H$$

γ-Keto-acids may also be synthesised by other methods.

Keto-acids with the oxo group in the δ- or a higher position are very conveniently prepared by the action of a dialkylcadmium on the half-ester acid chloride. On the other hand, δ-keto-acids may be prepared as follows from cyclohexane-1,3-dione (Stetter *et al.*, 1952); *e.g.*, 5-oxo-heptanoic acid:

$$\text{cyclohexane-1,3-dione} \xrightarrow[\text{(ii) MeI}]{\text{(i) KOH—MeOH}} \text{2-methylcyclohexane-1,3-dione} \xrightarrow[\text{(ii) } H_2SO_4]{\text{(i) } Ba(OH)_2} MeCH_2CO(CH_2)_3CO_2H$$

The cyclic dione, since it contains an active methylene group, forms the potassium salt which can be alkylated, and the product readily undergoes fission, thus behaving like the acyclic β-diketones.

The keto-acids may be reduced by the Wolff-Kishner method to the corresponding saturated acid to heptanoic acid. Alternatively, the alkylated ring may be opened and the carbonyl group reduced in one step step by refluxing with hydrazine in alkaline glycol solution.

SYNTHESIS OF KETONES

Bowman *et al.* (1952) have introduced the following general synthesis of ketones and b-keto-esters (R^2= tetrahydropyran-2-ol):

$$R^1CH(CO_2R^2)_2 \xrightarrow[\text{(ii) } R^3COCl]{\text{(i) NaH}} R^3COCR^1(CO_2R^2)_2 \xrightarrow[\text{reflux}]{CH_3CO_2H} R^3COCH_2R^1 \text{ (90-92\%)}$$

Johnson *et al.* (1952) have used t-butyl esters of malonic acid (R^2 = t-butyl):

$$R^1CH(CO_2R^2)_2 \xrightarrow[\text{(ii) } R^3COCl]{\text{(i) NaH}} R^3COCR^1\ (CO_2R^2)_2 \xrightarrow[CH_3CO_2H]{CH_3-C_6H_4-SO_3H} R^3COCH_2R^1 \text{ (56–85\%)}$$

Good yields of ketone are obtained by the above method only when anhydrous conditions are used throughout the preparation. Consequetly, insead of the usual method of hydrolysis (aqueous alkali), acidolysis is used. Acetic acid is satisifactory for this purpose, especially in the presence of a small amount of sulphuric acid or *p*-toluenesulphonic acid. In this way, the ester group is transferred from the malonic ester to the acetic acid.

However, for the transfer to be successful, the ester group must be very sensitive to acidolysis; the above two esters are particularly sensitive in this way.

Synthesis of higher ketonic acids. Sodiomalonic ester is treated with the acid chloride-ester derivative of a dicarboxylic acid, *e.g.*, e-ketoheptanoic acid:

$$[(C_2H_5O_2C)_2CH]^-Na^+ + ClCO(CH_2)_4CO_2C_2H_5 \longrightarrow$$

$$(C_2H_5O_2C)_2CHCH(CH_2)_4CO_2C_2H_5 \xrightarrow[\text{(ii) HCl}]{\text{(i) KOH}}$$

$$(HO_2C)_2CHCO(CH_2)_4CO_2H \xrightarrow{150\text{–}200°C} [HO_2CCH_2CO(CH_2)_4CO_2H] \xrightarrow{-CO_2} CH_3CO(CH_2)_4CO_2H$$

The intermediate b-keto-dicarboxylic acid is unstable and is readily decarboxylated. This might have been anticipated, since it may be regarded as a derivative of acetoacetic acid, which is readily decarboxylated on warming.

SYNTHESIS OF POLYCARBOXYLIC ACIDS

It has been pointed out previously that the monoalkyl derivatives of malonic acid are readily brominated, and that these bromo-derivative may be used to synthesise polycarboxylic acid, *e.g.*,

$$CH_2(CO_2C_2H_5)_2 \xrightarrow{Br_2} CHBr(CO_2C_2H_5)_2$$

$$CHBr(CO_2C_2H_5)_2 + [CH(CO_2C_2H_5)_2]^-Na^+$$

$$\longrightarrow NaBr + (C_2H_5O_2C)_2CHCH(CO_2C_2H_5)_2 \xrightarrow{Br_2}$$

$$\begin{matrix}(C_2H_5O_2C)_2C-Br \\ | \\ (C_2H_5O_2C)_2C-Br\end{matrix} \xrightarrow{2[CH(CO_2C_2H_5)_2]^-Na^+} \begin{matrix}(C_2H_5O_2C)_2C-CH(CO_2C_2H_5)_2 \\ | \\ (C_2H_5O_2C)_2C-CH(CO_2C_2H_5)_2\end{matrix}$$

Malonic ester may also be used for the preparation of unsaturated acids (Knoevenagel reaction), alicyclic compounds and heterocyclic compounds.

EXERCISES

1. Draw the energy profiles for tautomerism between A and B and for resonance between A and B.
2. Explain the following equation:

$$MeCOCH_2CO_2Et + MeCOCHR^1CO_2Et \xrightarrow[\text{(ii) 1}R^2X]{\text{(i) 1 EtONa}} MeCOCHR^2CO_2Et + MeCOCHR^1CO_2Et$$

3. How could you differentiate, by i.r. and NMR spectroscopy, between

   (A) $MeCOCH_2CH_2COMe$ and

   (B) $MeCOCH_2COCH_2CH_3$?

4. Starting with E.A.A. or D.E.M., give one synthesis (using equations) for:

   (i) 4-methylpentan-2-one;

   (ii) 2,3-dimethylbutanoic acid;

   (iii) methylpentanedioic acid;

   (iv) cyclopentate-1,2-dicarboxylic acid;

   (v) $PhCOCH_2CO_2Et$;

(vi) 3-carboxy-2-methylpentanedioic acid.

5. The b-keto-acid shown in decarboxylated with difficulty. Suggest an explanation.

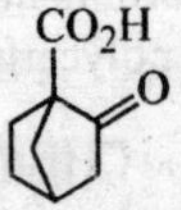

CO_2H

6. What two compounds will condense to give each product shown (after acidification)?

(i) A + B $\xrightarrow{\text{EtONa}}$ $MeCOCH_2COCO_2Et$

(ii) C + D $\xrightarrow{\text{EtONa}}$ $PhCH(CN)COCO_2Et$

(iii) E + F $\xrightarrow{\text{EtONa}}$ PhCH(CN)COMe

7. Starting with E.A.A., synthesise 3-methylpentan-2-ol.

8. Complete the equations:

(i) $CH_2(CO_2Me)_2$ + ? $\xrightarrow[\text{(ii) AcOH}]{\text{(i) Na}}$ $CH(CO_2Me)_3$

(ii) $MeCH(CO_2Et)COCO_2Et$ $\xrightarrow{\text{heat}}$?

(iii) $MeCH_2CO_2Et$ + ? $\xrightarrow[\text{(ii) AcOH}]{\text{(i) EtONa}}$ $MeCH(CHO)CO_2Et$

(iv) $MeCH_2CO_2Et$ + $(CO_2Et)_2$ $\xrightarrow[\text{(ii) AcOH}]{\text{(i) EtONa}}$? $\xrightarrow{\text{heat}}$?

9. Suggest a synthesis for ethyl 4-keto-3-mehylpentanoate.

10. A compound $C_4H_6O_2$, shows a very strong i.r. band at 1721 cm^{-1} and its N.M.R. spectrum shows only one singlet signal. What is the compound?

11. The starting material is an optically active form of s-BuOH, and is subjected to the following reactions:

s-BuOH $\xrightarrow{\text{Na}}$? $\xrightarrow{\text{MeI}}$?

Complete the equations and state the configuration of the s-Bu group in each product. Explain your answer.

12. Which of the two reactions would you use to prepare Me_3COMe, and why?

(*i*) $Me_3CBr + MeO^-K^+ \longrightarrow$;

(*ii*) $Me_3CO^-K^+ + MeBr \longrightarrow$.

13. Complete the following equations and comment:

    (*i*) MeOEt $\xrightarrow{\text{HI}}$?

    (*ii*) Et2O $\xrightarrow{\text{Na}}$?

    (*iii*) $Me_2C{=}CH_2 \xrightarrow[\text{pressure}]{H_2SO_4-H_2O}$? $\xrightarrow[\text{pressure}]{Me_3COH}$?

14. The compound $C_6H_{14}O$ showed two NMR signals. What is the compound?

15. The reaction between HI and C_2H_4 in EtOH gives predominantly EtI, whereas the reaction with HCl under the same conditions gives predominantly Et_2O. Explain.

16. Show the fragmentation paths whereby the ether, Et—O—CHMeEt, could produce ions *m/e*: 102, 87, 59, 57, 45, 31, 29.

4

ALDEHYDES AND KETONES

INTRODUCTION

Aldehydes and ketones both have the general formula $C_nH_{2n}O$ and both contain the **oxo (carbonyl) group,** $>C=O$ In **aldehydes,** the functional group is —CHO, *i.e.*, one of the available valencies of the carbonyl group is attached to hydrogen, and so the aldehyde group occurs at the *end* of a chain. In **Ketones** both available valencies are attached to carbon atoms, and so the keto group occurs *within* a chain (*cf.* ketens).

Nomenclature

Aldehydes. The lower members are commonly named after the acids that they from on oxidation. The suffix of the names of acids is *-ic* (the names of the trivial system are used; this suffix is deleted and replaced by *aldehyde, e.g.*,

$$\underset{\text{formaldehyde}}{HCHO} \xrightarrow{[O]} \underset{\text{formic acid}}{HCO_2H}$$

$$\underset{\text{isobutyraldehyde}}{(CH_3)_2CHCHO} \xrightarrow{[O]} \underset{\text{isobutyric acid}}{(CH_3)_2CHCO_2H}$$

The positions of side-chains or substituents are indicated by Greek letters, he α carbon atom being the one *adjacent* to the aldehyde group, *e.g.*,

$$\overset{\gamma}{C}H_3\overset{\beta}{C}H(OH)\overset{\alpha}{A}C_2CH \quad \beta\text{-hydroxybutyraldehyde}$$

According to the I.U.P.A.C. system of nomenclature, aldehydes are designated by the suffix *-al*, which is added to the name of the hydrocarbon from which they are derived. The longest carbon chain containing the aldehyde group is chosen as the parent hydrocarbon; the positions of side-chains or substituents are indicated by numbers, and the aldehyde group is given the number 1, which may be omitted from the name if it is the only functional group present in the compound, *e.g.*,

CH_3CHO	ethanal
$CH_3CH_2CH(CHO)CH(CH_3)CH_2CH_3$	2-ethyl-3-methylpentanal

Ketones. The lower members are commonly named according to the alkyl groups attached to the keto group, *e.g.*,

CH_3COCH_3	dimethyl ketone
$CH_3CH_2COCH(CH_3)_2$	ethyl isopropyl ketone

The positions of side-chains or substituents are indicated by Greek letters, the α-carbon atom being the one *adjacent* to the keto group, *e.g.*,

$CH_3CHClCOCH_2CH_2Cl$ α,β'-dichlorodiethyl ketone

If the two alkyl groups in a ketone are the same, the ketone is said to be simple or symmetrical; if unlike, mixed or unsymmetrical (*cf.* ethers).

According to the I.U.P.A.C. system of nomenclature, ketones are designated by the suffix *-one*, which is added to the name of the hydrocarbons from which they are derived. The longest carbon chain containing the keto group is chosen as the parent hydrocarbon; the positions of side-chains or substituents are indicated by numbers, and the keto group is given the lowest number possible, *e.g.*,

$CH_3COCH_2CH_2CH_3$	pentan-2-one
$(CH_3)_2CHCOCH(CH_3)CH_2CH_3$	2,4-dimethylehxan-3-one

The carbonyl or oxo group, $\rangle CO$, is prefixed in aldehydes as the oxo, aldo (for aldehydic O), or formyl (for CHO), and in ketones is prefixed as oxo or keto.

Since aldehydes and ketones both contain the carbonyl group, it might be expected that they would resemble one another. It is therefore instructive to compare their general methods of preparation and their general properties.

General Methods of Preparation of Aldehydes and Ketones

1. *Aldehydes*. By the oxidation or dehydrogenation of a *primary* alcohol.

t-Butyl chromate (prepared by adding chromium trioxide to t-butanol) oxidises primary alcohols to aldehydes almost quantitatively (Oppenauer *et al.*, 1949).

Ketones. By the oxidation or dehydrogenation of a *secondary* alcohol.

There is, however, a specific reagent for oxidising secondary alcohols to ketones, *viz.*, aluminium t-butoxide, $[(CH_3)_3CO]_3$ Al (the **Oppenauer oxidation,** 1937). The secondary alcohol is refluxed with the reagent, and then acetone (or cyclohexanone, etc.) is added. The mechanism of the reaction is the *reverse* of that of the Meerwein–Ponndorf-Verley reduction:

$$R^1CHOHR^2 + Me_3COAl/3 \rightleftharpoons R^1R^2CHOAl/3 + Me_3COH$$

$$R^1R^2CHOAl/3 + Me_2CO \rightleftharpoons R^1R^2CO + Me_2CHOAl/3$$

t-Butyl alkoxide is used as the regent, since the t-butyl alcohol produced is *not* oxidised under these conditions.

This reagent is particularly useful for oxidising *unsaturated* secondary alcohols because it does not affect the double bond. On the other hand, *primary* alcohols (particularly unsaturated alcohols) may also be oxidised to aldehydes if acetone is replaced by *p*-benzoquione. In general, quinones and aromatic ketones are better hydrogen acceptors than acetone.

Primary and secondary alcohols, both saturated and unsaturated, may be oxidised to the corresponding carbonyl compound by means of manganese dioxide in acetone solution (*inter alia*, Bharucha, 1956). Ruthenium tetroxide also oxidises saturated primary and secondary alcohols to the corresponding carbonyl compound (Berkowitz *et al.*, 1958). ON the other hand, *p*-toluensulphonates (tosylates) of n-alcohols can be oxidised to aldehydes by dimethyl sulphoxide in the presence of sodium hydrogen carbonate at 100°C (Kornblum *et al.*, 1959).

$$RCH_2OSO_2C_6H_4CH_3 + (CH_3)_2SO \xrightarrow{NaHCO_3} RCHO + (CH_3)_2S$$
$$+ CH_3C_6H_4SO_3Na \ (65\text{–}85\%)$$

Another method of oxidising primary and secondary alcohols to the corresponding carbonyl compound is that of Barton *et. al.* (1964). The alcohol is converted into its alkyl chloroformate, which is then dissolved in dimethyl sulphoxide, and finally triethylamine is added:

$$R_2CHOH \xrightarrow{COCl_2} R_2CHOCOCl \xrightarrow{Me_2SO} R_2CH{-}O{-}CO{-}O{-}\overset{+}{S}Me_2 \longrightarrow$$

$$CO_2 + R_2CH{-}O{-}\overset{+}{S}Me_2 \xrightarrow[-H^+]{Et_3N} R_2CO + Me_2S \ (57 - 80\%)$$

2. *Aldehydes.* By heating a mixture of the calcium salts of formic acid and any one of its homologues (yield: variable due to many side reactions):

$$(RCO_2)_2Ca + (HCO_2)CA \longrightarrow 2RCHO + 2CaCO_3$$

Ketones. By heating the calcium salt of any monocarboxylic acid other than formic acid.

$$(RCO_2)_2CA \longrightarrow R2CO + CaCO_3$$

The yields are usually low, but are high when the iron salts of n-acids are heated (Schultz *et al.*, 1962). The mechanism of the reaction is uncertain, but a possibility is one via the formation of an aldol and then a β-keto-acid (M = metal):

$$\text{(cyclic } RCH_2{-}CO{-}O{-}M{-}O{-}CO{-}CH_2R) \longrightarrow \text{(cyclic } RCH{-}CO{-}O{-}M{-}O{-}C(OH)(CH_2R)) \xrightarrow{-MO} RCH_2COCHRCO_2H \xrightarrow{-CO_2} RCH_2COCH_2R$$

No aldol formation is possible unless the acid contains an α-hydrogen also from ketones on heating:

$$(RCH_2CO)_2O \longrightarrow \text{(cyclic } RCH{-}CO{-}O{-}C(OH)(CH_2R)) \longrightarrow RCH_2COCHRCO_2H \xrightarrow{-CO_2} RCH_2COCHRCO_2H$$

If a mixture of calcium salts is used, mixed ketones are obtained:

$$(R^1CO_2)_2Ca + (R^2CO_2)_2Ca \longrightarrow 2R^1COR^2 + 2CaCO_3$$

R_2CO and RCHO are obtained as by-products, and the reaction probably proceeds via the manganous salt.

Ketones. By passing the vapour of any monocarboxylic acid other than formic acid over manganous oxide at 300°C:

$$2RCO_2H \xrightarrow{MnO} R_2CO + CO_2 + H_2O\,(g.)$$

A mixture of monocarboxylic acids gives mixed ketones:

$$R^1CO_2H + R^2CO_2H \longrightarrow R^1COR^2 + CO_2 + H_2O$$

R_2^1CO and R_2^2CO are obtained as by-products.

4. *Aldehydes.* By the reduction of an acid chloride with hydrogen in boiling xylene using a palladium catalyst supported on barium sulphate (**Rosenmund's reduction,** 1918):

$$RCOCl + H_2 \longrightarrow RCHO + HCl \ (g.\text{–}v.g.)$$

Aldehydes are more readily reduced than are acid chlorides, and therefore one would expect to obtain the alcohol as the final product. It is the barium sulphate that prevents the aldehyde from being reduced, acting as a poison (to the palladoium catalyst) in this reaction. Generally, when the Rosenmund reduction is carried out, a small amount of quinoline ad sulphur is added; these are very effective in poisoning the catalyst in the aldehyde reduction.

Acid chlorides are readily reduced to aldehydes by lithium tri-t-butoxyaluminium hydride and by tri-n-butyltin hydride. Esters also give aldehydes with sodium aluminium hydride.

Ketones. There is no analogous method for the preparation of ketones.

5. *Aldehydes.* By the ozonolysis of alkenes of the type $R^1CH = CHR_2$.

Ketones. By the ozonolysis or oxidation with the Lemieux reagent of alkenes of the type $R_2^1C = CR_2^2$.

6. *Aldehydes.* By the oxidation of 1,2-glycols of the type $R^1CH(OH)CH(OH)R^2$ with lead tetraacetate or periodic acid.

Ketones. By the oxidation of 1,2-glycols of the type $R_2^1C(OH)C(OH)R_2^2$ with lead tetraacetate or periodic acid.

7. *Aldehydes.* Acetylene, when passed into hot dilute sulphuric acid in the presence of mercuric sulphate as catalyst, is converted into acetaldehyde.

On the other hand, 1-alkynes give aldehydes by hydroboronation with disimaylborane followed by oxidation (Brown *et al.*, 1961).

$$RC \equiv CH + Si\alpha_2BH \longrightarrow RCH \equiv CHBSi\alpha_2 \xrightarrow[OH^-]{H_2O_2}$$

$$[RCH = CHOH] \longrightarrow RCH_2CHO \ (g.)$$

The addition to the *terminal* carbon atom is due to the fact that the borane is sterically hindered to addition at C-2.

Ketones. All homologues of acetylene, treated in the same way as acetylene, form ketones.

Hydroboronation of non-terminal alkynes, followed by oxidation, gives ketones:

$$RC \equiv CR \xrightarrow{Si\alpha_2BH} RCH = CRBSi\alpha_2 \xrightarrow[OH^-]{H_2O_2}$$

$$[RCH = CR(OH)] \longrightarrow RCH_2COR$$

8. *Aldehydes.* By Stephen's method (1925). An alkyl cyanide is dissolved in ether, or better, in ethyl formate or ethyl acetate (Stephen *et al.*, 1956), and reduced with stannous chloride and hydrochloric acid, and then steam distilled.

According to Hantzsch (1931), Stephen's reaction proceeds via the iminochloride or aldimine hydrochloride (which is present as the stannichloride):

$$RC \equiv N \xrightarrow{HCl} [RC \equiv NH]^+Cl^- \xrightarrow[HCl]{SnCl_2}$$

$$[RCH = NH_2]_2^+ SnCl_6^{2-} \xrightarrow{H_2O} RCH_2COR$$

Tolbert *et al.* (1963) have isolated these aldimine stannichlorides.

Cyanides and t-amides $R^1CONR^2_2$) are reduced to aldehydes by lithium triethoxyaluminium hydride.

Ketones. There is no analogous method for the preparation of ketones.

9. *Aldehydes.* By means of a Grignard reagent and formic ester.

Ketones. By means of a Grignard reagent and, for example, an alkyl cyanide.

Ketones are also formed by reaction between an alkyl cyanide and lithium-alkyls.

10. *Ketones.* By the ketonic hydrolysis of the alkyl derivatives of acetoacetic ester, and also from ethyl malonate derivatives.

Aldehydes. There is no analogous method for the preparation of aldehydes. However, the reduction of acylmalonic esters with sodium borohydride gives aldehydes (Muxfeldt *et al.*, 1965).

$$RCOCH(CO_2Et)_2 \xrightarrow{NaBH_4} RCHO + CH_2(CO_2Et)_2$$

11. *Aldehydes.* Many aldehydes may be prepared by the *oxo* process, *e.g.*, propionaldehyde (*q.v.*).

Ketones. There is no analogous method for the preparation of ketones.

12. *Aldehydes.* Aldehydes may be prepared (in yields up to 50

percent) by means of the modified **Sommelet reaction** (Angyal *et al.*, 1953; see benzaldehyde); *e.g.*, n-hexanal is produced by adding an aqueous solution of n-hexylamine hydrochloride to a solution of hexamine in acetic acid through which steam is passed.

$$CH_3(CH_2)_5NH_2{\cdot}HCl + (CH_2)_6N_4 \longrightarrow CH_3(CH_2)_4CHO\ (17\%)$$

Ketone. There is no analogous method for the preparation of ketones.

13. *Ketones.* Methyl ketones may be prepared from esters and dimethyl sulphoxide (Corey *et al.*, 1964). The latter compound is converted into a methylide by means of sodium hydride, and this is then treated as follows (THF = tetrahydrofuran):

$$2CH_3SOCH_3 \xrightarrow{2NaH} 2CH_3SOCH_2^- \xrightarrow{RCO_2Me}$$

$$MeO^- + CH_3SOCH_3 + [RCO\bar{C}HSOCH_3]$$

$$\xrightarrow{H^+} RCOCH_2SOCH_3 \xrightarrow[THF-H_2O]{Al-Hg} RCOCH_3\ (70\text{–}98\%)$$

14. *Aldehydes and ketones.* See the Darzens Glycidic Ester condensation.

General Properties of Aldehydes and Ketones

Dipole moment measurements of aldehydes and ketones have shown that the values are larger than can be accounted for by the inductive effect of the oxygen atom, but can be accounted for if carbonyl compounds are resonance hybrids:

$$>C=O \longleftrightarrow >\overset{+}{C}-O^-$$

Thus the carbon atom has a positive charge and consequently can be attacked by nucleophilic reagents. The carbonyl group also exhibits basic properties; it is readily protonated by strong acids to form oxonium salts. Since oxygen is more electronegative than carbon, the second resonating structure will make a larger contribution than the first.

$$>C=O + H^+ \rightleftharpoons >C=O^+-H \longleftrightarrow >C^+-O-H$$

Hence, protonation increases the electrophilic character of the carbonyl group and so it can be expected that nucleophilic additions will be catalysed by acids. It should also be noted that, because of the positive charge on the carbon a atom, the CO group has a strong –I effect.

Many addition reactions of carbonyl compounds may be represented by the general equation

$$\begin{matrix}R^1\\R^2\end{matrix}\!\!>C{=}O + H{-}Z \rightleftharpoons \begin{matrix}R^1\\R^2\end{matrix}\!\!>C<\!\!\begin{matrix}OH\\Z\end{matrix}$$

Thus, the hybridisation of the carbon atom changes from sp^2 to sp^3 This results in increased crowding and so reaction can be expected to be increasingly sterically hindered with increasing size of R^1 and R^2. For aldehydes. $R^2 = H$ (for formaldehyde, $R^1 = R^2 = H$), and hence it can be anticipated that aldehydes will be more reactive than ketones. This is observed in practices, reaction being faster and the equilibrium shifted more to the right.

The + I effect of an alkyl group also decreases the reactivity of the carbonyl group towards nucleophilic reagents, this being due to the partial neutralisation of the positive charge on the cabonyl carbon atom. Hence, ketones will be less reactive than aldehydes due to both steric and inductive factors.

If we consider the energetics of these addition reactions:

$$>C{=}O + H{-}Z \rightleftharpoons >C<\!\!\begin{matrix}OH\\Z\end{matrix}$$

$$(+694.5) \quad (+x) \qquad (-334.7-464.4-y)$$

$$\text{then } \Delta H^{\ominus} = (x - y - 104.6)\ \text{kJ}$$

where x and y are the bond energies of the H—Z and C—Z bonds, respectively. Suppose Z is OH, *i.e.*, H—Z is H—OH. In this case, $\Delta H^{\ominus} = (464.4 - 334.7 - 104.6) = 25.1$ kJ. Now suppose Z is CN, *i.e.*, HZ is H—CN. Then

$$\Delta H^{\ominus} = (414.2 - 347.3 - 104.6) = -37.7\ \text{kJ}$$

Aldehydes and ketones have low boiling points compared with those of the alcohols, and this is probably due to their inability to form intermolecular hydrogen bonds. On the other hand, carbonyl compounds have higher boiling points than the corresponding alkanes, and this is due to dipole-dipole interactions in the former. Since carbonyl compounds can form hydrogen bonds with water and also form hydrates (to various extents), they are soluble (or partially soluble) in water.

The infra-red absorption region (C = O stretch) for saturated aldehydes is 1740–1720 cm^{-1}(s) and for acyclic ketones is 1725–1700 cm^{-1}(s). It is therefore possible to distinguish between aldehydes and ketones. Once the presence of a carbonyl group has been established, aldehydes and ketones may be differentiated by the fact that there is the C—H (str.) in the aldehyde group that absorbs in the region 2880–2665 cm^{-1}(w-m). There are also bands due to C—H (def.) which are characteristic of a methyl or a methylene group directly attached to the carbonyl group in ketones : CH_3CO—, 1365–1355 cm^{-1}(s); —CH_2CO—, 1435–1405cm^{-1} (s).

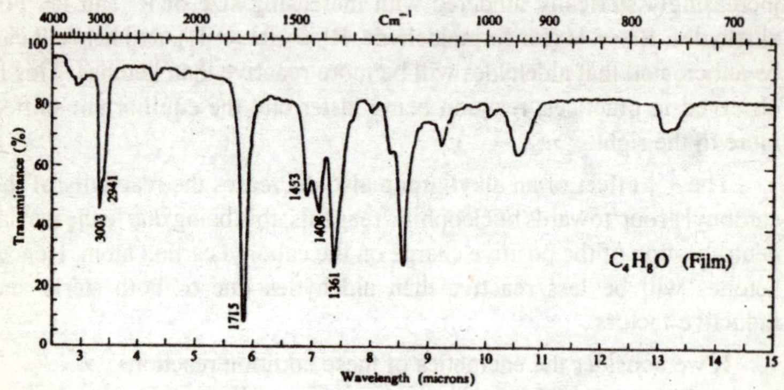

Fig. 4.1

1715 cm^{-1}: C═O (str..) in saturated ketones (and aldehydes?). 3003 and 2941 cm^{-1}: C—H (str..) in Me and/or CH_2. 1453 cm^{-1}: C—H (def.) in —CH_2 CO—. 1361 cm^{-1}: C—H (def.) in Me in MeCO—.

Fig. 4.1 is the infra-red spectrum of ethyl methyl ketone (as a film). The region of the CO absorption is in that for saturated ketone, but is very close to the lower frequency region for aldehydes. However, the absence of a band at 2880–2665 cm^{-1} is strongly in favour of the absence of an aldehyde.

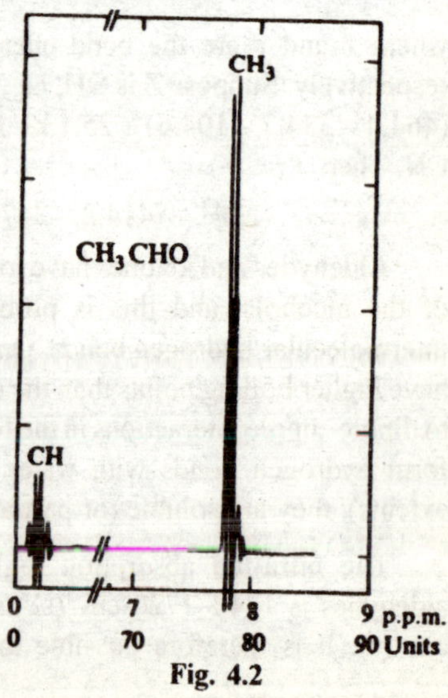

Fig. 4.2

The NMR spectra of carbonyl compounds are affected by the strong inductive effect (–I) of the carbonyl group and also by the magnetic anisotropy of the carbon-oxygen double bond. Both effects deshield an *aldehydic* proton, the result being a very low τ-value 0.1–0.7 p.p.m. On the other hand, protons in *Ketones* are deshielded mainly by the – I effect (the protons are at the 'edge' of the shielding cone) and consequently the shift downfield is much less (than for the aldehydic proton), *e.g.*, aliphatic ketones containing the MeCO group have a τ-value 7.8–8.1 for the methyl protons. Thus, infra-red spectroscopy will show the presence of a carbonyl group, and the NMR spectrum will distinguish between aldehydes and ketones. This is clearly shown by the NMR spectra of acetaldehyde (Fig. 4.2) and that of ethyl methyl ketone (Fig. 4.3; both in CCl_4 at 60 MHz).

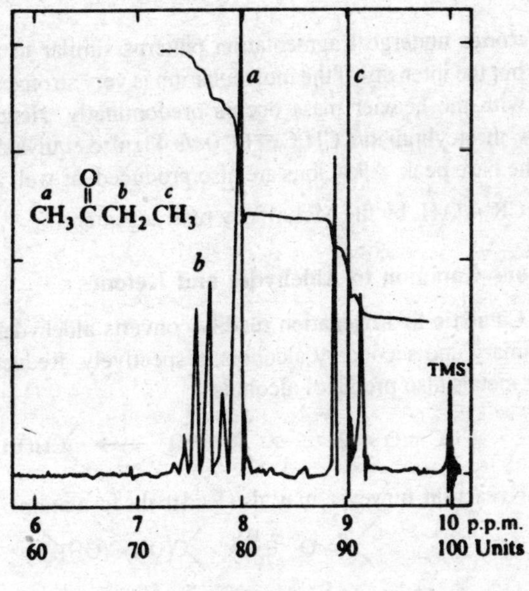

Fig. 4.3

Mass spectra. Aldehydes give molecular ions of low intensity and readily undergo α-cleavage of produce acylium ions:

$$H\cdot + RC{\equiv}O^+ \longleftarrow RCH{=}O^+ \longrightarrow R\cdot + HC{\equiv}O^+ (m/e\ 29)$$

The presence of ions $M-1$ and m/e 29 are usually characteristic of aldehydes (R^+ is also formed). It should be noted that the ion m/e 29 could also be $C_2H_5^+$, which is given by the higher aldehydes (the two ions may be distinguished by high resolution).

Aldehydes also undergo β-cleavage (the ion m/e 43 could also be $C_3H_7^+$).

$$[R—CH_2—CH=O]^+ \longrightarrow R\cdot(M-43) + [CH_2=CH—O]^+ (m/e\ 43)$$

When a γ-hydrogen atom is present, the McLafferty rearrangement also occurs:

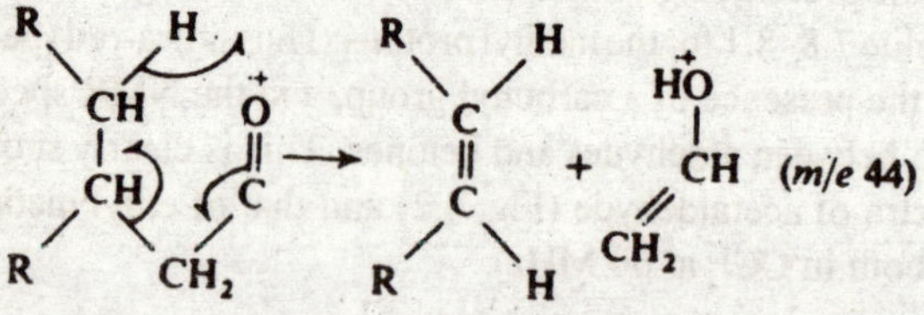

$$R\text{—}CH(\text{—}H)\text{—}CH(R)\text{—}CH_2\text{—}C=\overset{+}{O} \longrightarrow RCH=CHR + CH_2=CH\text{—}\overset{+}{O}H\ (m/e\ 44)$$

Ketones undergo fragmentation patterns similar to those of aldehydes, but the intensity of the molecular ion is very strong, and loss of the group with the heavier mass occurs predominatly. Hence, for methyl ketones, the acylium ion, $CH_3C\equiv O^+$ (*m/e* 43; also equivalent to $C_3H_7^+$) is often the base peak. Alkyl ions are also produced, as well as alkenes and $CH_2=CR—\dot{O}\overset{+}{H}$ by the McLafferty rearrangement.

Reactions Common to Aldehydes and Ketones

1. Catalytic hydrogenation readily converts aldehydes and ketones into primary and secondary alcohols, respectively. Reduction with dissolving metals also produces alcohols:

$$>C=O + 2e \longrightarrow >C^-—O^- \longrightarrow >CHOH$$

This reaction, however, may also lead to the formation of a 1,2-glycol:

$$2>C=O \xrightarrow{e;\ H^+} >C(OH)C(OH)<$$

Reduction of both aldehydes and ketones to alcohols may be effected by means of the **Meerwein-Ponndorf-Verley reduction** (1925, 1926). The carbonyl compound is heated with is removed from the equilibrium mixture by slow distillation:

$$R_2CO + (CH_3)_2CHOAl/_3 \rightleftharpoons R_2CHOAl/_3 +$$

$$CH3COCH_3 \xrightarrow[H_2SO_4]{dilute} R_2CHOH\ (g.)$$

This reducing agent is specific for the carbonyl group, and so may be used for reducible, *e.g.*, a double bond or a nitro-group.

The reduction occurs by hydride ion transfer from aluminium isopropoxide to the carbonyl compound via a cyclic transition state.

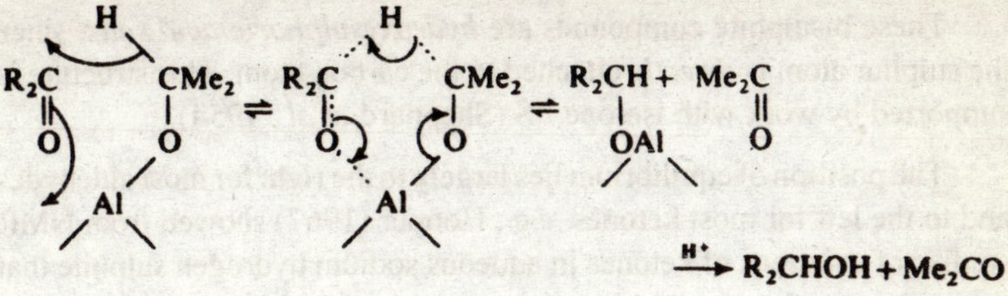

$$R_2C(=O)\cdots Al \;\; CMe_2(H)-O \rightleftharpoons R_2C\cdots O \cdots Al \cdots O-CMe_2(H) \rightleftharpoons R_2CH-OAl + Me_2C=O \xrightarrow{H^+} R_2CHOH + Me_2CO$$

Fig. 4.4

The equilibrium is displaced to the right by distillation of the acetone (which is the component with the lowest b.p. in the system). Of all the metal alkoxides that may be used for this to form coordination compounds.

Aldehydes and ketones are readily reduced to alcohols by various metallic hydrides.

Clemmensen reduction (1913): the carbonyl compound is reduced with *amalgamated zinc* and concentrated hydrochloric acid:

$$R^1COR^2 \xrightarrow{e;H^+} R^1CH_2R^2$$

The Clemmensen reduction does not appear to work well for aldehydes, but is reasonably good for many ketones.

Several mechanisms have been proposed for the Clemmensen reduction, *e.g.*,

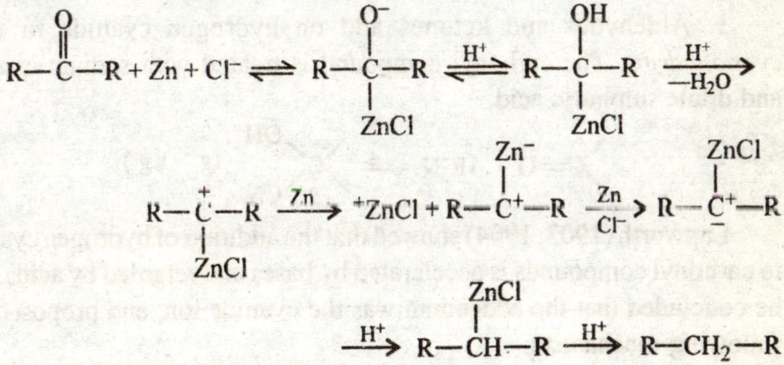

$$R-\overset{O}{\overset{\|}{C}}-R + Zn + Cl^- \rightleftharpoons R-\overset{O^-}{\underset{ZnCl}{C}}-R \underset{}{\overset{H^+}{\rightleftharpoons}} R-\overset{OH}{\underset{ZnCl}{C}}-R \xrightarrow[-H_2O]{H^+}$$

$$R-\underset{ZnCl}{\overset{+}{C}}-R \xrightarrow{Zn} {}^+ZnCl + R-\overset{Zn^-}{C^+}-R \xrightarrow[Cl^-]{Zn} R-\overset{ZnCl}{\underset{-}{C^+}}-R$$

$$\xrightarrow{H^+} R-\overset{ZnCl}{CH}-R \xrightarrow{H^+} R-CH_2-R$$

2. Aldehydes and ketones add on sodium hydrogen sulphite to form *bisulphite* compounds:

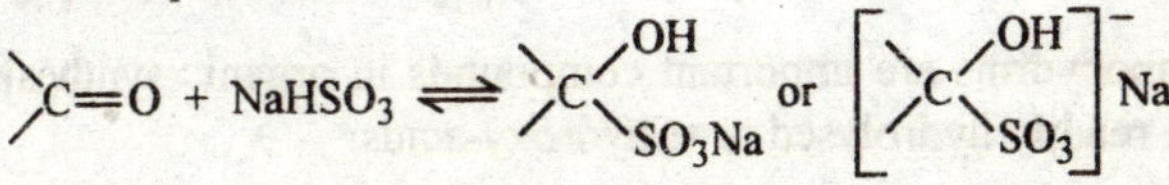

$$>C=O + NaHSO_3 \rightleftharpoons >C<^{OH}_{SO_3Na} \quad \text{or} \quad \left[>C<^{OH}_{SO_3}\right]^- Na^+$$

These bisulphite compounds are *hydroxysulphoric acid salts*, since the sulphur atom is *directly* attached to the carbon atom. This structure is supported by work with isotope ^{34}S (Sheppard *et al.*, 1954).

The position of equilibrium lies largely to the right for most aldehydes and to the left for most ketones. *e.g.*, Hooper (1967) showed from NMR studies of solutions of ketones in aqueous sodium hydrogen sulphite that the amounts of *unreacted* ketone were: Me_2CO, 24·6; MeCOEt, 71·8 MeCOisoPr, 100 per cent. These results may be partly explained in terms of steric effects.

Bisulphite compounds are usually crystalline solids, insoluble in sodium hydrogen sulphite solution. Since they regenerate the carbonyl compound when heated with dilute acid or sodium carbonate solution, their formation affords a convenient means of separating carbonyl compounds from non-carbonyl compounds.

The mechanism of bisulphite formation has been the subjected of much discussion. Hinshelwood *et al.* (1958), from a study of the *reverse* reaction, have proposed the following mechanism.

$$HSO_3^- \rightleftharpoons H^+ + SO_3^{2-}$$

$$\rangle C{=}O + SO_3^{2-} \rightleftharpoons \rangle C\langle^{OH}_{SO_3^{2-}} \xrightleftharpoons{HSO_3^-} \rangle C\langle^{OH}_{SO_3^-} + SO_3^{2-}$$

3. Aldehydes and ketones add on hydrogen cyanide to form *cyanohydrins*. The carbonyl compound is treated with sodium cyanide and dilute sulphuric acid:

$$\rangle C{=}O + HCN \rightleftharpoons \rangle C\langle^{OH}_{CN} \quad (g. - v.g.)$$

Lapworth (1903, 1904) showed that the addition of hydrogen cyanide to carbonyl compounds is accelerated by bases and retarded by acids, and he concluded that the addendum was the cyanide ion, and proposed the following mechanism:

$$R_2C{=}O + CN^- \xrightleftharpoons{slow} R_2C\langle^{O^-}_{CN} \xrightarrow{H^+;\ fast} R_2C\langle^{OH}_{CN}$$

Cyanohydrins are important compounds in organic synthesis since they are readily hydrolysed to α-hydroxy-acids:

$$RCH(OH)CN \xrightarrow{H^+} RCH(OH)CO_2H$$

All aldehydes form cyanohydrins; only the ketones acetone, butanone, pentan-3-one, and pinacolone form cyanothydrins.

4. Aldehydes and ketones combine with a variety of compounds of the type Z—NH_2 (Z = OH, NH_2, etc.). The general reaction may be written

$$>C=O + NH_2-Z \longrightarrow \left[>C\begin{matrix}OH\\NHZ\end{matrix}\right] \xrightarrow{-H_2O} >C=N-Z$$

Oximes are formed with hydroxylamine:

$$R_2CO + NH_2OH \longrightarrow \left[R_2C\begin{matrix}OH\\NHOH\end{matrix}\right] \xrightarrow{-H_2O} R_2C=NOH \quad (v.g.)$$

Oximes are usually well-defined crystalline solids, and may be used to identify carbonyl compounds.

The absorption region of oximes (C=N stretch) is 1690–1630 cm^{-1} (v) and that for the O—H stretch is 3650–3500 cm^{-1}(s) (*cf.* alcohols).

Aldoximes form cyanides when boiled with acetic anhydride, whereas ketoximes form the acetyl derivative of the oxime:

$$RCH=NOH \xrightarrow[-H_2O]{(CH_3CO)_2O} RCN$$

$$R_2C=NON \xrightarrow{(CH_3CO)_2O} R_2C=NOCOCH_3$$

Aldehydes and ketones combine with *N*-substituted hydroxylamines to form *nitrones*:

$$>C=O + RNHOH \longrightarrow H_2O + >C=\overset{+}{N}\begin{matrix}R\\O^-\end{matrix} \longleftrightarrow >\overset{+}{C}-\ddot{N}\begin{matrix}R\\O^-\end{matrix}$$

Nitrones readily undergo 1,3- addition reactions, *e.g.*,

$$\begin{matrix}R\\H\end{matrix}>C=\overset{+}{N}\begin{matrix}R\\O^-\end{matrix} + HCN \longrightarrow \begin{matrix}R\\H\end{matrix}>\underset{CN}{C}-N\begin{matrix}R\\OH\end{matrix} \xrightarrow{-H_2O} \begin{matrix}R\\NC\end{matrix}>C=N-R \quad \text{cyno-imine}$$

Hydrazones and *azines* are formed with hydraxine:

$$>C=O + NH_2NH_2 \longrightarrow \left[>C\begin{matrix}NHNH_2\\OH\end{matrix}\right] \xrightarrow{-H_2O} >C=NNH_2 \xrightarrow{>CO} >C=NN=C<$$

By using suitable derivatives of hydrazine, more well-defined crystalline products are obtained (and azine formation is avoided), *e.g.*, *phenylhydrazine* forms **phenylhydrazones**:

$$C_6H_5NHNH_2 + O{=}C\langle \longrightarrow H_2O + C_6H_5NHN{=}C\langle$$

Semicarbazide forms **semicarbazones**:

$$NH_2CONHNH_2 + O{=}C\langle \longrightarrow NH_2CONHN{=}C\langle + H_2O$$

The mechanism of the formation of semicarbazones and oximes has been studied in great detail.

These reactions (as are most reactions of this type, *i.e.*, $\rangle CO + NH_2{-}Z$) are catalysed by acids, and it has been found that the rates of formation of the products reach a maximum at some particular pH. According to Jencks *et al.* (1959, 1962), the pH optima observed are due to transitions in the rate-determining steps with changing concentration of acid. Let us consider the following mechanism ($Z = -NHCONH_2$ or $-OH$).

$$R_2C{=}O + NH_2{-}Z \longrightarrow R_2C\langle \longrightarrow R_2C\langle^{OH}_{NHZ} \; \text{(I)}$$

$$\longrightarrow R_2C{=}NH + H_2O \; \text{(II)}$$

Jencks showed, by means of ultraviolet and infra-red studies, that at pH 7 the formation of (I) is fast, and that as the acidity increases, the overall rate of the reaction increases, since the dehydration step, which is the r/d step, also increases (through protonation of the OH, etc.). However, as the acidity increases, the addition step becomes progressively shower because $Z{-}NH_2$ is decreasing in concentration due to its conversion into its conjugate acid, $Z{-}NH_3^+$. The latter no longer has a lone pair of electrons on the N atom and so is not a nucleophile. Hence, at a sufficiently high acidity, the addition step is slowed down to such an extent that this now becomes the r/d step in the reaction.

Oximes and hydrazones regenerate the carbonyl compound when refluxed with dilute hydrochloric acid. Regeneration from phenylhydrazones by this method is usually difficult. One fairly successful method is by exchange of the phenylhydrazino group with another oxo compound, *e.g.*, pyruvic acid. The most satisfactory method, however, appears to be the use of acetylacetone in faintly acid solution (Ried *et al.*, 19862); the product is 3,5-dimethyl-1-phenylpyrazole.

$$>C{=}NNHPh + MeCOCH_2COMe \longrightarrow >C{=}O + \text{(1-phenyl-3,5-dimethylpyrazole: Me, N, N, Me, Ph)}$$

Wolff-Kishner reduction (1912). When hydrazones (or semicarbazones) are heated with sodium ethoxide 180°C, nitrogen is eliminated, and a hydrocarbon is obtained, *i.e.*, by this means the carbonyl group is converted into the methylene group:

$$>C{=}NNH_2 \longrightarrow >CH_2 + N_2 \quad (g.\text{–}v.g.)$$

The yield of hydrocarbon is better than that obtained by the Clemmensen reduction, but both the Wolff-Kishner and the Clemmensen reduction are generally unsuccessful with sterically hindered ketones. Barton *et al.* (1954, 1955), however, have overcome this problem by using a modification of the improved technique of the Wolff-Kishner reduction introduced by Huang-Minlon (1946). All the methods described use elevated temperatures (180–200°C), but Cram *et al.* (1962) have shown that the reaction proceeds at room temperature if dimethyl sulphoxide is used as solvent.

According to Szmant *et al.* (1966), the mechanism of Wolff-Kishner reaction is (B is the base):

$$R_2C{=}NNH_2 + B \rightleftharpoons R_2C{=}N\bar{N}H + BH^+ \longrightarrow$$

$$R_2\overline{C}H + N_2 \xrightarrow{BH^+} R_2CH_2$$

With α, β-unsaturated carbonyl compounds, the Wolff-Kishner reduction may lead to migration of the double bond. In the above mechanism, the first step is the abstraction of a proton. Hence the use of a very strong base should enable the reaction to be carried out at a lower temperature. Henbest *et al.* (1963), using potassium t-butoxide, found that the reduction proceeded normally with certain compounds that had previously been reported to give abnormal products.

Girard's reagents ford carbonyl compounds (1934, 1936). Girard introduced two reagents for carbonyl compounds: Girard's reagents 'T' and 'P', which are respectively trimethylaminoacetohydrazide chloride: $[(CH_3)_3NCH_2CONHNH_2]^+Cl^-$, and pyridinium-acetohydrazide chloride, $[C_5H_5NCH_2CONHNH_2]^+Cl^-$. These reagents react with carbonyl compounds to form derivatives of the type $\left[(CH_3)_3NCH_3CONHN{=}C<\right]^+Cl^-$.
These compounds are *soluble* in water and have been found particularly useful for isolating certain ketonic hormones.

Aldimines, Schiff bases, or azomethines, are formed when aldehydes react with aliphatic *primary* amines;

$$R^1CHO + R^2NH_2 \longrightarrow R^1CH(OH)NHR^2 \xrightarrow{-H_2O} R^1CH{=}NR^2$$

In some aces, the product is a polymer of the aldimine.

Aldehydes and ketones with α-hydrogen from *enamines* with cyclic secondary amines.

5. Aldehydes and ketones condense with thioalcohols (mercaptans), in the presence of hydrochloric acid, to form *mercaptals* and *mercaptols*, respectively :

$$>C{=}O + 2RSH \longrightarrow >C(SR)_2 + H_2O$$

6. Aldehydes and ketones add on a molecule of a Grignard reagent, and the complex formed, when decomposed with acid, gives a secondary alcohol form an aldehyde (except formaldehyde, which gives a primary alcohol), and a tertiary alcohol from a ketone :

$$>C{=}O + RMgX \longrightarrow >C(OMgX)R \xrightarrow{H^+} >C(OH)R$$

7. Both aldehydes and ketones undergo the Wittig reaction of form alkenes.

8. Phosphorus pentachloride reacts with *simple* carbonyl compounds to form 1,1-dichlorides:

$$>C{=}O + PCl_5 \longrightarrow >CCl_2 + POCl_3$$

The mechanism of reaction is uncertain, but a possibility is attack by PCl_4^+ (*solid* PCl_5 is $PCl_4^+PCl_6^-$):

$$>C{=}O + PCl_4^+ \rightleftharpoons >\overset{+}{C}{-}OPCl_4 \overset{Cl^-}{\rightleftharpoons} >C(Cl)(OPCl_4) \xrightarrow{-OPCl_4^-}$$

$$>\overset{+}{C}{-}Cl \xrightarrow{Cl^-} >CCl_2$$

The chloride ions are probably derived from PCl_6^- (Newman *et al.*, 1959).

Good yields of 1,1-dichlorides are obtained by reaction between carbonyl compounds and 1,1-dichlorodimethyl ether (Rieche *et al.*, 1960).

Very little 1,1'-dibromides are obtained with phosphorus pentabromide; the main product from any carbonyl compound is the α-bromo-derivative. Phosphorus pentabromide dissociates more readily then phosphorus pentachloride, and it is probable that the halogenation in the α-position is brought about by the free halogen.

9. Chlorine or bromine replaces one or more α-hydrogen atoms in aldehydes and ketones, *e.g.*, acetone may be brominated in glacial acetic acid to give monobromoacetone:

$$CH_3COCH_3 + Br_2 \longrightarrow CH_3COCH_2Br + HBr \ (43\text{–}44\%)$$

The halogenation of carbonyl compounds is catalysed by acids and bases. Let us consider the case of acetone. In alkaline solution, tribromoacetone and bromoform are isolated. Thus, the introduction of a second and a third bromine atom is more rapid than the first. In aqueous sodium hydroxide, the rate has been shown to be independent of the bromine concentration., but first order with respect to both acetone and base (Lapworth, 1904), *i.e.*,

$$\text{rate} = k[\text{acetone}]\,[OH^-]$$

Thus, the reaction is zero order in bromine, and this suggests at least a two-stage process, the first being the rate-determining step of formation of an intermediate which then rapidly reacts with bromine to give the products. The problem is now to formulate this intermediate. A most likely one is the enolate ion, since it is known that both acids and bases catalyse enolisation of a ketone and that enols react extremely rapidly with bromine. The following mechanism is consistent with the facts mentioned:

$$CH_3COCH_3 + OH^- \underset{}{\overset{\text{slow}}{\rightleftharpoons}} H_2O + CH_3\overset{\overset{O}{\|}}{C}CH_2^- \leftrightarrow CH_3\overset{\overset{O^-}{|}}{C}{=}CH_2 \xrightarrow[\text{fast}]{Br_2} CH_3COCH_2Br + Br^-$$

There is much evidence to support this mechanism, *i.e.*, iodination of acetone proceeds at the same rate as bromination and is zero order in iodine (Bartlett, 1934). The above mechanism is independent of both the nature of the halogen and its concentration.

In the acid-catalysed bromination of acetone, it is possible to isolate the mono-, di-, and tribromo-derivatives. Hence, in contrast with the base-catalysed reaction, the introduction of a second and a third bromine atom is less rapid than the first.

Another point that we shall discuss here is: How is it possible for a C—H bond to be broken so readily? This has been explained as follows. Owing to the inductive effect of the carbonyl group, the electrons of the C—H bonds on the α-carbon atom are displaced towards the carbon atom, thus facilitating the release of a proton from C_α. At the same time, the anion produced has greater resonance stabilisation than the parent carbonyl compound, *i.e.*, the conjugate base is stabilised with respect to its acid (see above equation).

The inductive effect is very much weaker on a β-carbon atom since the I effect falls off rapidly from the source. Hence proton release is far less easy than on α-carbon. Also, if a proton were released from the β—C, this negative atom would be 'insulated' from the CO group by the intervening saturated carbon atom, and consequently such an anion would no have increased resonance stabilisation. Thus, because of the increased reactivity of hydrogen in a methylene group ($—CH_2—$) or in a methyne group (=CH—) when *adjacent* to a carbonyl group or any other strongly electron-attracting group, these groups are referred to as the 'active' methylene or methine group.

Aldehydes and ketones with α-hydrogen atoms readily react with sulphuryl chloride at room temperature in the absence of a catalyst, to replace-α-hydrogen atoms only, *e.g.*,

$$CH_3COCH_3 \xrightarrow{SO_2Cl_2} CH_3COCH_2Cl$$

A general method for preparing α-bromoaldehydes is as follows:

$$RCH_2CHO + (MeCO)_2O \xrightarrow{MeCO_2^-K^+} RCH{=}CHOOCMe \xrightarrow{Br_2}$$

$$RCHBrCHBrOOCMe \xrightarrow{EtOH} RCHBrCH(OEt)_2 \xrightarrow{HCl}$$

$$RCHBrCHO\ (v.g.)$$

The halogen atom in the group CXCO is very reactive by the S_N2 mechanism, but not by the S_N1. The reason for this is not clear, but a contributing factor is the – I effect of the carbonyl group which draws the covalent pair in the C—Cl bond away from the Cl and consequently makes its ionisation difficult.

10. Aldehydes and ketones with a methyl or methylene group adjacent to the carbonyl group are oxidised by selenium dioxide at room temperature to dicarbonyl compounds; *e.g.*, acetaldehyde forms glyoxal, and acetone forms methylglyoxal:

$$CH_3CHO + SeO_2 \longrightarrow OCHCHO + Se + H_2O$$

$$CH_3COCH_3 + SeO_2 \longrightarrow CH_3COCHO + Se + H_2O$$

The reaction is usually carried out in acetic acid, and the actual reagent is selenous acid. The mechanism of the oxidation appears to be far from settled, but a possibility is:

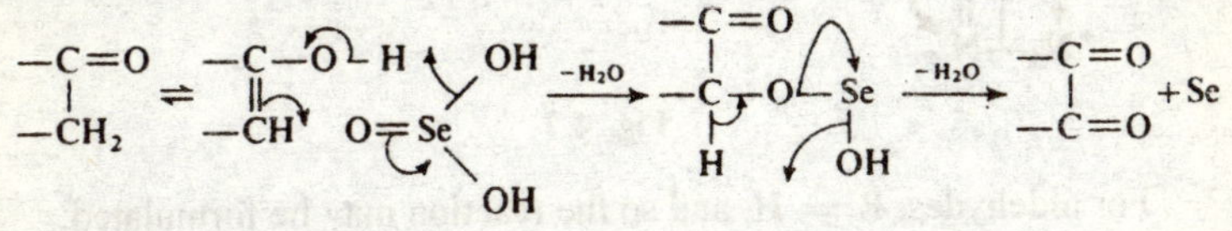

Fig. 4.5

11. Aldehydes and ketones undergo the **Schmidt reaction** (1924). This is a reaction between a carbonyl compound ad hydrazoic acid in the presence of *e.g.*, concentrated sulphuric acid. Aldehydes give a mixture of cyanide and formyl derivatives of primary amines, whereas ketones give amides.

$$RCHO + HN_3 \xrightarrow{H_2SO_4} RCN + RNHCHO + N_2$$

$$RCOR + NH_3 \xrightarrow{H_2SO_4} RCONHR + N_2$$

The mechanism of the reaction is uncertain. It has been shown to be intramolecular, and Smith (1948) has proposed the following mechanism, which is an example of the 1,2-shift (from carbon to nitrogen); for ketones:

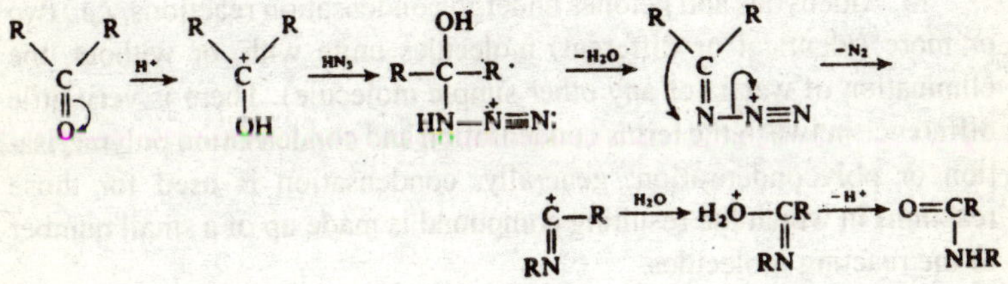

Fig. 4.6

In ketones, if the two groups are not identical, then two geometrical isomers of (I) are possible. It is also reasonable to suppose that the *anti* group (to the diazonium nitrogen) is the group that migrates. In this way it is possible to explain how steric factors may influence the isomer ratio of amides formed:

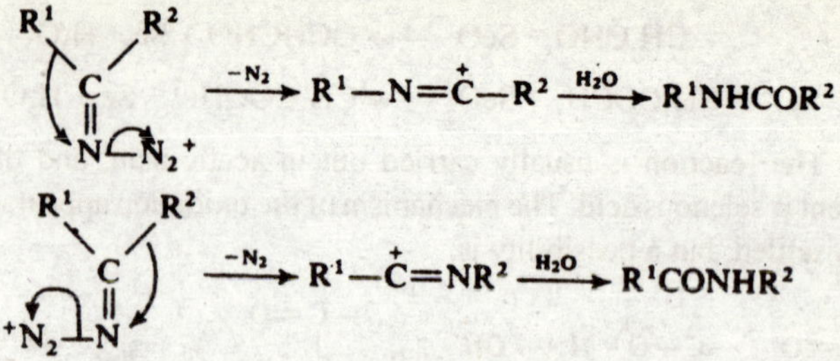

$$R^1(R^2)C{=}N{-}N_2^+ \xrightarrow{-N_2} R^1{-}N{=}\overset{+}{C}{-}R^2 \xrightarrow{H_2O} R^1NHCOR^2$$

$$R^1(R^2)C{=}N{-}N_2^+ \xrightarrow{-N_2} R^1{-}\overset{+}{C}{=}NR^2 \xrightarrow{H_2O} R^1CONHR^2$$

Fig. 4.7

For aldehydes, $R^1 = H$, and so the reaction may be formulated:

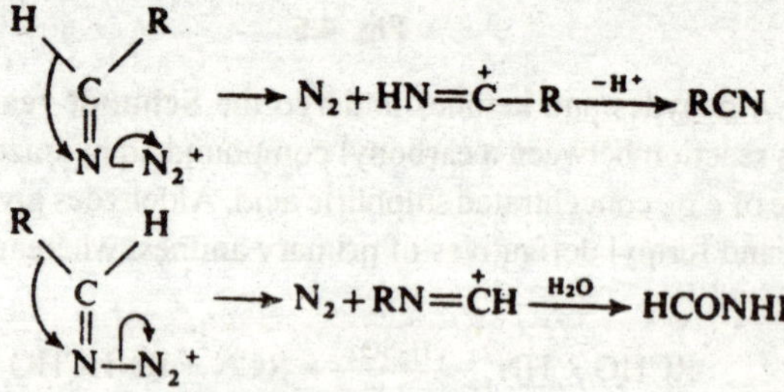

$$H(R)C{=}N{-}N_2^+ \longrightarrow N_2 + HN{=}\overset{+}{C}{-}R \xrightarrow{-H^+} RCN$$

$$R(H)C{=}N{-}N_2^+ \longrightarrow N_2 + RN{=}\overset{+}{C}H \xrightarrow{H_2O} HCONHR$$

Fig. 4.8

12. Aldehydes and ketones undergo cyclic additions with certain substrates to form heterocyclic compounds, *e.g.*, formaldehyde forms β-propiolactone with keten.

13. Aldehydes and ketones undergo condensation reactions, *i.e.*, two or more (identical or different) molecules unite with, or without, the elimination of water (or any other simple molecule). There is very little difference between the terms condensation and condensation polymerisation or polycondensation; generally, condensation is used for those reactions in which the resulting compound is made up of a small number of the reacting molecules.

Aldol condensation. Acetaldehyde, in the presence of dilute sodium hydroxide, potassium carbonate or hydrochloric acid, undergoes condensation to form a syrupy liquid known as *aldol*:

$$2CH_3CHO \xrightarrow{NaOH} CH_3CH(OH)CH_2CHO\ (50\%)$$

On heating , aldols eliminate water to form unsaturated compounds, *e.g.*, aldol forms crotonaldehyde:

$$CH_3CH(OH)CH_2CHO \longrightarrow CH_3CH{=}CHCHO + H_2O$$

In many cases it is the unsaturated compound that is isolated, and not the aldol, *e.g.*, mesityl oxide and phorone. This is generally the case with *acid* catalysts, since these can readily bring about dehydration of alcohols.

The aldol condensation can occur: (*i*) between two aldehydes (identical or different); (*ii*) between two ketones (identical or different); and (*iii*) between an aldehyde and a ketone. Whatever the nature of the carbonyl compound, *it is only the α-hydrogen atoms which are involved in the aldol condensation.*

(*i*) Generally, with two different aldehydes all four possible condensation products are obtained; but by using different catalysts one product may be made to predominate in the mixture, *e.g.*,

$$CH_3CHO + CH_3CH_2CHO \xrightleftharpoons{NaOH} CH_3CH_2CH(OH)CH_2CHO$$

$$CH_3CHO + CH_3CH_2CHO \xrightleftharpoons{HCl} CH_3CH(OH)CH(CH_3)CHO$$

(*ii*) Acetone, in the presence of barium hydroxide, gives diacetone alcohol:

$$2CH_3COCH_3 \xrightleftharpoons{Ba(OH)_2} (CH_3)_2C(OH)CH_2COCH_3$$

This equilibrium lies almost completely on the left, but the yield of diacetone alcohol may be increased (71 percent) by boiling acetone in a Soxhlet with solid barium hydroxide in the thimble. The acetone in the flask gets richer and richer in diacetone alcohol, since the boiling point of the latter is 164°C, and that of the former is 56°C.

When acetone is treated with hydrochloric acid, mesityl oxide and phorone are formed:

$$2CH_3COCH_3 \xrightleftharpoons{HCl} (CH_3)_2C{=}CHCOCH_3 \xrightleftharpoons{CH_3COCH_3} (CH_3)_2C{=}CHCOCH{=}C(CH_3)_2$$

(*iii*) When aldehydes condense with ketones, it is the α-hydrogen atom of the *ketone* which is involved in the condensation, *e.g.*,

$$CH_3CHO + CH_3COCH_3 \xrightleftharpoons{NaOH} CH_3CH(OH)CH_2COCH_3 \quad (25\%)$$

The yield of 4-hydroxypentan-2-one is low because aldol and diacetone alcohol are also formed.

It is generally accepted that the base-catalysed aldol condensation of acetaldehyde takes place in two steps, the first being the formation of the carbonaion (I), and the second the combination of this anion with a second molecule of acetaldehyde to form the anion (II) of the aldol. The simplest mechanism that embraces these facts is:

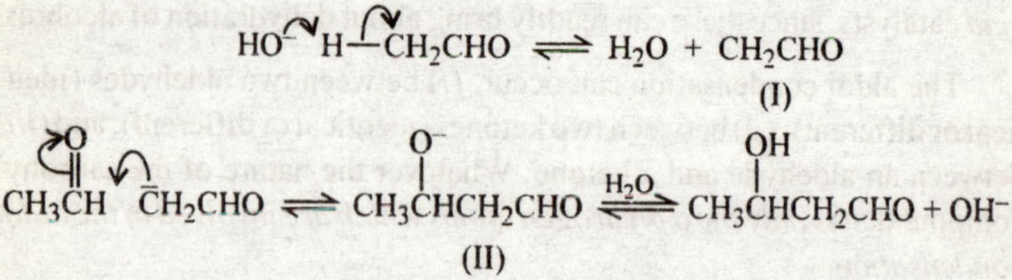

$$HO^- + H{-}CH_2CHO \rightleftharpoons H_2O + \underset{(I)}{\bar{C}H_2CHO}$$

$$CH_3CH{=}O + \bar{C}H_2CHO \rightleftharpoons \underset{(II)}{CH_3CH(O^-)CH_2CHO} \underset{}{\overset{H_2O}{\rightleftharpoons}} CH_3CH(OH)CH_2CHO + OH^-$$

On the basis of analogy with acetaldehyde, the mechanism of the base-catalysed condensation of acetone will be:

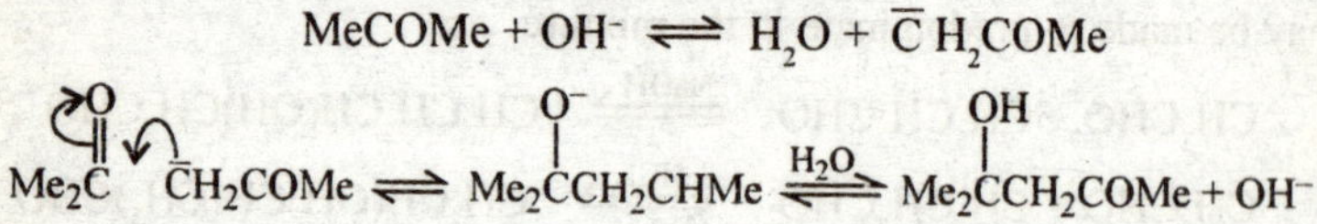

$$MeCOMe + OH^- \rightleftharpoons H_2O + \bar{C}H_2COMe$$

$$Me_2C{=}O + \bar{C}H_2COMe \rightleftharpoons Me_2C(O^-)CH_2CHMe \overset{H_2O}{\rightleftharpoons} Me_2C(OH)CH_2COMe + OH^-$$

There appear to be no detailed kinetic investigations on acid-catalysed aldol condensations, but it is generally assumed that the condensation proceeds by reaction between conjugate acid and the *enol form* of the carbonyl compound; *e.g.*, the formation of mesityl oxide:

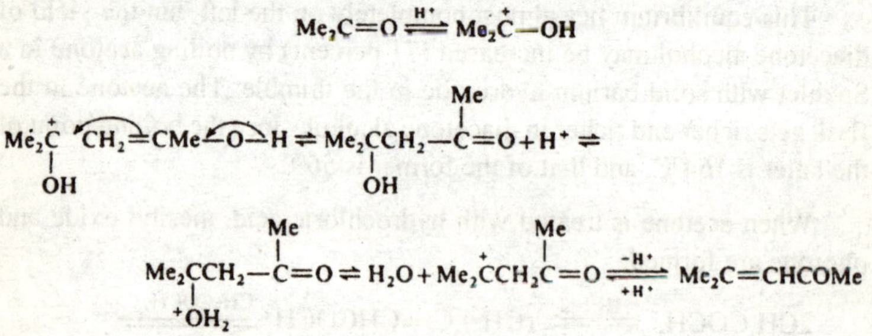

$$Me_2C{=}O \overset{H^+}{\rightleftharpoons} Me_2\overset{+}{C}{-}OH$$

$$Me_2\overset{+}{C}(OH) + CH_2{=}CMe{-}O{-}H \rightleftharpoons Me_2C(OH)CH_2{-}C(Me){=}\overset{+}{O}H \rightleftharpoons$$

$$Me_2C(\overset{+}{O}H_2)CH_2{-}C(Me){=}O \rightleftharpoons H_2O + Me_2\overset{+}{C}CH_2C(Me){=}O \underset{+H^+}{\overset{-H^+}{\rightleftharpoons}} Me_2C{=}CHCOMe$$

Fig. 4.9

Claisen condensation (1881). This is the condensation between an ester and another molecule of an ester or ketone.

Claisen–Schmidt reaction (also known as the **Claisen reaction**) is the condensation between an *aromatic* aldehyde (or ketone) and an aldehyde or ketone, *in the presence of dilute alkali* to form an α,β-unsaturated compound. This reaction is similar to the aldol condensation, ad is illustrated by the formation of cinnamaldehyde from benzaldehyde and acetaldehyde:

$$C_6H_5CHO + CH_3CHO \xrightarrow{NaOH} C_6H_5CH{=}CHCHO + H_2O$$

The mechanism of the Claisen-Schmidt reaction is (Noyce *et al.*, 1959; Stiles *et al.*, 1959):

$$HO^- \; H{-}CH_2CHO \rightleftharpoons H_2O + \bar{C}H_2CHO$$

$$C_6H_5CH{=}O \; \bar{C}H_2CHO \rightleftharpoons C_6H_5CH(O^-)CH_2CHO \underset{}{\overset{H_2O}{\rightleftharpoons}} C_6H_5CH(OH)CH_2CHO + OH^-$$

$$C_6H_5CH(OH){-}CH(H)CHO + OH^- \rightleftharpoons H_2O + C_6H_5CH(OH){-}\bar{C}HCHO \rightleftharpoons OH^- + C_6H_5CH{=}CHCHO$$

Other condensations involving aldehydes or ketones are the *Knoevengel* and *Perkin* reactions.

14. Darzens Glycidic Ester Condensation (1904). This reaction involves the condensation of an aldehyde or ketone with an α-halogeno-ester to produce an α,β-epoxy-ester (*glycidic ester*): the condesning agent is usually sodium ethoxide or sodamide:

$$ClCH_2CO_2Et + B \rightleftharpoons \bar{C}HClCO_2Et + BH \xrightarrow{R^1R^2CO}$$

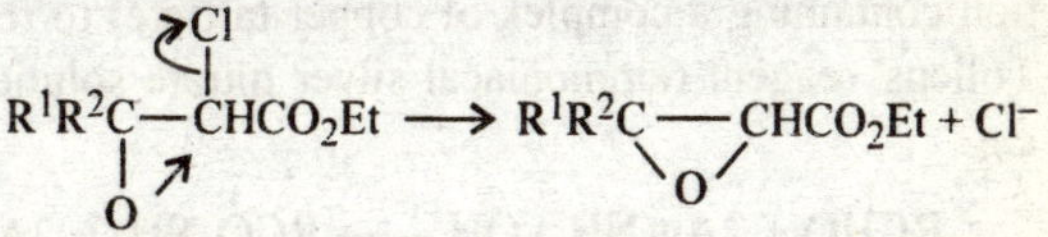

$$R^1R^2C(O^-){-}CHClCO_2Et \longrightarrow R^1R^2C{-}CHCO_2Et\,(\text{epoxide, }{-}O{-}) + Cl^-$$

This is an example of neighbouring group participation.

On alkaline hydrolysis, the ester gives the salt of the epoxy-acid, and when this is warmed with dilute acid, the product is an aldehyde if $R^3{=}H$, and a ketone if $R^3{=}$alkyl group. The mechanism is possibly (*cf.* acetoacetic acid):

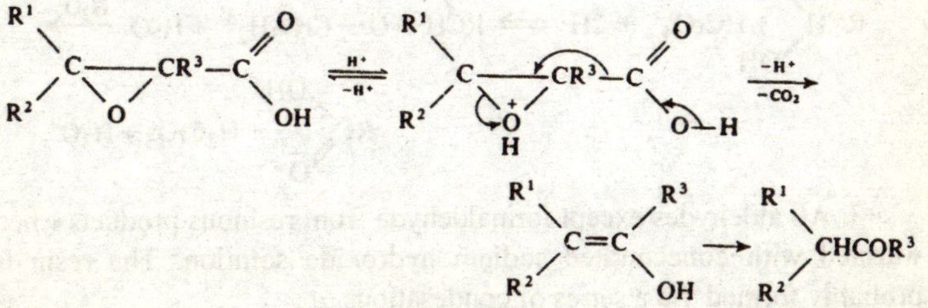

$$R^1R^2C{-}CR^3{-}C(=O)OH\,(\text{epoxide}) \underset{-H^+}{\overset{H^+}{\rightleftharpoons}} R^1R^2C{-}CR^3{-}C(=O){-}O{-}H\,(\text{protonated epoxide, }\overset{+}{O}H) \xrightarrow[-CO_2]{-H^+} R^1R^2C{=}C(OH)R^3 \longrightarrow R^1R^2CHCOR^3$$

Fig. 4.10

This synthesis can be shortened by using t-butyl glycidates, since the ester, on pyrolysis, eliminates isobutene to give, an aldehyde or ketone (Blanchard *et al.*, 1963):

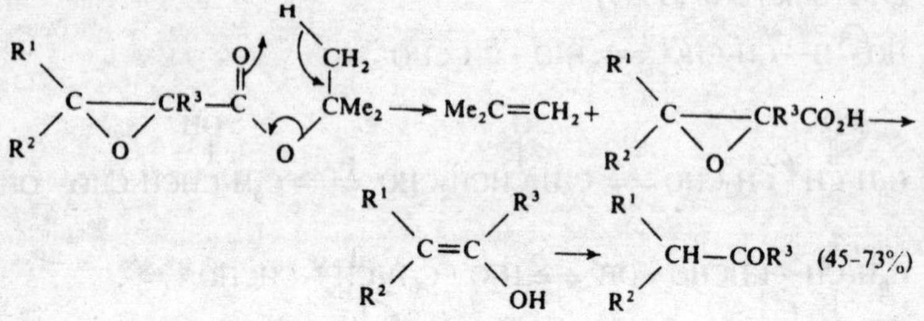

$$R^1R^2\underset{\diagdown O \diagup}{C\text{---}}CR^3\text{---}\overset{O}{\overset{\|}{C}}\text{---}O\text{---}CMe_2(CH_2\text{---}H) \longrightarrow Me_2C{=}CH_2 + R^1R^2\underset{\diagdown O \diagup}{C\text{---}}CR^3CO_2H \longrightarrow$$

$$R^1R^2C{=}C(R^3)OH \longrightarrow R^1R^2CH\text{---}COR^3 \quad (45\text{–}73\%)$$

Fig. 4.11

Reactions Given by Aldehydes Only

1. Aldehydes restore the magenta colour to Schiff's reagent (rosaniline hydrochloride is dissolved in water, and sulphur dioxide is passed in until the magenta colour is discharged). The mechanism of this reaction is uncertain.

2. Aldehydes are very easily oxidised, and hence are powerful reducing agents. Aldehydes reduce Fehling's solution (an alkaline solution containing a complex of copper tartrate) to red cuprous oxide; and Tollens' reagent (ammoniacal silver nitrate solution) to metallic silver, *e.g.*,

$$RCHO + 2Ag(NH_3)_2OH \longrightarrow RCO_2NH_4 + 2Ag + 3NH_3 + H_2O$$

Aldehydes are readily oxidised by acid dichromate, permanganate, etc. (*cf.* ketones, below)

The oxidation with dichromate is believed to proceed through the hydrate by a mechanism similar to that of prim.- and s-alcohols:

$$RCH(OH)_2 + HCrO_4^- + 2H^+ \rightleftharpoons RCH(OH)\text{---}O\text{---}CrO_3H_2^+ + H_2O \xrightarrow{H_2O} RC(OH){=}O + H_2CrO_3 + H_3O^+$$

3. All aldehydes except formaldehyde from resinous products when warmed with concentrated sodium hydroxide solution. The resin is probably formed via a series of condesations, *e.g.*,

$$2CH_3CHO \longrightarrow CH_3CH(OH)CH_2CHO \xrightarrow{-H_2O}$$

$$CH_3CH{=}CHCHO \xrightarrow{CH_3CHO} CH_3CH{=}CHCH(OH)CH_2CHO$$

$$\xrightarrow{-H_2O} CH_3CH{=}CHCH{=}CHCHO, \text{ etc.}$$

The Cannizzaro reaction is mainly applicable to aromatic aldehydes.

Although the Cannizzaro reaction is characteristic of aldehydes having no α-hydrogen atoms, it is not confined to them, *e.g.*, certain aliphatic α-monoalkylated aldehydes undergo quantitative disproportionation when heated with aqueous sodium hydroxide at 170–200°C (Hausermann, 1951):

$$2(CH_3)_2CHCHO + NaOH \xrightarrow{200^\circ C}$$

$$(CH_3)_2CHCO_2Na + (CH_3)_2CHCH_2OH \ (100\%)$$

The Cannizzaro reaction can take place between two different aldehydes, and is then known as a 'crossed' Cannizzaro reaction.

All aldehydes can be made to undergo the Cannizzaro reaction by treatment with aluminium ethoxide. Under these conditions the acid and alcohol are combined as the ester, and the reaction is then known as the **Tischenko reaction** (1906); *e.g.*, acetaldehyde gives ethyl acetate, and propionaldehyde gives propyl propionate.

The mechanism is uncertain, but a possibility is one that involves a hydride-ion shift as in the MPV reduction (p. 210):

$$RCH{=}O + Al(OEt)_3 \rightleftharpoons R\overset{+}{C}H{-}O{-}\overset{-}{Al}(OEt)_3 \underset{}{\overset{RCH=O}{\rightleftharpoons}} RCH(O{-}\overset{+}{C}HR){-}O{-}Al(OEt)_3$$

$$\longrightarrow R{-}\underset{\underset{O}{\|}}{C}{-}OCH_2R + Al(OEt)_3$$

4. Aldehydes (except formaldehyde) react with ammonia in ethereal solution to give a precipitate of *aldehyde-ammonia, e.g.*,

$$CH_3CHO + NH_3 \longrightarrow CH_3CH(OH)NH_2 \ (50\%)$$

The structure is actually more complicated than that given, and X-ray analysis has led to the suggestion that it is the trihydrate of trimethylhexahydrotraizine, which may be prepared anhydrous by allowing 'acetaldehyde ammonia' to stand over sulphuric acid. Acetaldimine is formed first and then trimerises.

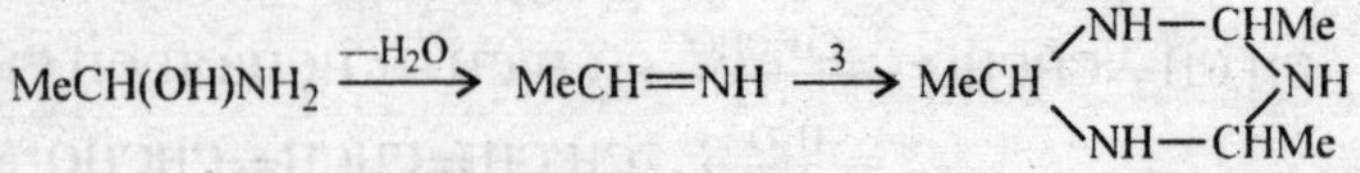

$$MeCH(OH)NH_2 \xrightarrow{-H_2O} MeCH{=}NH \xrightarrow{3} MeCH\begin{matrix} \diagup NH-CHMe \\ \quad\quad\quad \rangle NH \\ \diagdown NH-CHMe \end{matrix}$$

5. Aldehydes combine with alcohols in the presence of *e.g.*, dry hydrogen chloride, to form first the *hemi-acetal*, and then the *acetal*:

$$R^1CHO + R_2OH \underset{}{\overset{HCl}{\rightleftharpoons}} R^1CH(OH)OR_2 \overset{R^2OH}{\rightleftharpoons} RCH(OR_2)_2 + H_2O\ (g)$$

The hemi-acetal is rarely isolated since it very readily forms the acetal. Acetals are diethers of the unstable 1,1-dihydroxyalcohols, and may be named as 1,1-dialkoxyalkanes. Unlike the parent dihydroxyalcohols these acetals are stable. They are also stable in the presence of alkali, but are converted into the aldehyde by acid. Thus, acetal formation may be used to protect the aldehyde group against *alkaline* oxidising agents. On the other hand, the aldehyde group can be protected in *acid* solution by mercaptal formation.

The stability of acetals to *alkaline* media may be attributed to the fact that the alkoxyl group is a very poor leaving group, whereas the protonated group is a very good leaving group.

The mechanism of acetal formation has been elucidated by use of the principle of microscopic reversibility on acetals:

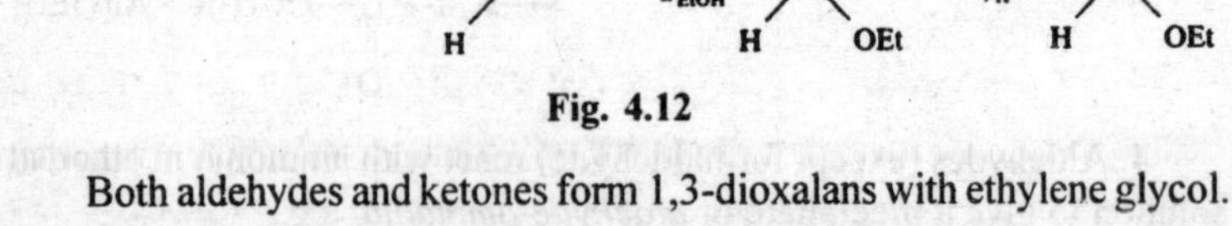

$$RCHO \underset{-EtOH}{\overset{+EtOH}{\rightleftharpoons}} RCH(OH)(OEt) \underset{-H^+}{\overset{+H^+}{\rightleftharpoons}} RCH(\overset{+}{O}H_2)(OEt) \underset{-H_2O}{\overset{-H_2O}{\rightleftharpoons}}$$

$$R(H)\overset{+}{C}-OEt \underset{-EtOH}{\overset{+EtOH}{\rightleftharpoons}} RCH(\overset{+}{O}(H)Et)(OEt) \underset{+H^+}{\overset{-H^+}{\rightleftharpoons}} RCH(OEt)(OEt)$$

Fig. 4.12

Both aldehydes and ketones form 1,3-dioxalans with ethylene glycol.

6. The lower aldehydes polymerise with great ease (see formaldehyde and acetaldehyde, below).

Reactions Given by Ketones Only

1. Ketones are not easily oxidised; they do not reduce Fehling's solution or ammoniacal silver nitrate.

α-Hydroxyketones, *i.e.*, compounds containing the group CH(OH)CO, readily reduce Fehling's solution and ammoniacal silver nitrate.

Strong oxidising agents, *e.g.*, acid dichromate, nitric acid, etc., oxidise ketones, only the carbon atoms adjacent to the carbonyl group being attacked, and the carbon atom joined to the smaller number of hydrogen atoms is preferentially oxidised:

$$CH_3COCH_2CH_3 \xrightarrow{[O]} 2CH_3CO_2H$$

If the adjacent carbon atoms have the same number of hydrogen atoms, the carbonyl group remains chiefly with the *smaller* alkyl group.

The mechanism of this oxidation is uncertain; a possibility is:

$$R^1COCH_2R^2 \rightleftharpoons R^1C(OH){=}CHR^2 \xrightarrow{[O]} R^1C(OH)_2CHOHR \xrightarrow{[O]} R^1CO_2H + R^2CO_2H$$

All ketones containing the acetyl group, CH_3CO, undergo the **haloform reaction.** This reaction is best carried out by dissolving the compound in dioxan, adding dilute sodium hydroxide, then a slight excess of iodine in potassium iodide solution, warming, and finally adding water. If the compound contains the acetyl group, iodoform is precipitated.

$$RCOCH_3 \xrightarrow{I_2,OH^-} RCOCCI_3 \xrightarrow{OH^-} RCO_2^- + CHI_3$$

The haloform reaction is very useful in organic degradations. A positive iodoform test is given by most compounds containing the acetyl group attached to either carbon or hydrogen, or by compounds which are oxidised under the conditions of the test to derivatives containing the acetyl group, *e.g.*, ethanol, isopropanol, etc.

Booth and Saunders (1950) have shown that a number of other compounds besides those mentioned above also give the haloform reaction *e.g.*, certain quinones, quinols and *m*-dihydric phenols. On the other, hand, acetoacetic ester, for example, does *not* give the test although it contains an acetyl and not the Me of the acetyl group.

An alternative method to the haloform reaction is to treat the monobrominated methyl ketone with pyridine, and then decompose the precipitated pyridinium salt with dilute ethanolic sodium hydroxide (Krohnke, 1933; King, 1944):

$$RCOCH_3 \xrightarrow{Br_2} RCOCH_2Br \xrightarrow{C_5H_5N} RCOCH_2{-}\overset{+}{N}C_5H_5Br^- \xrightarrow{NaOH/EtOH} RCO_2H + C_5H_5\overset{+}{N}Me\} Br^-$$

The advantage of this method is that the reaction can be carried out in non-aqueous solution.

Aliphatic ketones undergo the **Baeyer–Villiger oxidation** (1899–1900) to form esters or their hydrolysed products:

$$RCOR \longrightarrow RCO_2R \xrightarrow{H^+} RCO_2H + ROH$$

The authors used Caro's acid (permonosulphuric acid, H_2SO_5), but the reaction is now carried out with various organic per-acids, *e.g.*, perbenzoic, or monoperphthalic acid. The yields are reasonable, and so this reaction is useful in structure determination, and may also be used as method of preparing acids. According to Mateos *et. al.* (1964), the mechanism is:

$$R^1R^2C{=}O \rightleftharpoons R^1R^2C{=}\overset{+}{O}H \xrightarrow{R_3CO_3} R^1R^2C(O{-}O_2CR^3)(O{-}H) \longrightarrow$$

$$R^2{-}C({=}O){-}OR^1 + R_3CO_2^- + H^+$$

It was also shown that the rearrangement is intramolecular, and it has been established that the migratory aptitude of an alkyl group is t > s > prim.

$$(Ph)(Me)CHCOMe \xrightarrow{PhCO_3H} (Ph)(Me)CHOCOMe$$

3. Ketones react with ammonia to form complex condensation products, *e.g.*, if acetone is treated with ammonia and then followed by acidification of the reaction products, *diacetonamine* and *triacetonamine* are formed.

4. Ketones do not readily from *Ketals* when treated with alcohols in the presence of hydrogen chloride (*cf.* acetals, above). Ketals may, however, be prepared by treating the ketone with ethyl orthoformate (Helferich and Hauser, 1924):

$$R_2CO + HC(OC_2H_5)_3 \longrightarrow R_2C(OC_2H_5)_2 + HCO_2C_2H_5\ (g.)$$

In the presence of acids, however, the product is an enol ether.

$$RCOCH_2R + H(COEt)_3 \longrightarrow RC(OEt)_2CH_2R \xrightarrow{HCl} RC(OEt) = CHR + EtOH$$

Ketones readily react with 1,2-glycols to form dioxolans.

5. Ketones are reduced catalytically or by dissolving metals in alkaline solution to alcohols, but reduction by the latter reagent in neutral or acid solution gives 1,2-glycols as the main product; *e.g.*, with magnesium amalgam, acetone forms pinacol. This reaction is believed to proceed via free radicals:

$$2Me_2C{=}O + 2e \xrightarrow{Mg} [2Me_2C{-}O]Me^{2+} \longrightarrow$$

$$\begin{bmatrix} 2Me_2C{-}CMe_2 \\ | \quad\ \ | \\ O^- \ \ O^- \end{bmatrix} Mg^{2+} \xrightarrow[H^+]{H_2O} \begin{matrix} 2Me_2C{-}CMe_2 \\ | \quad\ \ | \\ OH \ \ OH \end{matrix} \qquad (43-50\%)$$

6. When ketones are treated with nitrous acid, the 'half oxime' of the α-dicarbonyl compound is formed, *e.g.*, acetone gives oximinoacetone:

$$CH_3COCH_3 + HNO_2 \xrightarrow{C_2H_5ONO/HCl} CH_3COCH = NOH + H_2O$$

All compounds containing the CH_2CO group form the oximino-derivative with nitrous acid, and this has been used to detect the presence of the CH_2CO group.

The mechanism of this reaction is uncertain. A possibility is via attack by the nitrosodium ion; the nitroso compound first formed then rearranges to the oxime:

$$HO{-}N{=}O + H^+ \rightleftharpoons H_2O^+{-}N{=}O \rightleftharpoons H_2O + NO^+$$

$$CH_3COCH_3 \rightleftharpoons CH_3C(OH){=}CH_2$$

$$CH_3\overset{\overset{O-H}{|}}{C}{=}CH_2 \ \ NO^+ \xrightarrow{-H^+} CH_3\overset{\overset{O}{\|}}{C}CH_2NO \xrightarrow{+H^+} CH_3CO\overset{\overset{H}{|}}{C}H{-}N{=}\overset{+}{O}H \xrightarrow{-H^+}$$

$$CH_3COCH{=}NOH$$

Benzaldehyde can also be used to detect the presence of the CH_2CO group:

7. Ketones condense with chloroform in the presence of potassium hydroxide to form chlorohydroxy-compounds:

$$CH_3COCH_3 + CHCl_3 \xrightarrow{KOH} (CH_3)_2C\begin{matrix} \diagup OH \\ \diagdown CCl_3 \end{matrix}$$

8. Ketones form *sodio-derivatives* when treated with sodium or sodamide in ethereal solution *e.g.*,

$$CH_3COCH_3 + NaNH_2 \longrightarrow \left[\begin{array}{c} O^- \\ | \\ CH_2C{=}CH_2 \end{array}\right]\overset{+}{Na} + NH_3$$

Formaldehyde (*methanal*) is prepared industrially:

(*i*) By the dehydrogenation or air oxidation of methanol in the presence of a silver catalyst at 400°C:

$$CH_2O + H_2 \xleftarrow{Ag} CH_3OH \xrightarrow[Ag]{air} CH_2O + H_2O$$

In the oxidation method, the condensate obtained is a mixture of formaldehyde, methanol and water. It is free from excess of methanol by distillation, and the resulting mixture is known as *formalin* (40 per cent formaldehyde, 8 per cent methanol, 52 per cent water).

(*ii*) By the air oxidation of methane or propane-butane mixtures (from natural gas) in the presence of various metallic oxides.

Formaldehyde is a colourless, pungent-smelling gas, p.b. – 21°C, extremely soluble in water. It is a powerful disinfectant and antiseptic, and so is used for preserving anatomical specimens. Its main uses are in the manufacture of dyes, the hardening of casein and gelatin, and for making plastics.

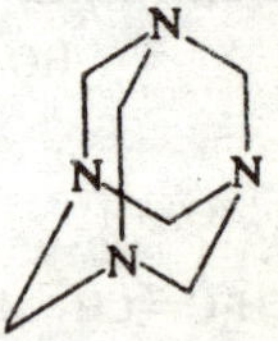

Formaldehyde undergoes many of the general reactions of aldehydes, but differs in certain respects. When treated with ammonia, it does not form an aldehyde-ammonia, but gives instead *hexamethylenetetramine* (m.p. 260°C).

N N N N

$$6HCHO + 4NH_3 \longrightarrow (CH_2)_6N_4 + 6H_2O \ (80\%)$$

The structure of hexamethylenetetramine consists of three fused chair conformations.

Formaldehyde, since i has no α-hydrogen atoms, readily undergoes the Cannizzaro reaction.

Primary ad secondary amines are methylated by heating with formaldehyde (formalin solution) ad an excess of formic acid at 100°C, *e.g.*, ethylmethylamine form ethylamine:

$$C_5HNH_2 + CH_2O + HCO_2H \longrightarrow C2H_2NHCH_3 + CO_2 + H_2O$$

This reaction is known as the **Eschweiler–Clarke methylation** (1905, 1933).

Polymers of formaldehyde. (*i*) In dilute aqueous solution, formaldehyde is almost 100 per cent hydrated to form methylene glycol (Bieber *et al.*, 1974). This is believed to be the reason for the stability of dilute formaldehyde solutions:

$$CH_2O + H_2O \rightleftharpoons CH_2(OH)_2$$

(*ii*) When a formaldehyde solution is evaporated to dryness, a white crystalline solid, m.p. 121–123°C, is obtained. This is known as *paraformaldehyde*, $(CH_2O)_n{\cdot}H_2O$, and it appears to be a mixture of polymers, *n* having values between 6 and 50. Paraformaldehyde reforms formaldehyde when heated.

Formaldehyde cannot be separated from methanol (in formalin) by fractionation; pure aqueous formaldehyde may be obtained by refluxing paraformaldehyde with water until solution is complete.

(*iii*) When a formaldehyde solution is treated with concentrated sulphuric acid, *polyoxymethylenes*, $(CH_2O)_n{\cdot}H_2O$ — *n* is greater than 100 — are formed. Polyoxymethylenes are white solids, insoluble in water, and reform formaldehyde when heated.

```
        CH2
       /   \
      O     O
      |     |
    H2C     CH2
       \   /
         O
```

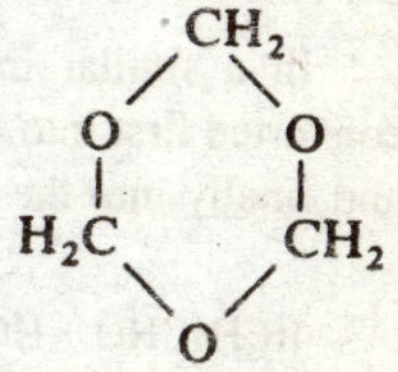

(*iv*) When allowed to stand at room temperature, formaldehyde gas slowly polymerises to a white solid, *trioxymethylene* (*metaformaldehyde, trioxan*) $(CH_2O)_3$, m.p. 61–62°C. Trioxan is prepared by distilling formaldehyde solution (60 per cent) containing a little sulphuric acid. This trimer is soluble in water, ad does not show any reducing properties. Hence it is believed to have a cyclic structure (in which there is no free aldehyde group).

Trioxan is very useful for generating formaldehyde since: (*a*) it is an anhydrous form of formaldehyde; (*b*) the rate of depolymerisation can be controlled; and (*c*) it is soluble in organic solvents.

(*v*) Formaldehyde polymerises in the presence of weak alkalis, *e.g.*, calcium hydroxide, to a mixture of sugars of formula $C_6H_{12}O_6$, which is known as *formose* or *α-acrose*.

Condensation reactions of formaldehyde. Formaldehyde can participate in the 'crossed' Cannizzaro reaction, and the nature of the final product depends on the structure of the other aldehyde. Aldehydes with *no* α-hydrogen atoms readily undergo the crossed Cannizzaro reaction; *e.g.*, benzaldehyde forms benzyl alcohol:

$$C_6H_5CHO + HCHO + NaOH \longrightarrow C_6H_5CH_2OH + HCO_2Na$$

Aldehydes with *one* α-hydrogen atom react as follows:

$$R^1R^2CHCHO + HCHO \xrightarrow{KOH} R^1R^2C\begin{cases} CH_2OH \\ CHO \end{cases}$$

This β-hydroxyaldehyde in the presence of excess of formaldehyde forms a substituted trimethylene glycol:

$$R^1R^2C\begin{cases} CH_2OH \\ CHO \end{cases} + HCHO + NaOH \longrightarrow R^1R^2C\begin{cases} CH_2OH \\ CH_2OH \end{cases} + HCO_2Na$$

Thus the first step in the above reaction is the replacement of the α-hydrogen atom by a *hydroxymethyl* group, CH_2OH, and the second step is the crossed Cannizzaro reaction.

In a similar manner, aldehydes with *two* α-hydrogen atoms are converted first into the hydroxymethyl, then into the bishydroxymenthyl, and finally into the trishydroxymethyl compound:

$$RCH_2CHO + HCHO \xrightarrow{KOH} RCH\begin{cases} CHO \\ CH_2OH \end{cases} \xrightarrow{HCHO}$$

$$RC\begin{cases} CHO \\ CH_2OH \\ CH_2OH \end{cases} \xrightarrow[NaOH]{HCHO} RC\begin{cases} CH_2OH \\ CH_2OH \\ CH_2OH \end{cases} + HCO_2Na$$

A special case is acetaldehyde, which has *three* x-hydrogen atoms. This reaction is best carried out by adding powdered calcium oxide to a suspension of paraformaldehyde in water containing acetaldehyde. Tetrakishydroxymethylmethane (tetramethylolmethane) or *pentaerythritol* is formed:

$$CH_3CHO + 4HCHO \xrightarrow{Ca(OH)_2} C(CH_2OH)_4 + HCO_2Ca/2 \quad (55\text{–}57\%)$$

Pentaerythritol is important industrially since its tetranitrate is a powerful explosive.

Acetaldehyde (*ethanal*) is prepared industrially:

(*i*) By the dehydrogenation or air oxidation of ethanol in the presence of a silver catalyst at 300°C (*cf.* formaldehyde):

$$CH_3CHO + H_2 \xleftarrow{Ag} CH_3CH_2OH \xrightarrow[Ag]{air} CH_3CHO + H_2O$$

(*ii*) By the hydration of acetylene (p. 136) This method is decreasing in importance.

(*iii*) By passing a mixture of ethylene and oxygen, under pressure, into an aqueous solution of palladium and cupric chlorides at 50°C

$$CH_2 = CH_2 + PdCl_2 + H_2O \longrightarrow CH_3CHO + Pd + 2HCl$$

$$2CuCl_2 + Pd \longrightarrow 2CuCl + PdCl_2$$

$$2CuCl + 2HCl + \frac{1}{2}O_2 \longrightarrow 2CuCl_2 + H_2O$$

Acetaldehyde is a colourless, pungent-smelling liquid, p.b. 21°C, miscible with water, ethanol and ether in all proportions. It is used in the preparation of acetic acid, ethanol, paraldehyde, rubber accelerators, phenolic resins, synthetic drugs, etc.

Polymers of acetaldehyde. (*i*) When acetaldehyde is treated with a few drops of concentrated sulphuric acid, a vigorous reaction takes place and the trimer *paraldehyde*, $(CH_3CHO)_3$, is formed. This is pleasant-smelling liquid, b.p. 128°C, and is used in medicine as an hypnotic. When paraldehyde is distilled with dilute sulphuric acid, acetaldehyde is regenerated. Paraldehyde has no reducing properties, and its structure is believed to be (I):

```
       CH
      /  \
     O    O
     |    |
  CH3CH   CHCH3
      \  /
       O
      (I)
```

```
CH3CHOCHCH3
   |   |
   O   O
   |   |
CH3CHOCHCH3
    (II)
```

(ii) When acetaldehyde is treated with a few drops of concentrated sulphuric acid at 0°C, the tetramer *metaldehyde* $(CH_3CHO)_4$, is formed. This is a white solid, m.p. 246°C, and regenerates acetaldehyde when distilled with dilute sulphuric acid. Its structure may be (II).

Chloral (*trichloroacetaldehyde*) is prepared industrially by the chlorination of ethanol. Chlorine is passed into cooled ethanol, and then at 60°C, until no further absorption of chlorine takes place. The final product is chloral alcoholate, $CCl_3CH(OH)OC_2H_5$, which separates out as a crystalline solid which, on distillation with concentrated sulphuric acid, gives chloral.

Chloral is a colourless, oily, pungent-smelling liquid, p.b. 98°C, soluble in water, ethanol and ether. When heated with concentrated potassium hydroxide, it yields pure chloroform:

$$CCl_3-\overset{O}{\overset{\|}{C}}H \;\; \bar{O}H \rightleftharpoons CCl_3-\overset{O^-}{\overset{|}{C}}H-OH \longrightarrow :\bar{C}Cl_3 + H\overset{O}{\overset{\|}{C}}-OH$$

$$:\bar{C}Cl_3 \xrightarrow{H_2O} CHCl_3 + OH^- \qquad HCOOH \xrightarrow{OH^-} HCO_2^- + H_2O$$

Chloral is oxidised by concentrated nitric acid to trichloroacetic acid and is reduced by aluminium ethoxide to trichloroethanol.

$$CCl_3CO_2H \xleftarrow{HNO_3} CCl_3CHO \xrightarrow{Al(OC_2H_5)_3} CCl_3CH_2OH$$

Chloral undergoes the usual reactions of an aldehyde, but its behaviour towards water and ethanol is unusual. When chloral is treated with water or ethanol, combination takes place with the evolution of heat, and a crystalline solid is formed, chloral hydrate, m.p. 57°C, or chloral alcoholate, m.p. 46°C, respectively. *These compounds are stable*, and the water or ethanol can only be removed by treatment with concentrated sulphuric acid. Thus, in chloral hydrate the water is present as *water of constitution, i.e.*, the structure of chloral hydrate is $CCl3CH(OH)_2$; similarly, the structure of chloral alcoholate is $CCl_3CH(OH)OC_2H_5$.

Carbonyl compounds are hydrated by water, the reaction being catalysed by acids and bases.

Acid $$>C=O + H^+ \rightleftharpoons >C=\overset{+}{O}-H \longleftrightarrow >\overset{+}{C}-O-H \xrightleftharpoons{H_2O} >C\begin{matrix}OH\\OH_2^+\end{matrix} \underset{+H^+}{\overset{-H^+}{\rightleftharpoons}} >C\begin{matrix}OH\\OH\end{matrix}$$

Base $$>C=O + OH^- \rightleftharpoons >C\begin{matrix}\bar{O}\\OH\end{matrix} \xrightleftharpoons{H_2O} >C\begin{matrix}OH\\OH\end{matrix} + OH^-$$

Since the reactions are reversible, any factor that stabilises the hydrate will shift the equilibrium to the right. In practice, aldehydes are more hydrated than ketones. One contributing factor is the steric effect Another factor is the + I effect of alkyl groups. Both mechanisms involve the addition of a nucleophile in the r/d step (H_2O or OH^-), and the greater the + I effect, the less easy is the addition of the nucleophile and the easter is its removal. Ketones, with two alkyl groups, will therefore exert a greater

+ I effect than will aldehydes (with one alkyl group), and consequently the equilibrium will be more to the left with ketones than with aldehydes.

Hooper (1967) has examined these equilibria by N.M.R. spectroscopy ad found that the amounts of *unreacted* compound in dilute aqueous solution were: MeCHO, 58; EtCHO, 63; Me_3CCHO, 100; Me_2CO, 100; $MeCOCH_2Cl$, 66 per cent.

The unusual stability of chloral hydrate has been attributed to the – I effect of chlorine and to the formation of intramolecular hydrogen bonds, the presence of which has been shown by Davies (1940) from infra-red studies. At the same time, the and u.v. spectroscopy, by the presence of the carbonyl group in chloral itself and its absence in the hydrate.

```
      H
   .·  \
 Cl     O
   \   /
Cl—C—CH
   /   \
 Cl     O
   `·  /
      H
```

Propionaldehyde, b.p. 49°C, **butyraldehyde,** b.p. 75°C, and **isobutyraldehyde**, p.b. 61°C, are prepared industrially by the *oxo process.*

Acetone (*dimethyl ketone, propan*-2-*one*) is prepared industrially:

(*i*) By the catalytic dehydrogenation (Cu or ZnO) or catalytic oxidation (Ag) of isopropanol.

(*ii*) By passing a mixture of propene and oxygen, under pressure, into an aqueous solution of palladium and cupric chlorides (*cf.* acetaldehyde).

(*iii*) Acetone is also manufactured by the oxidation of natural gas, and is obtained as a by-product in the oxidation of cumene to phenol.

Acetone is a colourless, pleasant-smelling liquid, b.p. 56°C, miscible with water, ethanol and ether in all proportions. Acetone is used as a solvent for acetylene, cellulose acetate and nitrate, celluloid, lacquers, etc., and for the preparation of keten, sulphonal, etc.

Although acetone does not exist as a hydrate in water (see above), it nevertheless undergoes reversible hydration. On the other hand, Wilson *et al.* (1963) have shown that acetone hydrate exists in the *solid* state.

Ketones do not polymerise, but readily undergo condensation reactions. Acetone readily forms mesityl oxide, phorone and diacetone alcohol, but in addition to these condensations, acetone forms mesitylene when distilled with concentrated sulphuric acid.

Ethyl mehyl ketone, butan 2-one, is prepared industrially by the catalytic dehydrogenation or air oxidation of s-butanol, and form but-1-ene and palladium chloride, etc. (*cf.* acetone).

Butanone is a pleasant-smelling liquid, b.p. 80°C, may be prepared in several ways, but the best method is by treating t-pentanol with bromine and hydrolysing the dibromo-derivative, whereupon a molecular rearrangement takes place (see pinacone, below):

$$(CH_3)_2C(OH)CH_2CH_3 \xrightarrow{Br_2} (CH_3)_2CBrCHBrCH_3 \xrightarrow{H_2O}$$

$$(CH_3)_2C(OH)CH(OH)CH_3 \xrightarrow[\text{rearranges}]{-H_2O} (CH_3)_2CHCOCH_3$$

t-Butyl methyl ketone, pinacone, p.b. 119°C, may be prepared by distilling pinacol or pinacol hydrate with sulphuric acid:

$$(CH_3)_2C(OH)C(OH)(CH_3)_2 \xrightarrow{H^+} CH_3COC(CH_3)_3 \ (65\text{–}75\%)$$

This conversion of pinacol into pinacone is an example of the 1,2-shift, and is known as the *pinacol-pinacone rearrangement.*

Pinacolone is oxidised by alkaline sodium hypobromite to trimethylacetic acid:

$$CH_3COC(CH_3)_3 \xrightarrow{Br_3/NaOH} (CH_3)_3CCO_2H \ (71\text{–}74\%)$$

The pinacol-pinacone rearrangement is general for 1,2-glycols under acid conditions; the migrating group may be alkyl or aryl. The detailed mechanism of this arrangement is not certain, but it is generally accepted that the first stage is the addition of a proton to one of the hydroxy-groups, followed by loss of this as water. This leaves a carbonium ion and it is the nature of this ion and the steps leading to the product that are uncertain. According to Bunton *et al.* (1959), the mechanism for the rearrangement of pinacol itself is:

$$\underset{\text{OH}\ \ \text{OH}}{Me_2C\text{—}CMe_2} + H^+ \rightleftharpoons \underset{\text{OH}\ \ \overset{+}{O}H_2}{Me_2C\text{—}CMe_2} \quad \text{(I)}$$

$$H_2O + Me\text{—}\underset{\underset{\text{OH}}{|}}{\overset{\overset{\text{Me}}{|}}{C}}\text{—}\overset{+}{C}Me_2 \longrightarrow Me\text{—}\underset{\underset{\text{H—O}}{|}}{\overset{+}{C}}\text{—}CMe_3 \longrightarrow H^+ + MeCOCMe_3$$

(II)

Kinetic measurements have shown that the conversion of (I) into (II) is the rate-determining step.

Whitmore *et al.* (1939), using a related pinacolic deamination, showed that when an optically active amine containing one asymmetric centre was used, this asymmetric centre was predominatly inverted in the product (*cf.* Walden inversion; Ph = C_6H_5):

$$\mathrm{Ph{-}C(Ph)(OH){-}CHMe(NH_2)} \xrightarrow{HNO_2} \mathrm{Ph{-}C(Ph)(OH){-}CHMe(N_2^+)} \longrightarrow N_2 + \mathrm{Ph{-}\overset{+}{C}(OH){-}CHMe(Ph)} \longrightarrow H^+ + \mathrm{PhCOCHMe(Ph)}$$

These results mean that the classical carbonium ion, which is flat, if formed at all, must rearrange so fast that racemisation cannot occur. On the other hand, the inversion may be explained by a synchronous migration of a phenyl group and breaking of the C—N bond. In this case, the intermediate could be the *phenonium ion, i.e.*, a bridged ion in which the configuration of the asymmetric carbon atom is held in the inverted position:

$$\mathrm{Ph{-}C(C_6H_5)(OH){-}CHMe(N_2^+)} \longrightarrow \mathrm{Ph{-}C(OH){-}CHMe\ (bridged\ by\ C_6H_5^+)} \longrightarrow \mathrm{Ph{-}\overset{+}{C}(OH){-}CHMe(C_6H_5)}$$

A pair of electrons of the benzene ring is used to form the bridge, and the rest of the benzene molecule is represented as a pentadienyl cation.

When dealing with unsymmetrical pinacols, there are several other points to be considered, *e.g.*, which hydroxyl group is removed. Since the OH group is removed as water, any polar effect in the molecule which weakens one C—O bond more than the other will facilitate the release of this OH group. Consider the two groups phenyl and methyl. The phenyl group has a powerful conjugative effect, whereas the methyl group has a weak inductive effect in comparison:

$$\mathrm{C_6H_5{\rightarrow}C(OH){-}C(OH){\leftarrow}Me}$$

Therefore the OH group lost should be the one on the phenyl side, and consequently it will be the methyl group that migrates. This is borne out in practice, *e.g.*,

$$Ph_2C(OH)C(OH)Me_2 \xrightarrow{H_2SO_4} Ph_2MeCCOMe$$

Another way of explaining which hydroxyl group is removed is based on the relative stabilities of the alternative carbonium ions, the more stable of the two being the preferred one. In the example given here, (III) is stabilised by resonance, whereas (IV) is stabilised by the + I simply spreading the charge by the inductive effect, (III) is far more stable than (IV).

$$C_6H_5-\overset{+}{C}(Ph)-C(OH)Me_2 \longleftrightarrow C_6H_5{=}\overset{+}{C}(Ph)-C(OH)Me_2 \longleftrightarrow \cdots$$

(III)

$$Ph_2C(OH)-\overset{+}{C}(Me)_2$$

(IV)

Another problem is that of 'migration aptitude', *i.e.*, which of the two groups migrates when the groups are different. A great deal of work has been done with pinacols of the type (Ar = aryl) $Ar^1Ar^2C(OH)C(OH)Ar^1AR^2$, and it has been shown that the aryl group with the greater electron-releasing property is the one that migrates (Bachmann *et al.*, 1932–34). This is understandable on the basis of the formation of an intermediate phenonium ion. Any polar factor that helps to release a pair of electrons from a benzene ring will therefore facilitate the formation of the bridged ion with *this* benzene ring. It has also been found that migration aptitudes depend on steric factors.

ALKANES

Introduction : The aliphatic compounds are open-chain or *acylic* compounds, and the name aliphatic arises from the fact that the first compounds of this class to be studied were the fatty acids (Greek : *aliphos* fat) Carbon forms a large number of compounds with hydrogen only and these are known collectively as hydrocarbons. The alkanes or the paraffins are the saturated hydrocarbons.

Many occur naturally, and the chief source of the alkanes is mineral oil or petroleum, which occurs in many parts of the worlds. The simplest organic compounds are hydrocarbons containing only carbon-carbon single bonds in their molecules and hydrogen only. In nature hydrocarbons occur chiefly in coal, natural gas and petroleum. All organic compounds can be considered as the derivative of hydrocarbons, obtained by substituting hydrogen both an appropriate functional group.

The compounds of carbon and hydrogen are called hydrocarbons based upon. The nature of bonds reactivity, hydrocarbon can be broadly classified into two main categories (i) Aliphatic (i) Aromatic aliphatic hydrocarbons are be broadly classified into saturated (alkanes) and in saturated hydrocarbon. C Alkenes and alkanes, and acyclic (cycloalkanes) hydrocarbons.

Structural Formula

The simplest alkane is methane, and its molecular formula is CH_4. Assuming the quadrivalency of carbon and the univalency of hydrogen, we find that there is only one structure possible for methane, viz., (1)

```
  H           Cl          Cl
  |           |           |
H—C—H       H—C—H       H—C—Cl
  |           |           |
  H(I)        Cl(II)      H(III)
```

Study of the reactions of methane shows that all four hydrogen atoms are equivalent, *e.g.*, methylene dichloride, CH_2Cl_2. prepared by totally different methods is always the same,. Thus, (II) and (III) are different ways of writing the same structure. At first sight it may appear that these two structural formula are different. They are different if the molecule is two-dimensional, but saturated compounds the four valencies of car arranged tetrahedrally. Examination of the two structures as tetrahedral figures shows that they are identical. The chief disadvantage of the two-structural formula is that it does not show the spatial arrangement of the atoms. On the other hand, the three dimensional structural formula is cumber some, and for many complicated molecules cannot easily be drawn on paper. Hence we usually adopt the plane-formula when dealing with compounds from the point of view of their structure, only when we which to stress the spatial when we wish to stress the spatial arrangement of the atoms or groups in molecule do we resort to solid diagram. How we show the relative positions of the various atoms or groups in the plane-structural formula of a given compound usually depends on our selves, but where possible the simplest method of writing the structure should be chosen.

Atomic models' are very useful to the organic. Sets of these models may be bought.

Consider 1.5-dichloropentane $C_5H_{10}Cl_2$. Thus is a 'straight' chain compound, and its structural formula is usually written $ClCH_2$ CH_2 CH_2

CH_2 CI. Now, 1.5-dichloropentane gives cyclopentane when treated with zinc. Cyclopentane is a ring compound, and to show its formation from 1.5-dichloropentane, we write the structure of the latter as follows.

$$CH_2\begin{matrix} \diagup CH_2CH_2Cl \\ \diagdown CH_2CH_2Cl \end{matrix} + Zn \longrightarrow CH_2\begin{matrix} \diagup CH_2 \diagdown CH_2 \\ | \\ \diagdown CH_2 \diagup CH_2 \end{matrix} + ZnCl_2$$

Thus 'straight' chains may be 'bent' to stress a particular point we may have in mind.

Substitution reactions of methane. Chlorine has no action on methane in the dark. In bright sunlight the reaction is explosive, and hydrogen chloride and carbon are formed.

$$CH_4 + 2CI_2 \longrightarrow C + 4HCI$$

In diffused sunlight no explosion occurs, but a series of reactions takes place whereby the four hydrogen atoms in methane are successively replaced by chlorine atoms.

$CH_4 + Cl_2 \longrightarrow CH_3\ Cl + HCl$ methyl chloride

$CH_3CI + Cl_2 \longrightarrow CH_2\ Cl_2 + HCl$ methylene dichloride

$CH_2\ Cl_2 + Cl_2 \longrightarrow CHCl_3 + HCl$ chloroform

$CHCl_3 + CI_2 \longrightarrow CCl_4 + HCl$ carbon tetrachloride.

In methane the four carbon valencies are satisfied by combination with four hydrogen atoms. Carbon never exhibits a valencies of more than four, and so cannot combine with more than four hydrogen atoms or four other univalent atoms or groups.

Hence in the reaction with chlorine, the hydrogen atoms are displaced and chlorine atoms take their place. This type of reaction is known as substitution, and is the direct replacement of hydrogen by some other atom or group.

The product so formed are known as substitution products. The atom or group that has replaced the hydrogen atom is called the substituent, and when a substituent atom or group is replaced by some other atom or group, the reaction is referred to as replacement (or displacement) reaction. It should be noted that in substitution or replacement reactions there is no change is structure. The spatial arrangement of the molecule, however, may have changed.

Structural Isomerism in Alkanes

All the alkanes can be derived formally from methane by substituting hydrogen atoms by methyl groups (CH_3). When the alkane contains three or more carbon atoms, substitution now gives rise to isomerism, *e.g.*, (i) Propane, C_3H_5, can give rise to two butanes. C_4H_{10}, (1) by substitution at a terminal carbon atom, and (1) by substitution at the central carbon atom.

```
                                            H
                                            |
  H   H   H   H                    H   H—C—H   H
  |   |   |   |                    |      |     |
H—C—C—C—C—H                H—C———C———C—H
  |   |   |   |                    |      |     |
  H   H   H   H                    H      H     H
```

$CH_3CH_2CH_2CH_3$ (I)

$CH_3CH(CH_3)CH_3$ — written as CH_3CHCH_3 with CH_3 on the central carbon, or $(CH_3)_3CH$ (II)

(1) Has a straight chain, and (II) a branched chain. Both isomers are knows, and thus 'butane' is the first alkane to exhibit structural isomerism. This is an example of chain (or nuclear isomerism, and is characterised by the manner of linking of the carbon chain. (1) is known as normal butane, and (II) as isobutane.

(ii) Three pentanes. C_5H_{12}, may be derived from the butanes.

$\underset{\text{normal pen tan e}}{CH_3CH_2CH_2CH_2CH_3}$ $\underset{\text{isopen tan e}}{(CH_3)_2CHCH_2CH_2CH_3}$ $\underset{\text{neopen tan e}}{(CH_3)_4C}$

The bond-diagram way of writing structural formula uses up a lot of a space. Hence it is general practice to use contracted' structural formula, the method is to write the formula in a 'straight' line and enclosing in parentheses any side-chain or substituent group. The atoms or groups in parentheses are joined directly to the preceding carbon atom in the chain not placed in parentheses *e.g.*,

In many cases where there is no ambiguity, the parentheses may be omitted, *e.g.*., $CH_3CH(OH)CH_3$ is often written $CH_3CHOHCH_3$.

As the number of carbon atoms in the alkane increases, the number of possible isomers increases rapidly, *e.g.*, the alkane $C_{15}H_{32}$ can exists in 4347 isometric forms. The number of isomers of a given alkane may be calculated by means of mathematical formula, in most cases very few have actually been prepared.

Examination of the two butane structures shows that all the carbon atoms are not equivalent, and also that all the hydrogen atoms are not equivalent. A primary carbon atom is one that is joined to one other carbon atom, a secondary carbon atom to two other carbon atoms, and a tertiary carbon atom to three other carbon atoms. Hydrogen atoms joined to primary, secondary and tertiary carbon atoms are known as primary, secondary, or tertiary hydrogen atoms, respectively. Thus normal butane contains two primary carbon atoms, two secondary carbon atoms, six primary and four secondary hydrogen atoms. Isobutane contains three primary carbon atoms, one tertiary carbon atom, nine primary hydrogen atoms, and one tertiary hydrogen atom. It will also be seen that normal pentane contains two primary and three secondary carbon atoms, six primary and six secondary hydrogen atoms, and isopentane contains one secondary, one teritary, and three primary carbon atoms, and one tertiary, two secondary, and nine primary hydrogen atoms. Neopentane, however, contains four primary carbon atoms and one quaternary carbon atom, *i.e.*, a carbon atom joined to four other carbon atoms also all the twelve hydrogen atoms are primary. As we shall see later the behavior of these various types of hydrogen atoms differs considerably.

Nomenclature of the Alkanes

Whenever a new branch of knowledge is opened up, there is always the problem of introducing a system of nomenclature. The early chemists usually named a compound on the basis of its history, *e.g.*., methane. This is the parent hydrocarbon of methyl alcohol, CH_3OH. Methyl alcohol was originally obtained by the destructive distillation of wood, and was named 'word-spirit'. From this arose the word methyl, which is a combination of two Greek words, methu (wine) and hule (wood). Other examples of this way of naming compounds are acetic acid, which is the chief consituent of vinegar (Latin : *actum,* vinegar), malic acid, which was first isolated from apples (Latin : *malum,* apple), and so on. Thus grew up a system of common or trivial names, but in many cases the origin of the name has been forgotten. One advantage of the trivial system is that the names are usually short and easily remembered, but a disadvantage is that a particular compound may have a number of names.

As the number of knows organic compounds increased, it became apparent that it was necessary to systematise the method of nomenclature. This task was originally begun in 1992 by an international committee of chemists a Geneva, and hence is referred to as the Genera system of nomenclature. The work as carried on by the *International Union* of Chemists (I.U.C) by committee appointed in 1922, and in 1931 these drew up a report which is often referred to as the I.U.C, system.

Nomenclature is always undergoing revision, and the latest rules are those recommended in 1969 by the Commission on the Nomenclature of Organic Chemistry of the International Union of Pure and Applied Chemistry (I.U.P.A.C). The reader should always consult the *Chemical Society Handbook* is the wished to publish work in the *Journal* , and any mistakes he makes in nomenclature will soon be put right by the Editors!

(1) *In the trivial system :* of nomenclature the straight chain compounds are always designated as normal compounds, and the word normal is usually abbreviated to n. If it contains a quaternary carbon atom, the compound is known as the neo-compound, It is impossible to name many of the more complex alkanes by the trival system (later for examples of the trival system).

Univalent groups that are formed by the removal of one hydrogen atom from an alkane are knows alkyl or alkyl groups. The name of each individual group is obtained by changing the suffix -one of the parent hydrocarbon. The first four alkyl groups are often represented by a shorthand notation. The group derived from pentanes was named amyl. This name is now abandoned and pentyl and isopentyl used.

The alkanes, originally and still known as the paraffins, derived this name from the fact that an alkyl group plus one hydrogen atom gives a paraffin. *i.e.*, alkyl H = alkane.

In chemical equations if we are dealing with alkyl compounds as group and we do no wish to specify any particular member, we use the symbol R to represent the unspecified alkyl group, *e.g.*, RCI represents any alkyl chloride.

(2) In this system of nomenclature the hydrocarbon, except the n-compound, is regarded as a substitution product of methane.

The most highly branched carbon atom in the compound is named as the methane nucleus and the alkyl groups attached to this carbon atom are named in alphabetical order. Hydrogen atoms, if joined to the carbon atom chosen as the methane nucleus are not named.

This system of nomenclature is fairly good, since the name indicates the structure of the compound. It is impossible, however to name of the complex alkanes by this system.

(3) In the I.U.P.A.C. system of nomenclature the longest chain possible is chosen and the compound is named as a derivative of this n-hydrocarbon. The carbon chain is numbered from one end to the other by arabic numerals and the positions of side-chains are indicated by numbers, the direction of numbering being so chosen as to give the lowest numbers possible to the side- chains. When series of locants containing the same number of terms are compound term by term, that series is 'lowest' which contains the lowest number on occasion of the first difference, and this principle is used irrespective of the nature of the substituents *e.g.*,

$$\overset{10}{C}H_3\overset{9}{C}H_2\overset{8}{C}H(CH_3)\text{—}\overset{7}{C}H(CH_3)\overset{6}{C}H_2\overset{5}{C}H_2\overset{4}{C}H_2\overset{3}{C}H_2\overset{2}{C}H(CH_3)\overset{1}{C}H_3$$

When two sets of numbers are equally possible, then the order of the prefixes in the name decides which shall be used, *e.g.*, I-bromo3-chloropropane and not 3-bromo 1-chloropropane. It should also be noted that the names of prefixes are arranged alphabetically, regardless of the number of each *e.g.*, 5 ethyl 2,3-dimethyl octane.

The I.U.P.A.C. system of nomenclature is undoubtedly superior to the other two, since it permits the naming of any alkane on sight. N.B. The following names are retained for unsubstituded alkanes only, isobutane, isopentane, neopentane, isohexane.

When several chains are of equal length, that chain chosen goes in series to:

(a) The chain which has the greatest number of side-chains,

(b) The chain whose side chains have the lowest numbered locants,

(c) The chain having the greatest number of carbon atoms in the smaller side-chains,

Homologous Series

If we examine the formula of the various alkanes, we find that the formula of each individual differs from that of its 'neighbour' by CH_2, *e.g.*, CH_4, C_2H_6, C_3H_6, C_3H_5, C_4H_{10}, C_5H_{12}, ... A set of compounds, such as the alkanes, in which the members differ in composition from one another by CH_2, is known as an homologous series, the individual members begin known as homologues.

Through organic chemistry we find homologous series, each series being characterised by the presence of a functional group. The functional group is an atom or a group of atoms that causes a compound to behave in a particular way, *i.e.*, it is the functional group that gives rise to homologous series.

It is also possible for a compound to contain two (or more) identical or different functional groups and this gives rise to polyfunctional compounds.

If we examine the formula of the various alkanes, we find that the formula C_nH_{2n+2} will represent any particular homologue when n is given the appropriate value, *e.g.*, for pentane n is 5, therefore the formula of pentane is C_5H_{12}. The formula C_nH_{2n+2} is know as the general formula of the alkanes. The composition of any homologous series can be expressed by means of a general formula.

When we study the methods of preparation of the different alkanes, we find that several methods are common to all, *i.e.*, similar methods may be used for the preparation of all the homologues. This gives rise to the general methods of preparation of a particular homologous series.

Examination of the properties of the alkanes shows that many properties are, more or less, common to all the alkane homologues. This rise to the general properties of an homologous series.

The occurrence of homologous series facilitates the study of organic chemistry, since it groups together compounds having many resemblances. If we know the properties of several of the lower homologues we can obtain a fair idea of the properties of higher homologues, *i.e.*, we can forecast (within limits) the properties of a compound that we have not yet prepared.

The reason for these limitation is that the behaviour of a functional group in an homologous series depends on structural differences. We can now, however, explain many of these differences in behaviour in terms of polar and steric effects, and we are therefore in a much stronger position to make our predictions.

Only when one functional group is present do the properties of the compounds containing it show many similarities. If two (or more) functional groups are present, the properties depend on the 'distance' between them. When the groups are widely separated, each generally behaves in the normal way, but when they are close to each other, the properties of the compound may be very different from those of the individual functional groups.

There are various ways of studying organic chemistry, *e.g.*, as homologous series, or treating as a class compounds containing a particular functional group, irrespective of the 'rest' of the molecule. In this book, the treatment is the discussion of homologous series.

The General Methods of Preparation of the Alkanes

(1) *Grignard reagents* : Alkyl halides in ether react with magnesium to form alkyl, gnesium halides or Grignard reagents which, on treatment with water or dilute acid, are decomposed to alkanes.

$R - MgX + H_2O$—boil $\longrightarrow$ R—H + Mg CO + DBr for *ex.*

1. $C_2H_5\ Mg\ Br + H_2O \xrightarrow{\text{Boil}} C_2H_6 + Mg\,(OH)$ for ethane.

2. $CH_3\ MgI + H_2O \xrightarrow{\text{Boil}} CH_4 + Mg\,(OH)\,I$ methane.

3. $RI + \xrightarrow{\text{either}} R - Mg - I \xrightarrow{H^+} RH$ (*e.g.*)

4. By heating a mixture of the sodium salt of a carboxylic acid and soda lime.

$$RCO_2Na + NaOH\,(Ca) \longrightarrow RH + Na_2CO_3$$

This process of eliminating carbon dioxide from a carboxylic acid is know as decarboxylate. Soda-lime is a very useful reagent fro this purpose, but cases tested propionate, butyrate and caporate various products were obtained, *e.g.*, with sodium propionate,

$$C_2H_5CO_2Na \xrightarrow{NaOH} \underset{(44\%)}{C_2H_6} + \underset{(20\%)}{CH_4} + \underset{(33\%)}{H_2} + \text{Unsaturated compounds.}$$

This is therefore not useful general method for the preparation of alkanes, since the separation of the products is usually difficult.

When dealing with a preparation, there are at least three important points that must be considered, (i) the yield of crude product, (ii) the yield of pure product, (iii) the accessibility of the starting materials. In certain cases the yield of crude material is high, but the nature of the impurities is such that purification causes a large loss of material, resulting in a poor yield of pure product. On the other hand, it often happens that the product of one reaction is to be used as the starting materials for some other compound which can readily be freed (*i.e.*, purified with very little loss) from the original impurity. Provided, then that this impurity does not interfere with the second reaction, the crude material of the first step can be used as the starting material for the second. Thus the yield alone of a particular reaction cannot decide the usefulness of that method of preparation, the subsequent history of the product must also be taken into consideration. Furthermore, all things being equal, the more accessible materials, *i.e.*, readily prepared or purchased, are used as the starting materials.

(2) By the catalytic reduction of unsaturated hydrocarbons, *e.g.*, reduction of ethylene using nickel. Formerly, the nickel catalyst was usually prepared by reducing with hydrogen nickel

$$R-CH=CH_2+H_2 \xrightarrow{Ni/573k} RCH_2CH_3 \quad \text{alkane}$$

$$C_2H_4 + H_2 \xrightarrow{Ni} C_2H_6(ex.)^*$$

$$CH_3-CH_3=CH_2\ H_2 \xrightarrow{Ni-573k} CH_3-CH_2 \quad \text{propane}$$

oxide deposited on a suitable inert, porous support, *e.g.*, kieselguhr. The support is impregnated with a nickel salt, treated with sodium hydroxide, washed and dried, and the resulting nickel oxide reduced with hydrogen at 300 – 450°C. Many organic compounds may be reduced by passing their vapours mixed with hydrogen over nickel heated at 200 – 300°C. Any reduction that is carried out in this manner is referred to as the Sabatier-Senderens reduction, in honour of the workers who first introduced this method. It is quite a common feature in organic chemistry to name a reaction after its discoverer or, in certain cases, after a worker who investigated the reaction and extended its application. The reader should always make himself familiar with the reaction associated with a particular name.

A common nickel catalyst is that prepared by the method introduced by Raney (1927). An alloy containing equal amounts of nickel and aluminium is digested with sodium hydroxide, the aluminum, is dissolved away, and the residual very finely divided nickel is washed and stored under water, ethanol, or any other suitable liquid. Raney nickel is more reactive than the supported nickel catalyst, and is usually effective at lower temperatures, often at room temperature

Alkenes may also be reduced by a variety of chemical reagents.

(3) *Reductional of alkyl halides* : This may be done is different ways.

(i) Reduction by dissolving metals, *e.g.*, zinc and acetic or hydrochloric acid, zinc and sodium hydroxide, zinc-copper couple and ethanol , etc. It is no longer considered that 'nascent' hydrogen is the reducing agent. The belief now is that there is an electron-transfer from the metal to the substrate, followed by the addition of protons from the solvent. Thus, reduction with a zinc-copper couple may be formulated as (where e is an electron).

$$Zn \longrightarrow Zn^{2+} + 2e$$

$$RX + e \longrightarrow X^- + R. \longrightarrow R^-$$

$$R{:}^- + C_2H_5OH \longrightarrow R - H + {}^-OC_2H_5 \text{ (g, vg.)}$$

(ii) Primary and secondary alkyl halides are readily reduced to alkanes by lithium aluminium, hydride ($LiAlH_4$), tertiary halides give mainly alkenes. On the other hand, sodium borohydride ($NaBH_4$) reduces secondary and tertiary halides, but not primary, whereas triphenyltin hydride (Ph_3SnH) reduces all three types of halides.

(iii) Catalytic hydrogenolysis

(iv) Iodides may be reduced by heating with concentrated hydriodic acid at 150°C.

Reduction with concentrated hydriodic acid is usually carried out in the presence of a small among of red phosphorus which regenerates the hydroids acid from the iodine formed. The hydriodic acid red phosphorus mixture is one of the most powerful reducing agents used in organic chemistry.

Alcholos, aldehydes, ketones, and carboxylic acids are also reduced to alkanes by hydrogen iodide and red phosphorus. On the other hand, aldehydes and ketones may be reduced to alkane by the Clammensen.

(4) *Wurtz reaction (1854) :* An ethereal solution of an alkyl halide (preferably the bromide or iodide) is treated with sodium, *e.g.*,

$$R^1X + R^2X + 2Na \longrightarrow R^1 - R^2 + 2NaX$$

As previously pointed out, when we do not wish to specify a particular alkyl group, we use the symbol R. When we deal with two unspecified alkyl groups which may, or may not, be the same, we can indicate this by R^1 and R^2, also when dealing with compounds containing a halogen atom, and we do not wish to specify the halogen, we can indicate the presence of the unspecified halogen atom by means of X.

Consideration of the equation given above shows that in addition to the desired alkane $R^1 - R^2$, there will also be present the alkanes $R^1 - R^1$ and $R^2 - R^2$. Unsaturated hydrocarbons are also obtained. Obviously, then the best yield of an alkane will be obtained when R^1 and R^2 are the same, *i.e.,* when the alkane contains an even number of carbon atoms and is symmetrical. It has been found that the Wurtz reaction gives good yields only for 'even carbon' alkanes of high molecular weight, and that the reaction generally fails with tertiary alkyl halides (q, v).

Sodium is used in the Wurtz reaction. Other metals, however, in a finely divided state, may also be used, *e.g.*, Ag Cu.

Tow mechanisms have been suggested for the Wurtz reaction.

(i) The intermediate formation of an organo-metallic compound, *e.g.*, the formation of n-butane from ethyl bromide.

$$C_2H_5 - Br + 2Na \longrightarrow C_2H_5 - Na^+ + NaBr$$

$$C_2H_5^- Na^+ + C_2H_5Br \longrightarrow C_2H_5 - C_2H_5 + NaBr$$

(ii) The intermediate formation of free radicals, *e.g.*,

$$C_2H_5 - Br + Na \longrightarrow C_2H_5 + NaBr$$

$$C_2H_5 + C_2H_5 \longrightarrow C_2H_5 - C_2H_5$$

On of the properties of free radicals is disproportionation, *i.e.*, intermolecular hydrogenation, one molecule acquiring hydrogen at the expense of the other, *e.g.*,

$$C_2H_5 + C_2H_5 \longrightarrow C_2H_6 + C_2H_4$$

This would account for the presence of ethane and ethylene in the products. According to Morton *et al.* (1942). However, ethane and ethylene may be produced as follows:

$$Na^*CH_3CH_2^- H—CH_2—CH_2—Br \longrightarrow CH_3CH_3 + CH_3{=}H_2 + NaBr$$

The is a bimolecular elimination mechanism. On the other hand, the mechanism of alkane formation is generally accepted as usually being a bimolecular nucleophilic substitution.

$$Na^*CH_3CH_2^- H—CH_2—CH_2—Br \longrightarrow CH_3CH_3 + CH_3{=}H_2 + NaBr$$

The free- radical mechanism, however, has been shown to operate in some cases (Bryce Smith, 1963).

(5) *Kolbe's electrolytic method (1849)* : A concentrated solution of the sodium or potassium salt of a carboxylic acid or mixture of carboxylic acids is electrolysed. *e.g.*,

$$R^1CO_2K + R^2CO_2K + 2H_2O \longrightarrow R^1–R^2+2CO_2+H_2+2KOH$$

$$R - COO\,Na\,(ag) \xrightarrow{\text{dissiociation}} Na^+ + R - COO^-$$

At Anode

$$R\,COO^- \longrightarrow R\,COO^+ + C$$

$$R\,COO^- \longrightarrow R^+ + CO_2$$

$$2R^+ \longrightarrow R - R \text{ higher altitude}$$

At Cathode

The Na– ions carry the current Through the solution but do not get reduced at cathode Instead H^+ ions produced due to one, ionsection of water are invalued in two electrode reaction at cathode. The reaction taking place at cathode are.

$$2H_2O \rightleftharpoons 2H^+ + 2OH^-$$

$$2H^+ + 2e^- \longrightarrow H_2(g)$$

Not reaction at cathode. $2H_2O + 2e^- \longrightarrow H_2\,(g) + 2OH^-$

If R^1 and R^2 are different, then hydrocarbons $R^1 - R^1$ and $R^2 - R^2$ are also obtained (of. Wurtz reaction). Such mixtures can often be separated readily. Yields of 50–90 per cent have been obtained with straight chain acids containing 2–18 carbon atoms. Alkyl groups in the a-position decrease the yield (usually below 10 percent). The by products are alkenes, alcohols (particularly in alkaline solution) and esters. It is also interesting to note that the yields of the alkanes are increased when dimethyl formamide is used as solvent (Finkelstien *et al.*, 1960).

The Kolbe's electrolytic method now has application in the synthesis of natural compounds, particularly lipids.

Several mechanisms have been proposed for the Kolbe's reaction. The free radical theory is the on now favoured, and strong support for it has been provided by Lippincott *et al.* (1956), *e.g.*, where sodium propionate is electroysed, n butane, ethane, ethylene and ethyl propionate are obtained. The propionate ion discharges at the anode to form a free radical.

$$C_2H_5CO_2 \longrightarrow C_2H_5CO_2 + e$$

This propionate free radical then breaks up into the ethyl free radical and carbon dioxide

$$C_2H_5CO_2 \longrightarrow C_2H_5 + CO_2$$

Then (i) $2C_2H_5 \longrightarrow C_4H_{10}$

(ii) $C_2H_5 + C_2H_5 \longrightarrow C_2H_5 + C_2H_4$

(iii) $C_2H_5 + C_2H_5CO_2 \longrightarrow C_2H_5CO_2C_2H_5$

Reaction (i) given n-butane, (ii) gives ethane and ethylene by disproportionation (ef. Wurtz reaction), and (iii) gives ethyl propionate.

(5) A very useful method for preparing long-chain alkanes involves the coupling of alkylboranes by means of silver nitrate in the presence of sodium hydroxide at 25°C, *e.g.*, 2-methylpentene gives 4,7-dimethyldecane (Brown et at., 1961).

$$6CH_3(CH_2)_2C(CH_3)=CH_2 \xrightarrow{2B_2H_4} [2CH_3(CH_2)_2$$

$$B \xrightarrow[NaOH]{AGNO_5} 3CH_3(CH_2)_2CH(CH_3)CH_2CH_2$$

$$CH(CH_3)(CH_2)_2CH_3 \qquad (75\text{-}80\%)$$

Coupling may also be effected between unlike boranes.

Conformation. It has already been pointed out that the valencey angle at the central carbon atom of a regular tetrahedron is 109°28'. It was originally postulated (van't Hoff). Another postulate was fixed but it is now known that it may deviate from this value (± 5°). Another postulate was the principle of free rotation about a single bond. Let us consider the ethane molecule, $CH_3 - CH_3$, and let us imagine that one methyl group is rotated about the C – C bond as axis with the other group at rest. Suppose we use, as the starting point, the position in which two C – H bonds are parallel, *i.e.*, these four atoms lie in a plane (Fig. 4.13c).

In this position, the *dihedral angle* (*angle of rotation or angle of torsion*) is zero, and if the rotation is free, then as the dihedral angle changes, the energy content of the molecule will remain constant (the plot of energy content against the dihedral angle will be a horizontal line). In this situation, the two haves can assume with complete freedom, and infinite number of positions relative to each other. Thus, the entropy of the molecule will be a maximum.

Pitzer *et al.* (1936) calculated the entropy of ethane based on the assumption that the internal rotation was free, and found that calculated value was greater then the observed value. This means that there is less freedom to assume all possible dihedral angles than was expected on the principle of free rotation. Pitzer therefore assumed that the internal rotation is hindered by a potential energy barrier, and calculated the change in entropy with increasing barrier height.

When he assumed a barrier of 12.55kj mol^{-1} (3k cal), the calculated and observed entropies were brought into good agreement. The potential energy curve obtained for ethane is shown in (Fig. 4.13a) and by convention the potential energy is measured relative to the energy of the most stable form. Because of the existence of potential energy barriers, the result is *mutual oscillation* (*libration*) about the conformations with minimum potential energy, there by producing 'more order' (more restriction) than had there been no barriers.

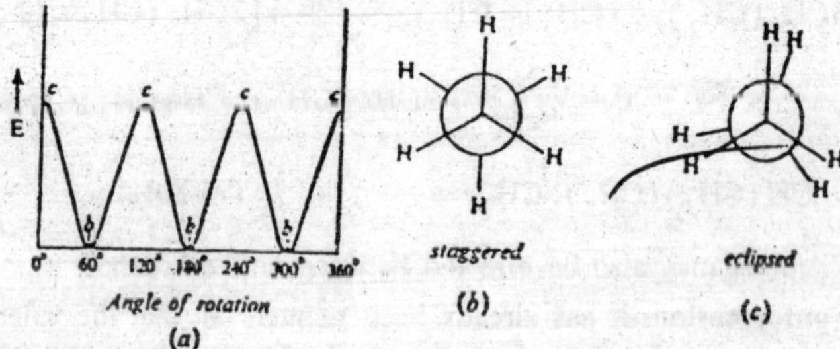

Fig. 4.13

Fig. 4.13b is the Newman projection formula. This is obtained by viewing the molecule along the bonding line of the two carbon atoms, with the carbon atom nearer to the eye being designated by equally spaced radii, and the carbon atom further from the eye by a circle with three equally spaced radial extensions.

The *staggered* (or transoid) conformation (in which the hydrogen atoms are as far apart as possible), and Ic the *eclipsed* (or cisoid) conformation (in which the hydrogen atoms are as close together as possible). It can be seen from that the energy of the eclipsed conformation is greated than that of the staggered conformation.

staggered (transoid)　　*gauche* or *skew*

fully eclipsed　　*eclipsed*

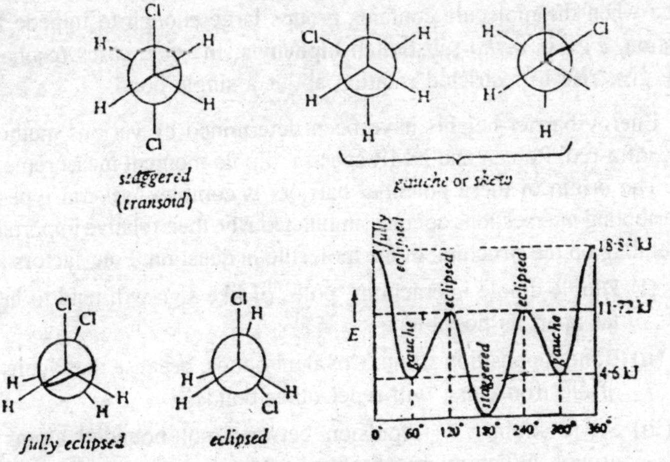

Fig. 4.14

The potential energy barrier (11.92 kj mol^{-1}) is much too small for either conformation to remain stable, *i.e.*, the eclipsed and staggered forms are readily inter convertible and hence neither can be isolated as such. However the staggered conformation is the preferred from, *i.e.*, its population is greated than that of the eclipsed form.

Now let us consider ethylene dichloride. According to Bernstein (1949), the potential energy undergoes the changes shown in Fig. 4.14.

There are two positions of minimum energy, but the staggered conformation is the more preferred one, *i.e.*, the one in which the molecule largely remains. Dipole moment studies show that this is so in practice, and according to Mizushima *et al* (1038), only the staggered from is present at low temperatures.

Thus, in theory, there is no free rotation about a single bond, but in practice it may occur the various forms do not differ very much in energy content (usually between 4.18 and 41.84kj mol^{-1}). Molecules which can form isomers by rotation about single bonds are called flexible molecules and the different forms taken up are different conformations or rotational isomers. Usually, the staggered conformation is the most stable eclipsed form is always avoided where possible.

Free rotation about a single bond is generally accepted in *simple* molecules. *Restricted rotation* about a single bond, however, may take place when the molecule contains groups large enough to impede free rotation, *e.g.*, in *ortho*-substituted biphenyls. In some cases resonance may give rise to restricted rotation about a single bond.

Energy barrier heights have been determined by various methods, *e.g.*, infra-red, Raman and NMR spectra, dipole moment measurements, etc. The origin of these potential barriers is complex, several types of contributing intersections operate simultaneously, their relative importance depending on the structure of the molecule in question. Four factors are:

(i) Dipole-dipole interactions, poles of like sign will tend to be as far apart as possible.

(ii) Bond opposition strain, Covalent bonds, because they 'consist' of electron pairs, will repel other bonds

(iii) Steric strain, *i.e.*, repulsion between non-bounded atoms or groups. When approach closely together, forces of repulsion come into play, and it has been found that the distance of closest approach is greater than the sum of the covalent radii of the atoms concerned. This distance of closest approach for each atom (or group) is known as the can der *Walls radius*, *e.g.*, H, 1.2Å, O 1.4Å, Cl, 1.8Å, Br, 1.95Å, I 2.13Å.

(iv) Bond angle strain, *i.e.*, distortion of bond angles.

When the stability of a molecule is decreased by internal forces product by interaction between constituent parts, that the molecule is said to be under strain. The study of the existence of preferred conformations in molecules, and the relating of physical and chemical properties of a molecule to its preferred conformation, is known as conformation analysis.

A particularly useful method for studying the conformations of molecules is NMR spectroscopy. However, before dealing with this, let us first consider the chemical shifts of the protons of CH_3, CH_2, and the CH group in acyclic compounds and the chemical shift of a proton attached to O, N, etc.

As we have seen (p, 11), the more electronegative Z is in the groups – CH – Z and – Z – H, the more is the proton desheilded and consequently the lower is the value lists the values of some of the more important groupings, and also lists some coupling constants.

We return now to the use NMR spectroscopy in conformational analysis. Let us first consider ethyl chloride, and since the most stable conformation is the staggered one, the molecule may be represented as (I), (II), and (III), In (I), the environments of H_a and H_c are identical,

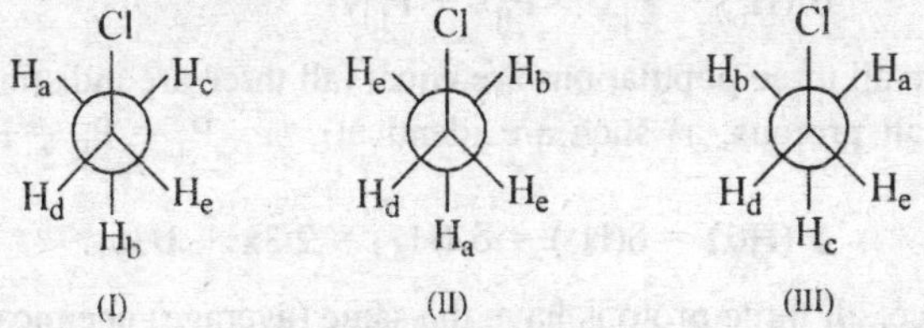

(I) (II) (III)

Fig. 4.15

but differ from that of H_b. This can be seen to be the case by replacing each proton, one at a time, by Z. This procedure for H_a and H_c produces mirror images, and so the two protons are chemically equivalent but for H_b the substituted product is different because of the different dihedral angles, and similarly for H_b. Thus, H_a and H_c are chemically but not magnetically equivalent.

By using similar arguments, it can be shown that H_a and H_b are also chemically but not magnetically equivalent. Hence, if the population of ethyl chloride conformation were completely represented by (1), protons Ha and H_b would give one signal and H_a another signal, provided that the chemical shifts were sufficiently different (0.1 – 0.2 p. pm.), *i.e.*, the methyl group gives would two signals. In practice, the methyl group gives only one signal a (a triplet), and therefore protons H_a, H_b, and H_b must be equivalent.

This is explained on the basis that there is free rotation about the carbon-carbon single bond. When the methyl group rotates (with respect to the CH_2CI group), it can take up the other two stable staggered conformations, (II) and (III), the result being that each proton has an average environment. Since these average environments are identical, rotation result in the equivalence of H_a, H_b, and H_c.

This equivalence may be demonstrated as follows. Let P_1, P_{II}, P_{III} be the population in conformations (I), (II), and (III), respectively, where $P_1 + P_{II} + P_{III} = 1$. Also let δ (the chemical shift) of any proton trans (anti) to another proton be x, and y if *trans* to Cl. Now, the observed δ of a given proton is the sum of the δ-contributions in each conformation, and so it follows that,

$$\delta\ (H_a) = P_1x + P_{II}y + P_{III}x$$

$$\delta\ (H_b) = P_1y + P_{II}x + P_{III}x$$

$$\delta\ (H_c) = P_1x + P_{II}x + P_{III}y$$

Since all three populations are equal (all three are indistinguishable because all protons, as such are identical), *i.e.*, $P_1 = P_{II} = P_{III} = 1/3$, therefore

$$\delta\ (H\alpha) = \delta(H_b) = \delta\ (H\chi) = 2/3x + 1/3y.$$

Hence, all three protons have the same (average) chemical shift (in the freely rotating state) and so there is no observable splitting for the methyl group.

Application of this method to protons H_d and H_e gives.

$$\delta\ (H_d) = P_1x + P_{II}x + P_{III}x = x$$

$$\delta\ (H_e) = P_1x + P_{II}x + P_{II}Ix = x$$

From the above discussion, it can be seen that if ethyl chloride were undergoing slow rotation, it would be an ABB.CC spin system, but when undergoing fast rotation, it rotation, it is an A_3B_2 spin system (see also below).

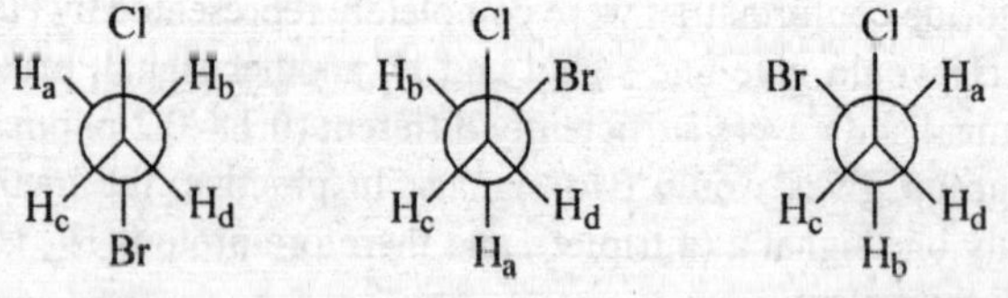

Cl, H_a, H_b, H_c, H_d, Br (IV) — Cl, H_b, Br, H_c, H_d, H_a (V) — Cl, Br, H_a, H_c, H_d, H_b (III)

(IV) (V) (III)

Fig. 4.16

Now let us consider the case of 1-bromo 2-chloroethane, and if we use the same arguments as before then in (IV), H_a and H_b are chemically but not magnetically equivalent, and this is also true for Hc and H_d. Also if δ for a proton trans to a proton is x, trans of Cl is y, and trans to Br is z, then,

$$\delta\,(H_a) = P_{iv}x + P_{v}y + P_{vi}x$$

$$\delta\,(H_b) = P_{iv}x + P_{v}x + P_{vi}y$$

In this molecular, although $P_v = P_{vi} \neq P_{iv}$, it still follows that $\delta\,(h_a) = \delta\,(H_b)$. In the same way, it can be shown that $\delta\,(H_c) = \delta\,(H_d)$.

We can extend the argument by also by also considering the possible coupling constants. If J_t and J_t represent, respectively, the vicinal trans and skew coupling, then,

$$J_{ad} = P_{iv}J_t + P_vJ_x + P_{vi}J_x$$

$$J_{bd} = P_{iv}J_x + P_vJ_t + P_{vi}J_x$$

Since $P_v = P_{vi} \neq P_{iv}$ therefore $J_{ad} \neq J_{bd}$. Hence H_a and H_b (and Hc and H_d) are chemically but not magnetically equivalent. Application of this method will show that the three protons in the methyl group in ethyl chloride are magnetically equivalent, as are also the two protons in the CH2CI group, for the freely rotation molecule.

Finally, let us consider a molecule of the type $R^1R^2CHCH^3R^4$. Each carbon atom is asymmetric and the environments of H_a are different in all three conformations, and this is also true for H_d.

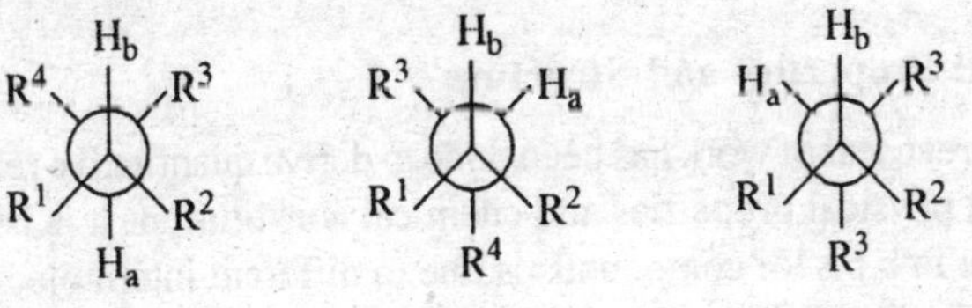

(VII) (VIII) (IX)

Fig. 4.17

Furthermore, because of the different steric effects, the three populations will be different (*i.e.* $P_{vii} \neq P_{viii} \neq P_{ix}$). Hence, the chemical shifts of H_a (and those of H_b) are different in each conformation. If rotation were slow, the NMR spectrum would be a composite of all three spectra (three AB spin systems for each pair of enantiomers). If the temperature is raised, the rate of rotation is increased, and the result is an average of chemical shifts and coupling constants to given one AB spin system (for each pair of enantiomers). These averages, however, will be weighted in favour of the most stable conformation, *i.e.*, the one with the highest population.

The name paraffin arose through contracting the two Latin words parum affinities, which means little affinity. The name was suggested because these hydrocarbons were apparently very reactive. It is difficult to define the terms 'reactive' and unreactive, since a compound may be reactive under one set of conditions and unreactive under another. Under 'ordinary' conditions, the alkanes are inert towards reagents such as acids, alkalis, oxidising reagents, reducing reagents, etc, but are reactive if the 'right' conditions are uses (see below). One possible reason for this lack of reactivity is that the C – H bond is very strong.

PHYSICAL ALKANES

The normal alkanes from C_1 to C_4 are colorless gases, C_3 to C_{17}, colorless liquids, and from C_{18} onwards, colorless solids. The b.p.s. rise fairly regularly as the number of carbon atoms in the compound increases. This holds good only for the normal compounds, and the difference in b.p.s. decreases as the higher homologues are reached. Other physical properties, such as m.p., density, viscosity, also increases in the same way as the b.p.s. (of the normal alkanes), *e.g.*, the density of the normal alkane increases fairly steadily for the lower members, and eventually tends to a maximum value of about 0.79.

Physical Properties and Structure

A great deal of work has been done to derive quantitative relationships between physical properties and chemical constitution. It is believed the variation in b.p.s. of compounds is due to different intermolecular forces such as hydrogen bounding, diploe moments, etc. Hydrogen bounding may produce association, and this will cause the b.p., to be higher than anticipated. The greater the dipole moment of the compound, the higher is the b.p., since owing to the charges, more work is required to separate then molecules, *e.g.*, nitro-compounds, RNO_2, which have large dipole moments, have much higher b.p.s. Than the alkanes in which the dipole moment is absent or very small. As far as the alkanes are concerned, the only attractive forces are the weak van der Waals forces.

Observation has shown that in group of isomeric compounds (acyclic), the normal compound always has the highest b.p., and m.p., and generally, the greater the branching the lower is the b.p. In the n-compound, owing to its large surface area due to the linear nature of the chain, the van der Waals attractive forces are relatively large. As branching increases, the molecule tends to become spherical, thereby resulting in a decreased surface area available for intermolecular attractions. Furthermore, because intermolecular forces are small in alkanes, the b.p.s. of these compounds are used as reference values for given molecular weights, *e.g.*, propane (M.W. 44, b.p. – 44.5°C), ethanol (M.W. 46, b.p. 78.1°C). The high b.p., of the latter is attributed to hydrogen bounding.

Solubility

The alkanes are almost insoluble in water, but readily soluble in ethanol and ether, the solubility diminishing with increase in molecular weight.

It is believed that solubility depends on the following intermolecular forces, solvent/solute, solute/solute, solvent/solvent. If the solvent/solute attractive forces are greater than either of the other two, solution will be expected to occur fairly readily. A non-electrolyte dissolves readily in water only if it can form hydrogen bonds with the water. Thus alkanes are insoluble or almost insoluble, in water. Methane is more soluble than any of its homologues, hydrogen bonding with the water is unlikely, and so other factors possibly molecular size–must also play a part. Useful rule in organic chemistry with respect to solubility is that like dissolves like. *e.g.*, if a compound contains a hydroxyl groups, then the best solvents usually contain hydroxyl groups.

X-ray analysis of solid alkanes has shown that the carbon chains are fully extended, i.e, zigzag. In the liquid state this extended form is also one of the stable conformations provided that the carbon chain is not very long. When the number of carbon atoms is sixteen or more, the extended form is no longer present in the liquid state.

Since the dipole moment of methane is zero, the dipole moment of the methyl group is equal to that of the fourth C – H bond (in methane) and is directed along this axis. Thus replacement of hydrogen by a methyl group will not be expected to change the dipole moment, *i.e.*, the dipole moment of all alkanes, whether straight -or branched-chain, will be zero. Thus has been found to be so in practice. This will always hold good whatever conformation is taken up by the alkane provided that no deformation of the normal carbon valencey angle (of 109°.28') is produced in the twisting, since all methyl groups will be balanced by a C – H bond. It therefore follows that the electron negativity of all alkyl groups is equal to that of hydrogen, namely zero. As soon as one hydrogen atom is replaced by another atom of group (other than alkyl), the resultant molecule will now be found to possess a dipole moment.

The position of the infra-red absorption region of the CH group (stretch) in alkanes depends on whether the carbon atom is primary, secondary or tertiary. Thus we have the following regions: —CH_3, 2975 – 2950 and 2885 – 2860 cm^{-1}(m); —CH_2—, 2940 – 2915 and 2870 – 2845 cm^{-1}(m); $\rangle$CH, 2900 – 2880 cm^{-1}(w). SOme C—H deformation absorption frequencies are: CH_3, 1470–1435 (m) and 1385 – 1370) cm^{-1} (s), – CH_2 – 1480 – 1440 cm^{-1} (m).

Two useful skeletal vibrations are is, Me_2CH – 1175–1165cm^{-1} (s), neo, Me_3C – 1255–1245 cm^{-1}(s). It is thus possible to detect the presence of these groups in a molecule. The infra-red spectrum of 3-methylpentane (as a film). These bands are characteristic of all types of alkanes containing methyl and methylene groups.

Mass Spectra

Since the stability of carbonium ions is in the order t > s > prim, fission of bonds in alkanes occurs preferentially at branched carbon atoms. When alternative fissions can occur, it is the heaviest side-chain that is eliminated preferentially. Since alkyl carbonium ions are formed, all of them (with ^1H and ^{12}C) will give peaks of odd masses. In particular, n-alkanes give a series of peaks separated by 14 mass units (CH_2). The relative abundance of these peaks is usually greatest for $C_3H_7^+$ (43), C (57), and $C_5H_7^+$ (71), and decreases fairly regularly for the larger masses. Furthermore, each peak is generally accompanied by peaks of mass 1 and 2 units lower, corresponding to the loss of 1 and 2 hydrogen atoms, respectively.

The molecular ion is always present for n-alkanes, its intensity decreasing with increasing molecular weight. On the other hand, the greater the branching in the alkane, the less is the likelihood of the appearance of the molecular ion and, if this does appear, its intensity is usually low.

Physical Properties of Alkanes

Some of the physical properties that we will study in this section are boiling point, melting point density and solubility. All these properties depend upon the intermolecular forces between the molecules. These forces are of the van der Walls' type. The van der Walls' forces depend upon the shape and structure of the molecules.

A straight chain molecule has more are of contact with an adjacent molecule as compared to a branched chain molecule which is more compact. Therefore, *the van der Waals' forces are greater in straight chain molecules than in branched chain molecules.*

Major physical properties of alkanes are described below.

1. *Physical state :* Lower alkanes containing upto four carbon atoms (C_1 to C_4) are colorless, odourless gasses. The alkanes with five to sixteen carbon atoms ($C_5 - C_{16}$) are colourless and odourless liquids, and the higher members ($> C_{18}$) are colourless (white) waxy solids at near room temperature.
2. *Boiling point :* Alkanes are non-polar molecules. The molecules of alkanes are bond to each other by weak van der Waals' forces, and therefore, have lower boiling points.

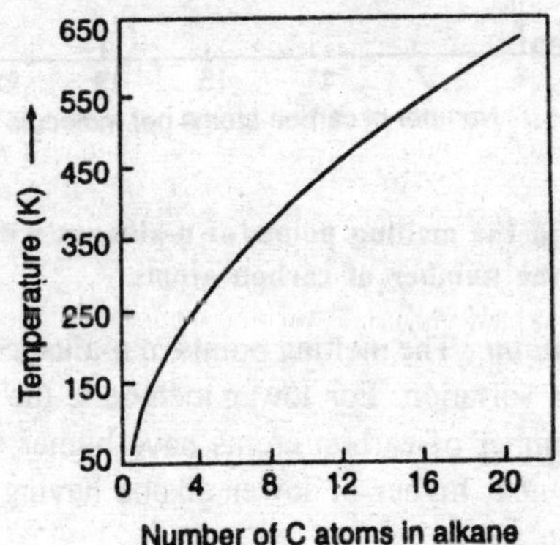

Fig. 4.18 : The Variation of the boilling points of n-alkanes.

(a) The boiling points of alkanes show a regular increases with the number of carbon atoms in the alkane. The boiling point increases by about 20–30 degrees for each addition of a $- CH_2$ group to the molecular formula. Variation of the boiling points of some n-alkanes is shown in Fig. 4.18.

Explanation. With an increases in the molecular size/mass, the strength of the intermolecular forces (van der Waals' forces) increases. The boiling point of a compound depends upon the strength of the intermolecular forces between the molecular. Therefore, the boiling points of alkanes increase with an increase in the chain length.

(b) *The branched chain isomers of an alkane have lower boiling points than the straight chain isomer.*

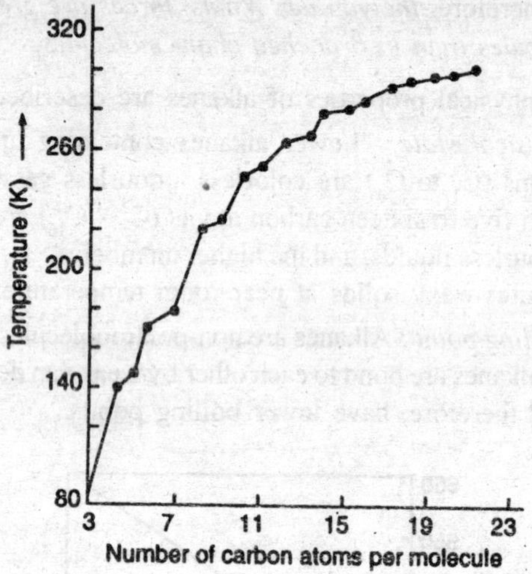

Fig. 4.19 : Variation in the melting points of n-alkanes with increase in the number of carbon atoms.

3. *Melting point :* The melting points of n-alkanes do not show a regular variation. For lower members, the alkanes with even number of carbon atoms have higher melting point than the next higher or lower alkane having odd number of carbon atoms.

However, four alkanes with four or more carbon atoms, the increase in the melting point in going from odd to even alkane, is mush higher than that observed in going from even to odd alkane. The variation of melting point with number of carbon atoms in the chain of some n-alkanes is shown in Fig. 4.19.

Explanation : The carbon chains in alkanes are zig-zig because the carbon atoms do not lie in the same vertical plane Fig. 4.20.

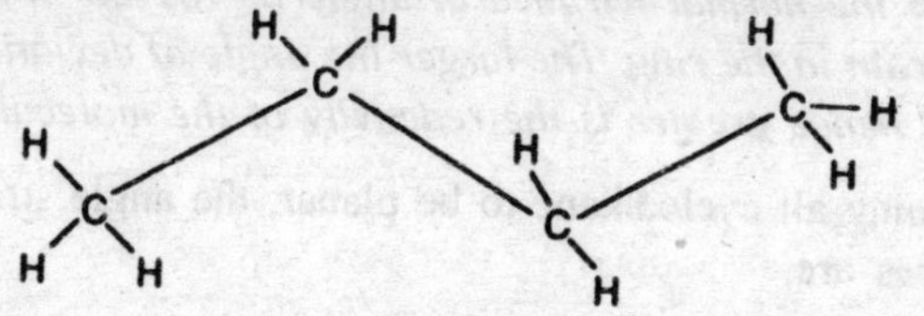

Fig. 4.20 : The zig-zag chain of carbon atoms in alkanes.

For an alkane consisting of even number of carbon atoms the terminal methyl groups lie on the opposite side. For alkanes consisting of odd number of carbon atoms, the terminal methyl groups (hence carbon atoms) line on the same side of the molecule. Such an arrangement leads to a better and more effective packing in the crystal lattice of alkanes having even number of carbon atoms. Such a close packing maximises the intersection forces and , therefore, leads to higher melting points for alkanes containing even number of carbon atoms.

4. *Densities :* Alkanes have very low densities. The densities increase gradually in going from lower to higher alkanes, and reaches a limiting value at about 0.8 g/mL (800 kg/m^3). The alkanes are lighter than water.
5. *Solubility :* Alkanes are non-polar compounds (dipole moment is zero) and hence insoluble in water and other polar solvents. However, they are soluble in non-polar solvents, such as, either benzene, carbon tetrachloride etc. The solubility of alkanes in any solvent, decreases with the increases in molecular mass of alkanes.

CONFORMATIONS IN CYCLOALKANES

In cycloalkanes, the reactivity decreases as the number of carbon atoms in the ring increases. Thus, the order of reactivity shown by cycloalkanes is

Cyclopropane > Cyclobutane > Cyclopentane > Cyclohexane

A. Von Baeyer explained the decreasing reactivity of cycloalkanes with increasing number of carbon atoms in the ring in terms of deviation of the $C\hat{C}C$ angle in any cycloalkane from the tetrahedral angle of 109°28. According to *Baeyer's Strain, Theory, any deviation of bond angle from the normal tetrahedral angle of 109°28′ would cause an internal strain in the ring. The larger the angle of deviation, larger the strain and hence greater is the reactivity of the molecule.*

Assuming all cycloalkane to be planar, the angle strain in various cycloalkanes are,

Cycloalkane:	Cyclopropane	Cyclobutane	Cyclopentane	Cyclohexane	Cycloheptane
Angle strain:	+ 24°44′	+ 9°44′	+ 44′	–5°16′	–9°33′

The + sign indicates that the $C\hat{C}C$ angle is compressed to satisfy the geometry of the ring. The –ve sign indicates widening of the $C\hat{C}C$ angle to satisfy the geometry of the ring. *The strain in the molecule and, therefore, its reactivity is determined by the magnitude of the angle strain irrespective of its sign. From the angle strain values given above, we would expect cyclopropane to the most reactive of all the cycloalkanes listed, and cyclopentane to be the least reactive. This is found to be so.* However, from the angle strain values, cyclohexane and higher cycloalkanes should show greater reactivity than cyclopentane. But, cyclohexane and higher cycloalkanes are found to be quite stable.

Conformations in cyclohexane : To explain the stability (or reactivity) of higher cycloalkanes, Sachse and Mohr (1998) pointed out that the rings of cyclohexane and higher cycloalkanes can become free of strain if all the carbon atoms in the ring are not forced into one plane. Thus, the molecules of cyclohexane and higher cycloalkanes avoid the angle strain by adopting conformations in which all the bond angles are close to 10°28' and are therefore, quite stable and unreactive. The two most important conformations for cyclohexane are the *chair* and the *boat forms* shown in Fig. 4.20(a).

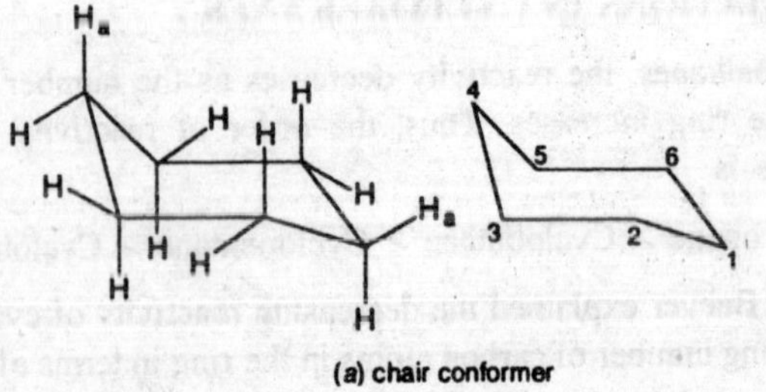

(a) chair conformer

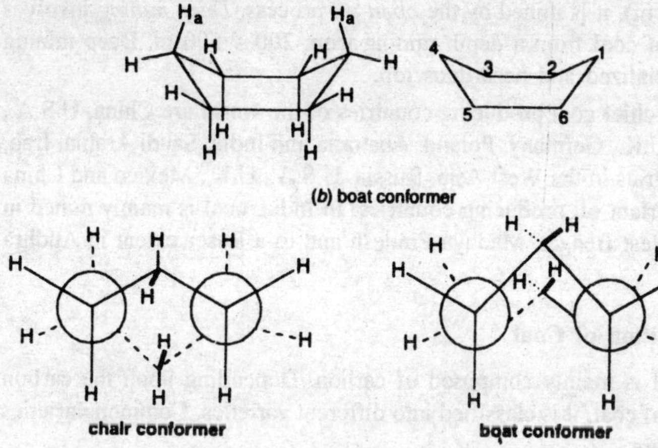

(*b*) boat conformer

Fig. 4.20(a)

Of the two, the chair form is more stable than the boat form, the energy difference being about 44kJ mol^{-1}, *i.e.*, the boat form has 44kJ/mol more energy than the chair form of cyclohexane. Therefore, *cyclohexane exists mainly in the chair form.* This is because in the boat form, many hydrogens on the adjacent carbons correspond to the unfavourable eclipsed conformation of ethane. Also, the two hydrogens marked Ha (commonly called flag pole hydrogens) in the boat form are quite close and repel one another. On the other hand, the chair form does not have these unfavourable interactions and all the hydrogens correspond to the more stable staggered conformation of ethane.

COAL AND PETROLEUM

Until the early years of the nineteenth century, only the plants and animals were the known sources of organic compounds. With the advent of the Industrial Revolution in Europe, fossil fuels like coal and petroleum gained prominence as the sources of organic compounds, particularly hydrocarbons. Today, hydrocarbons are mostly obtained from petroleum. However, in recent years coal has started gaining prominence, probably due to the uncertain conditions in the world oil market.

Coal and petroleum have organic (plant and animal) origin.

Coal

Coal is one of the earliest used fossil fuels. It occurs at varying depths under the upper crust of the earth. When found at low depths

(20 – 50 m), it is mined by the *open-pit* process. *Deep mining* involves mining of coal from a depth raining from 200 – 500 m. Deep mining is a specialized and hazardous job.

The chief coal producing countries of the world are China, U.S.A., Russia, U.K., Germany, Poland, Australia and India. Saudi Arabia, Iraq, the countries in the West Asia, Russia, U.S.A., U.K., Mexico and China are important oil producing countries. In India, coal is mainly mined in Bihar. West Bengal, Madhya Pradesh and to a lesser extent in Andhra Pradesh.

Composition of Coal

Coal is mainly composed of carbon. Depending upon the carbon content of coal, it is classified into different varieties. Common varieties of coal are,

Variety of coal:	Peat	Lignite	Bituminous	Anthracite
Carbon content:	10-15%	40%	60-70%	80-90%

Coal also contains arenes (aromatic hydrocarbons) like benzene, toluene, xylene, naphthalene and anthracene besides some organic compounds of sulphur and nitrogen.

Formation of Coal in Nature

It is believed that coal in nature was formed from the remains of the trees buried inside the earth some 500 million years ago. Due to the bacterial and chemical action on the wood, it got converted into **peat.** Then, as a result of high temperature and high pressure inside the earth peat got transformed into coal.

$$\begin{pmatrix}\text{dead plants}\\ \text{and trees}\end{pmatrix} \xrightarrow[\text{and chemical action}]{\text{bacterial}} \text{Peat} \xrightarrow[\text{pressure}]{\Delta} \text{Lignite}$$

$$\xrightarrow[\text{pressure}]{\Delta} \text{Bituminous coal} \xrightarrow[\text{pressure}]{\Delta} \text{Anthracite}$$

Aromatic Compounds From Coal

Coal is a complex mixture of hydrocarbons. It also contains some organic compounds containing nitrogen and sulphur in small amounts. It can be approximated to the formula $(C_3H_4)_n$

Coal is hydrogen-deficient substance. It is assumed that the basic structure of coal is probably built up of a large number of interlocked benzene rings, upto thirty in bituminous/anthracite coal. Hydrogen is present in the aliphatic sidechains. Approximate composition of some common varieties of coal are given below.

Table 4.1 : Approximate composition of some common varieties of coal

Element	*Peat (Yellow to light brown)*	*Lignite (Black)*	*Bituminous (Bright black)*	*Anthracite (Brown)*
Carbon	54.0	64.0	84.0	93.0
Hydrogen	5.5	5.0	4.5	3.3
Oxygen	35.0	26.0	8.0	2.0
Nitrogen	2.0	1.3	1.3	1.3
Sulphur	3.5	3.3	1.5	0.5

For obtaining organic compounds from coal, bituminous coal is thermally decomposed. This process is called destructive distillation of coal or pyrolysis of coal.

Destructive Distillation of Coal

Degradation (breakdown) of coal at high temperature in the absence of air is termed destructive distillation of coal. For this purpose, coal is heated in iron retorts in the absence of air, when the volatile material evolved is collected as distillate/condensate.

The process can be carried out under different temperature conditions, depending upon the nature of the products required.

Low Temperature Pyrolysis

Coal when heated in the absence of air at bout 723-973 K, gives soft solid, smokeless coke called *coalite*. The yield of coat tar and coal gas is doubled. Coal tar so obtained contains more of aliphatic compounds.

High Temperature Pyrolysis

In this process coal is heated in the absence of air upto about 1273-1473 K. The major products of high temperature pyrolysis are,

The yield per ton of coal on destructive distillation at higher temperature is shown in Fig. 4.21.

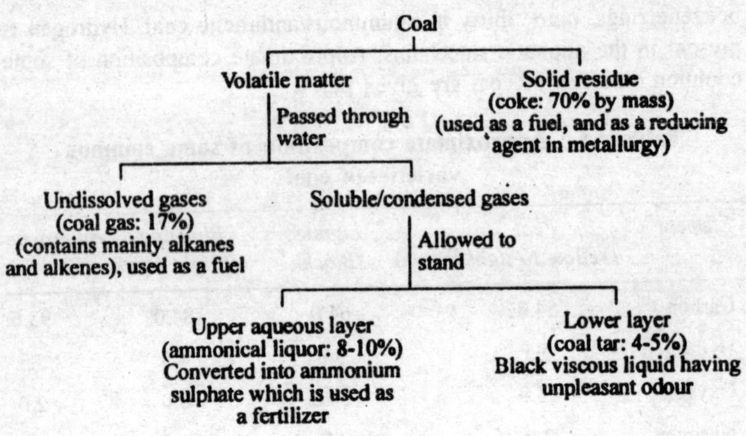

Fig. 4.21

Coke	*Coal gas*	*Coal tar*	*Ammonium sulphate*	*Light oil*
680 kg	30m^3	15-22 kg	1 kg	5-6 kg

Coal tar obtained as a heavy viscous dark liquid with an unpleasant odour is a valuable by-product in the destructive distillation of coal. Its composition depends upon the nature of coal used during distillation. However, coal tar generally contains the following compounds

Acid compounds.	Phenol and cresols etc.
Basic compounds.	Pyridine, etc.
Neutral compounds.	Benzene, toluene, naphthalene and anthracene etc.

FRACTIONAL DISTILLATION OF COAL TAR

Coal-tar obtained from the destructive distillation of coal is preheated in a preheater heated by the outgoing hot gases. Here coal tar loses most of the water it contains, along with some low boiling hydrocarbons.

Then, it is sent to the main iron still (a side retort) heated directly in a furnace. The vapours are condensed to obtain various fractional distillation of coal-tar.

PETROLEUM

Petroleum occurs deep inside the earth at a depth ranging from a few hundred to few thousand meters, Fig. 4.22. The name petroleum has been given to the crude oil because it occurs under the earth's crust entrapped in rocks (petra = oleum = oil). Crude oil (petroleum) is pumped out of earth by drilling oil wells.

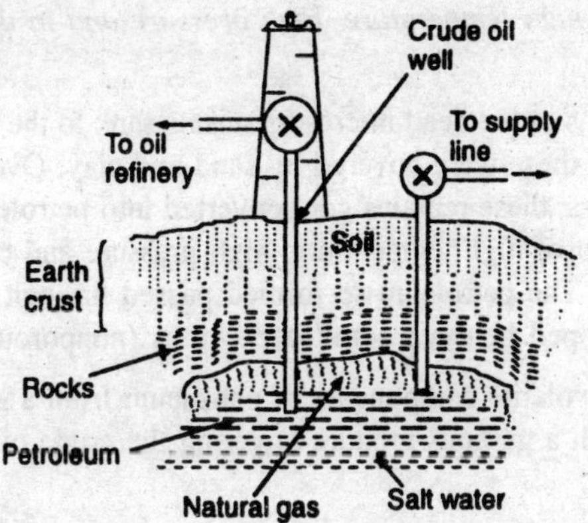

Fig. 4.22 : Occurrence of petoleum under the earth crust.

Type chief petroleum producing countries are, U.S.A., Russia, Saudi Arabia, Kuwait, Iraq, Libya, Nigeria, Algeria, Venezuela and Mexico. In India, oil fields are located in Assam., Gujarat, and Maharashtra. For the last few years, we have started producing oil from the high seas off Mumbai.

Petroleum is obtained by drilling a hole through the crust till it reaches the oil reserve. The oil gushes out of its own due to high pressure of the gas over its surface. However, when the pressure falls, it is pumped out with the help of pumps.

Large quantity of natural gas is generally associated with petroleum. Petroleum is often referred to as *liquid gold*, due to its importance in commerce and industry. Prosperity of any country depends upon its petroleum reserves.

COMPOSITION OF PETROLEUM

Crude oil (or petroleum) is a dark brown (sometimes with greenish tinge) viscous liquid. This crude oil is a mixture of about 150 different organic compounds.

Crude oil is largely made up of aliphatic hydrocarbons with much lesser amounts of aromatic compounds and organic compounds of sulphur and nitrogen.

FORMATION OF PETROLEUM IN NATURE

It is believed that *petroleum (or crude oil) was formed due to the decomposition of the prehistoric vegetable micro-organisms and sea animals under high temperature, high pressure and in the absence of the air.*

The remains of the dead micro-organisms sank to the bottom of the sea. Gradually, these were covered by sand and clay. Over a period of millions of years, these remains got converted into petroleum under the combined effect of high temperature, high pressure and bacteria in the absence of air. The petroleum so formed passed through porous rocks until it got trapped between some impervious (nonporous) rocks.

The more volatile components of petroleum from a gaseous cover petroleum. Such a gaseous mixture covering the crude oil is known as *nature gas.*

CRUDE OIL

The crude oil or petroleum as pumped out of the oil well is a viscous and complex mixture of several hydrocarbons and small amounts of other compounds. Crude oil (petroleum) cannot be used as such. Crude petroleum, however, can be separated into many useful fractions by fractional distillation. The process of separating crude oil into useful fractionals is called refining of petroleum.

In India, oil refineries are located at Koyali, Mumbai, Chennai, Kochi, Visakhapatnam,, Haldia, Barauni, Guwahati, Digboi and Mathura.

Refining of Crude Oil

The crude oil is first washed with acidic or basic solution as needed. Then, the crude oil is heated in a France to 650 K and the pressure reduced.

The vapours of the crude oil are then fed into a fractionating column through an inlet near its bottom. Fractionating tower is a tall cylindrical steel structure with horizontal trays fitted in it. Each tray has a number of holes and a short tube with a bubble cap. High boiling fraction condenses in the lower section of the tower, while the vapours move up. Due to a regular temperature gradient along the height of the column, the fraction with different boiling points get liquefied at different heights, and can be tapped out (Fig. 4.23).

Major products obtained from the fractional distillation of crude oil are listed.

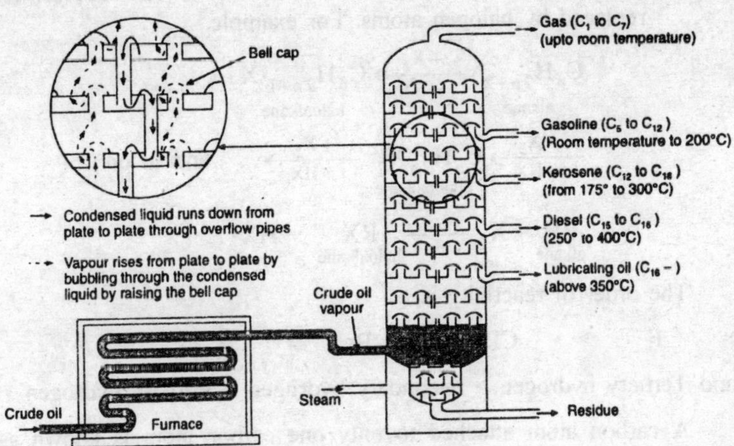

Fig. 4.23 : Fractional distillation of crude oil. The cutway view shows how ractionation of nore volatile components takes place.

GENERAL CHEMICAL PROPERTIES OF ALKANES

Under ordinary conditions, alkanes are quite inert that is, alkanes do not react with]many regents. This is because,

(*i*) They do no have any reactive functional group.

(*ii*) C – C and C – H sigma bonds are strong and require large amount of energy to break.

(*iii*) C – C and C – H bonds are essentially non-polar and hence electrophilic or nucleophilic substitution reactions do not take place.

However, alkanes undergo substitution reactions in which one or more hydrogen atoms of the alkane molecule are replaced by an equal number of other atoms or groups. The products so formed are termed as the substitution products. Some typical reactions of alkanes are described below.

Substitution Reactions

Alkanes given the following substitution reactions,

(*i*) *Halogenation :* When alkanes react with halogens (chlorine or bromine) in the presence of sunlight or in the presence of a halogen carrier (iron), the H-atoms of alkanes are successively replaced by halogen atoms. For example,

$$\underset{\text{alkane}}{C_nH_{2n+2}} \xrightarrow[-HX]{+X_2} \underset{\text{haloalkane}}{C_nH_{2n+1}X}$$

$$\xrightarrow[-HX]{+X_2} \underset{\text{dihaloalkane}}{C_nH_{2n}X_2} \xrightarrow[-HX]{+X_2} \ldots \quad \text{and so on}$$

$$\underset{\text{alkane}}{RH} + X_2 \longrightarrow \underset{\text{haloalkane}}{RX} + HX$$

The order of reactivity is,

F > Cl > Br > I

and Tertiary hydrogen > Secondary hydrogen > Primary hydrogen

A carbon atom attached to only one carbon atom is known as primary carbon atom and its three hydrogen atoms are known as primary hydrogen atoms. Similar, carbon atoms attached to two or three carbon atoms are known as secondary and tertiary carbon atoms respectively. The hydrogen atoms attached to these secondary and tertiary carbon atoms are respectively known as secondary and tertiary hydrogen atoms.

In the case of higher alkanes a mixture of all possible isomeric haloalkanes (monohalids) is obtained. The proportional amounts of the isomers formed depend upon the order of reactivity of H atoms.

The effect of halogens on the process of halogenation of alkanes is described below.

(a) *With Fluorine : Fluorine* reacts with alkanes violently and yield hydrogen fluoride and carbon. For example, with methane

$$\underset{\text{methane}}{CH_4} + 2F_2 \longrightarrow C + \underset{\text{hydrogen fluride}}{4HF}$$

However, controlled fluorination in the presence of cobalt moderator, fluoro derivatives are formed.

(b) *With Chlorine : Chlorine* in the presence of bright sunlight reacts almost explosively with alkanes, *e.g.*, with methane

$$CH_4 + 2Cl_2 \xrightarrow[\text{sunlight}]{\text{bright}} C + 4HCl$$

However, in the presence of diffused sunlight, hear or halogen carriers, chlorine reacts with alkanes to given chloro derivatives of alkanes. The final product depends upon the amount of chlorine used. For example, when methane is allowed to react with chlorine in the presence of diffused sunlight, the four H atoms of methane are successively replaced by chlorine atoms.

$$\underset{\text{methane}}{CH_4} + Cl_2 \xrightarrow{\text{sunlight}} \underset{\text{chloromethane}}{CH_3Cl} + HCl$$

$$CH_3Cl + Cl_2 \longrightarrow \underset{\text{dichloromethane}}{CHCl_2} + HCl$$

$$CH_2Cl_2 + Cl_2 \longrightarrow \underset{\substack{\text{trichloromeahtne} \\ \text{(chloroform)}}}{CHCl_3} + HCl$$

$$CHCl_3 + Cl_2 \longrightarrow \underset{\substack{\text{tetrachloromethane} \\ \text{(carbon tetrachloride)}}}{CCl_4} + HCl$$

Chlorination of isobutane at 300°C gives a mixture of two isomeric haloalkanes.

$$CH_3-\underset{\underset{CH_3}{|}}{CH}-CH_3 + Cl_2 \xrightarrow[-HCl]{300^\circ} CH_3-\underset{\underset{CH_3}{|}}{CH}-CH_2Cl + CH_3-\underset{\underset{CH_3}{|}}{CCl}-CH_3$$

(c) *With bromine :* Bromine reacts with alkanes in the presence of diffused light/halogen carriers to give bromo derivatives of alkanes similar to chlorine but less readily.

(d) *With iodine :* Iodine reacts with alkanes slowly and reversibly. For example, methane with iodine gives,

$$\underset{\text{methane}}{CH_4} + I_2 \rightleftharpoons \underset{\text{iodomethane}}{CH_3I} + HI$$

Iodination, therefore, is carried out in the presence of an oxidising agent such as, iodic acid (HIO_3), nitric acid or mercuric oxide etc., which oxidise the hydrogen iodide formed during the reaction and permits the reaction to go in the forward direction.

$$\underset{\substack{\text{produced in} \\ \text{the iodination step}}}{5HI} + HIO_3 \longrightarrow 3I_2 + 3H_2O$$

Mechanism of halogenation : Chlorine and bromine react with alkanes through the chain *mechanism*. It involves the follcwing steps.

(a) *Chain initiation step.* In this step, halogen molecule absorbs a photon of suitable radiation from the sunlight (visible region of the electromagnetic radiation) and undergoes homolytic split producing free chlorine atoms.

$$Cl_2 \xrightarrow{hv} \underset{\text{free chlorine atoms}}{2Cl^{\cdot}}$$

or $$CH_3 + {}^*Cl_2 \longrightarrow \underset{\text{chloromethane}}{CH_3Cl} + Cl^*$$

(b) *chain propagation step. The* free halogen atoms produced in step (a) attack the alkane molecule (say CH_4 molecule) to produce hydrogen halide and an alkyl free radical. The alkyl radical so produced is highly reactive and immediately reacts with another halogen molecule to produce haloalkane and releasing free halogen atm. The free halogen atoms so produced then react with another molecule of alkane and the process continuous in a cyclic (chain-like manner).

$$CH_4 + {}^*Cl \longrightarrow \underset{\text{free aradical}}{CH_3} + HCl$$

$$CH_3 + {}^*Cl_2 \longrightarrow \underset{\text{chloromethane}}{CH_3Cl} + Cl^*$$

These reactions proceed to produce more and more of the product. However, depending upon the amount of chlorine in the reaction mixture higher halogen derivatives may also be produced during the halogenation. Thus,

$$\underset{\text{chloromethane}}{CH_3Cl} + \underset{\text{from atom of chlorine}}{Cl} \longrightarrow \underset{\text{chloromethyl radical}}{CH_2Cl} + HCl$$

$$CH_2Cl + Cl_2 \longrightarrow \underset{\text{dichloromethane}}{CH_2Cl_2} + HCl$$

$$CHCl_2 + {}^*Cl \longrightarrow \underset{\text{dichloromethyl radical}}{CHCl_2} \quad HCl$$

$$^{\cdot}CHCl_2 + Cl_2 \longrightarrow \underset{\text{trichloromethane (chloroform)}}{CHCl_3} \quad {}^*Cl$$

$$CHCl_3 + {}^*Cl \longrightarrow \underset{\text{dichloromethyl radical}}{{}^*CCl_3} \quad HCl$$

$$^{*}CCl_3 + Cl_2 \longrightarrow \underset{\substack{\text{tetrachloromethane}\\(\text{carbon tetrachloride})}}{CCl_4} + {}^{*}Cl$$

(*c*) *Chain termination step.* The above mentioned chain reactions do not continue indefinitely. The chain reaction comes to an end when either the free radicals get deactivated or combine with each other to from a stable (relatively non-reactive) molecule. Such reactions are termed *chain-termination steps.*

In this reaction, the following reactions may lead to the chain termination.

$$Cl^{\cdot} + CH_3 \longrightarrow \underset{\text{chloromethane}}{CH_3Cl}$$

$$^{\cdot}CH_3 + {}^{\cdot}CH_3 \longrightarrow \underset{\text{ethane}}{CH_3 - CH_3}$$

$$Cl^{\cdot} + Cl^{\cdot} \longrightarrow Cl_2$$

2. *Nitration :* Alkanes when made to react with nitric acid under suitable conditions give substitution product in which the nitro $(-NO_2)$ group replaces the H-atom in the alkane molecule, viz.,

$$\underset{\text{alkane}}{RH} + HNO_3 \rightarrow \underset{\text{nitroalkane}}{R - NO_2} + H_2O$$

The compounds RNO_2 so obtained are termed **nitroalkanes**. The process of introducing NO_2 group in the molecule in called **nitration**. Higher members can be nitrated by heating with con. HNO_3. The lower members are attacked only by nitric acid vapours at higher temperatures (150 – 475°C) producing a mixture of all possible mononitro derivatives. For example, propane forms a mixture of mononitro derivatives, *e.g.*,

$$\underset{\text{propane}}{CH_3.CH_2CH_3} \xrightarrow[420°C]{NHO_3} \underset{\text{1-nitropropane}}{CH_3.CH_2CH_2NO_2}$$

$$+ \underset{\text{2-nitropropane}}{CH_3 \cdot \underset{|}{\overset{NO_2}{CH}} \cdot CH_3} + \underset{\text{nitroethane}}{C_2H_5NO_2} + \underset{\text{nitromethane}}{CH_3NO_2}$$

The liquid-phase nitration of alkanes is slow and generally polynitro compounds are obtained.

Mechanism of nitration : Vapour-phase nitration is believed to proceed by a free-radical mechanism. The details are, however uncertain.

A probable mechanism for the nitration of ethane is,

$$CH_3 - CH_3 + NO_2 \longrightarrow CH_3 - CH_2 \quad + HNO_2$$

$$\underset{\text{ethyl free radical}}{C_3 - CH_2} + NO_2 \longrightarrow \underset{\text{nitroethane}}{CH_3 - CH_2NO_2} + CH_3 - CH_2ONO_2$$

$$CH_3 - CH_2ONO \longrightarrow CH_3 - CH_2O + NO$$

$$CH_3 - CH_2O \longrightarrow CH_3 + CH_2O$$

$$CH_3 + NO_2 \longrightarrow \underset{\text{nitromethane}}{CH_3NO_2}$$

(3) *Sulphonation :* Conc H_2SO_4 does not react with alkanes even at elevated temperature. Higher alkanes, from hexane onwards, react with fuming sulphuric acid at high temperature to form alkane sulphonic acids.

$$\underset{\text{alkane}}{RH} + \underset{\text{sulphuric acid}}{HOSO_3H} \xrightarrow{400^\circ} \underset{\text{alkanesulphonic acid}}{R.SO_3H} + H_2O$$

$$\underset{\text{hexane}}{C_6H_{14}} + \underset{\text{oleum}}{H_2SO_4} \xrightarrow{400^\circ} \underset{\text{hexanesulphonic acid}}{C_6H_{13}.SO_3H} + H_2O$$

The sulphonation of n-alkanes is slow, but the branched chain alkanes undergo supohonation readily. The order of reactivity is,

Tertiary hydrogen > Secondary hydrogen > primary hydrogen

Oxidation

Alkanes undergo oxidation yielding products which depend upon the reaction conditions. Some of the oxidation reactions are described below.

(a) *Combustion :* Alkanes are highly combustible and burn readily with a non-luminous flame in excess of air (or oxygen) to give CO_2 and H_2O. The combustions being highly exothermic. It is due to this reaction that the alkanes are used as good fuels (LPG used for cooking purposes is in fact a mixture of lower alkanes). For example,

$$C_nH_{2n+2} + \left(\frac{3n+1}{2}\right) O_2 \longrightarrow nCO_2 + (n+1)\, H_2O + \text{heat}$$

For methane and ethane, the reactions are,

$$CH_4(g) + 2O_2 \longrightarrow CO_2(g) + 2H_2O(g) + 802.5 \text{ kJ}$$

$$2C_2H_6(g) + 7O_2 \longrightarrow 4CO_2(g) + 6H_2O(g) + 2856.5 \text{ kJ}$$

The heat liberated per unit mass of the fuel (the substance undergoing combustion) is termed as the calorific value of the fuel. The commercial value of a fuel is based on its calorific value.

(b) *Slow combustion* : Lower alkanes, such as, methane and ethane when treated with oxygen at high pressure and low temperature undergo slow combustion yielding alcohol, aldehyde, acid and finally carbon dioxide and water. For example, methane gives,

$$\underset{\text{methane}}{CH_4(g)} \xrightarrow{[O]} \underset{\substack{\text{methanol} \\ \text{(methyl alcohol)}}}{CH_3OH} \xrightarrow{[O]} \underset{\text{unstable}}{[CH_2(OH)_2]}$$

$$\xrightarrow[-H_2O]{} \underset{\substack{\text{methanal} \\ \text{(formaldehyde)}}}{HCHO} \xrightarrow{[O]} \underset{\substack{\text{methanoic acid} \\ \text{(formic acid)}}}{HCOOH} \xrightarrow{[O]} CO_2(g)\ H_2O(l)$$

(c) *Catalytic oxidation* : Lower alkanes on oxidation in the presence of a catalyst give alcohols, and aldehydes. Higher alkanes give higher fatty acids. For example, methane gives different products under different conditions.

(i) $$\underset{\text{(9 : 1 mixture)}}{CH_4(g) + O_2} \xrightarrow[\text{Cu, 200°C}]{\text{100 atm}} \underset{\text{methanol}}{CH_3OH}$$

(ii) $$CH_4(g) + O_2 \xrightarrow[\text{400°C, 200 atm}]{\text{molybdenum oxide}}$$

$$\underset{\text{methanol}}{CH_3OH} \longrightarrow \underset{\text{methanal (formaldehyde)}}{HCHO}$$

(iii) $$\underset{\text{higher alkane}}{2RCH_3} + \underset{\text{from air}}{3O_2} \xrightarrow[\text{100--160°C}]{\text{Mn-acetate}} \underset{\text{fatty acid}}{2RCOOH} + 2H_2O$$

(d) Alkanes are normally resistant towards oxidising agents like $KMnO_4$. However, such oxidising agents oxidise a tertiary hydrogen atom to a hydroxyl group. For example,

$$\underset{\text{isobutane}}{CH_3-\underset{\underset{CH_3}{|}}{\overset{\overset{CH_3}{|}}{C}H}} + [O] \longrightarrow \underset{\text{tert. butyl alcohol}}{CH_3-\underset{\underset{CH_3}{|}}{\overset{\overset{CH_3}{|}}{C}}-PH}$$

Thermal Decomposition

Breaking down of higher alkanes into lower alkanes by heating is termed thermal decomposition. This process is also called as pyrolysis or cracking. In actual practice, the vapours of higher alkanes are passed through a hot metal tube maintained at 500-700°C. For example, n-butane gives.

$$CH_3—CH_2—CH_2—CH_3 \xrightarrow[500\text{-}700°C]{heat} \begin{cases} CH_3—CH{=}CH—CH_3 + H_2 & \text{(but-2-ene)} \\ CH_4 \text{ (methane)} + C_3H_6 \text{ (propane)} \\ C_2H_6 \text{ (ethane)} + C_3H_4 \text{ (ethane)} \end{cases}$$

n-butane

Propane gives,

$$CH_3—CH_2—CH_3 \xrightarrow{heat} \begin{cases} C_3H_6 \text{ (propane)} + H_2 \\ CH_4 \text{ (methane)} + C_2H_4 \text{ (ethane)} \\ 2CH_4 + C \end{cases}$$

propane

The process of cracking of great industrial importance. Large quantities of high-boiling fractions (not of much use as such) of petroleum are converted into low-boiling gasoline by cracking.

Isomerisation

The straight chain alkanes when heated with aluminum chloride in the presence of dry HCl at 300°C given branched chain isomers. For example, n hexane gives,

$$CH_3—CH_2—CH_2—CH_2—CH_2—CH_3 \xrightarrow[300°C]{AlCl_3/HCl} \begin{cases} CH_3—CH(CH_3)—CH_2—CH_2—CH_3 & \text{(2-methylpentane)} \\ CH_3—CH_2—CH(CH_3)—CH_2—CH_3 & \text{(3-methylpentane)} \end{cases}$$

n-hexane

This reaction is of vital importance in petroleum industry. The branched chain alkanes have higher octane number (anti-knock properties).

AROMATISATION

Alkanes containing six or more carbon atoms, when heated under pressure in the presence of suitable catalysts get cyclised to give aromatic compounds. For example, n-hexane give benzene,

$$\underset{n\text{-hexane}}{CH_3{-}CH_2{-}CH_2{-}CH_2{-}CH_2{-}CH_3} \xrightarrow[500°C,\ 10\text{-}20\ atm]{Cr_2O_3/Al_2O_3} \underset{\text{benzene}}{C_6H_6} + 4H_2(g)$$

(benzene drawn as ring: C–H, HC, CH, HC, CH, C–H)

n-heptane gives toluene,

$$\underset{n\text{-heptane}}{CH_3{-}(CH_2)_5{-}CH_3} \xrightarrow[500°C]{Cr_2O_3/Al_2O_3} \underset{\text{toluene}}{C_6H_5CH_3} + 4H_2(g)$$

(toluene drawn as ring: CH_3–C, HC, CH, HC, CH, C–H)

Uses of Alkanes

Alkanes are the simplest organic compounds containing carbon and hydrogen. Some major uses of these compounds are,

(i) Lower alkanes occurring as natural gas and lighter petroleum fractions are used as fuels.

(ii) Low-boiling liquid alkanes *e.g.*, hexane, are used as solvents.

(iii) Heavy petroleum fractions are used as lubricants (grease), and for obtaining waxes and vaseline.

(iv) The products of cracking process are generally used for producing linearalkyl benzenes, (LAB) used as the raw material for manufacturing detergents.

Conformations of Alkanes

The carbon-carbon single bond (sigma bond) in alkanes is cylindrically symmetrical about the internuclear axis (line joining the two carbon nuclei). *The symmetry of the bond permits free rotation of one methyl group relative to others about the bond axis without breaking the bond.* The rotation about a sigma bond is not completely free. In fact, it is hindered by an energy barrier of 1–20 kJ/mol. The repulsive interaction between the neighboring bounding electron pairs is called *torsional strain.* As a result, a number of different spatial arrangements of the atoms or groups attached to the carbon atoms are possible.

The different structures (arrangements) which can be produced by the rotation about the carbon-carbon single bond are called *conformers.* The molecular geometry corresponding to a conformer is known as *conformation.*

Conformers are usually represented using either the *Sawhorse projection* or the *newman projection.*

Conformations of Ethane

In ethane ($CH_3 - CH_3$) molecule, the two carbon atoms are bounded together by a single bond. Due to the free rotation about the carbon-carbon bond an infinite number of arrangements of hydrogen atoms attached to one carbon atom with respect to the hydrogen atoms on the other carbon atom are possible. Thus, in ethane an infinite number of conformations are possible. However, only two conformations in between these two extreme conformations are called *gauche* or *skew*. These are shown in Fig. 15.4, and described below.

(i) *Eclipsed from :* The arrangement in which the hydrogen atoms of one carbon lie exactly behind those of the other is called eclipsed from.

(ii) *Staggered from :* The arrangement in which the hydrogen atoms of the nearer methyl group are seen exactly between those of the other methyl group, *i.e.*, the hydrogen atoms are staggered with respect to each other, is called staggered from.

These conformations are represented by Sawhorse, and Newman projections. The Sawhorse projections of the eclipsed and the staggered forms of ethane.

Newman projection of the conformations : Newman projection is easier to draw and visualise. To convert a conformer from a Sawhorse (or structural) from to the Newman from, the following procedure is adopted :

(a) The two carbons bonded by a single bond are represented by a small circle, the front carbon covering the second carbon lying exactly behind it.

(b) The C—H bonds of the front (nearer) carbon are shown by solid lines starting from the centre of the circle.

(c) The C—H bonds of the real carbon (carbon behind the front carbon) are shown by dotted lines starting from the circumference of the circle.

The Newman projection for the eclipsed and the staggered forms of ethane molecule are shown in Fig. 4.24.

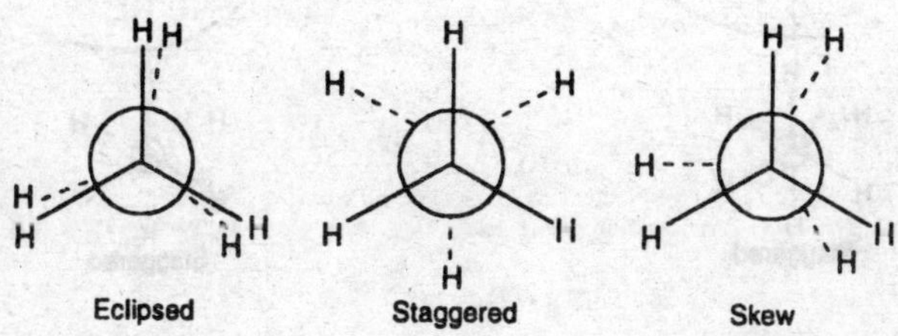

Fig. 4.24 : Eclipsed and Staggered conformations of ethane.

Relative Stabilities of Eclipsed and Staggered Conformations of Ethane

The various conformations of a compound posses identical physical characteristics but differ in energy.

In the eclipsed conformation, all the C—H bonds of one methyl group are aligned with those of the second, and in the staggered from the two methyl groups are rotated as far away as possible from one another.

Thus, the H-atoms on the two carbon atoms are closer in the eclipsed from than in the staggered from. So , the repulsive force acting between the two hydrogen atoms in eclipsed from is more than that in the staggered form.

As a result, the energy of the eclipsed form is higher than that of the staggered form of ethane. Thus, the staggered conformation of ethane is more stable than the eclipsed form. The order of stability for different conformations is

Staggered > Skew (or Gauche) > Eclipsed

The staggered conformer of ethane is more stable than the eclipsed conformer by about 12.6 kJ mol^{-1}. As the energy difference is small, there is no effective energy barrier to the rotation about a sigma bond. Thus, we can say that in ethane, there is free rotation about a C—C sigma (s) bond.

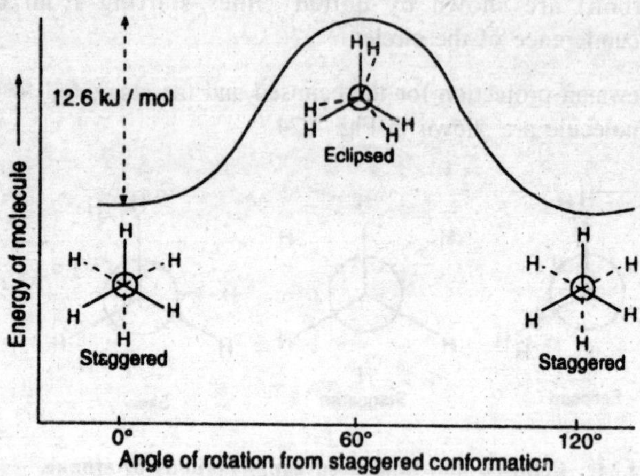

Fig. 4.25 : Degree of rotation from the Staggered form and energy of the molecule of ethane.

At room temperature, the eclipsed and staggered forms of ethane interconvert rapidly and cannot be isolated as separate conformers.

CRACKING OF HYDROCARBONS

The process by which higher hydrocarbons having high boiling points are broken into a variety of lower hydrocarbons, which are more volatile (low boiling), is called **cracking** (or sometimes as **pyrolysis).** For example, a higher hydrocarbon $C_{10}H_{22}$ splits according to the reaction

$$\underset{\text{higher hydrocarbon}}{C_{10}H_{22}} \xrightarrow[770\text{ K}]{\text{heat}} \underset{\text{lower hydrocarbons}}{C_8H_{18} + C_2H_4}$$

The process of cracking, thus i cracking, carbon-carbon bonds get its are formed during cracking, carbon-carbon bonds get broken in random manner. So various kinds of products are formed during cracking.

Creaking is generally carried out in the following ways.

Thermal cracking : Breaking down large molecules by heating at high temperature and pressure termed thermal cracking. Thermal cracking can be further classified into the following classes.

(*a*) *Liquid phase thermal cracking* : In this process, the higher boiling lubricating oil are converted into low boiling fractions by temperature of 750 K, and under a pressure of about 10.

(*b*) *Vapour phase thermal cracking* : In this process, low boiling fraction, *e.g.*, kerosene is cracked in the vapour phase at a temperature of about 875 K and under a atmosphere.

(*i*) *Catalytic cracking* : Higher hydrocarbons can be cracked at lower (600-650 K) and lower pressure (2 atm.) in the presence of a suitable catalyst. A typical catalyst used for this purpose is a mixture of silica (SiO_2), 4 parts; alumina (Al_2O_3), 1 part, and manganese dioxide (MnO_2), 1 part. Catalytic cracking produces gasoline of higher octane number and, therefore, this method is used for obtaining better quality gasoline.

(*ii*) *Steam cracking* : In this process, higher hydrocarbons in the vapour phase are mixed steam, heated for a short duration to about 900°C, and cooled rapidly. The process is suitable for obtaining lower unsaturated hydrocarbons.

APPLICATIONS OF CRACKING

Of the various products obtained hi straight-run refining, the most important are petrol, diesel and kerosene. The demand for these products is more than what is obtained during refining of the petroleum. On the other hand, high-boiling fractions find lesser use as such. So,

(i) In petroleum industry, the yield of the low boiling (lower molecular mass) fractions, such as gasoline is increased by the cracking of less useful high boiling fractions.

For example, large molecules hi the lubricating oil range (C_{18} to C_{21}) can be converted into more useful molecules in the petrol (gasoline) range (C_6 to C_9) by cracking process.

$$\underset{\text{hexadecane}}{C_{16}H_{34}} \xrightarrow{\text{cracking}} \underset{\substack{\text{octane} \\ \text{(alkanc)}}}{C_8H_{18}} + \underset{\substack{\text{oct-1-ene} \\ \text{(alkane)}}}{C_8H_{16}}$$

(ii) Cracking always yields low boiling alkenes as the by-products. These unsaturated hydrocarbons called petrochemicals, are converted into a variety of useful compounds such as, polyethylene etc.

REFORMING, OR AROMATISATION

The process of converting open chain (aliphatic) hydrocarbons and/or cycloalkanes into aromatic hydrocarbons (arenes) containing the same number of carbon atoms in the presence of a catalyst is called **reforming** or **aromatisation**. Reforming (or aromatisation) involves reactions of the type, n, cyclisation, and isomerisation.

In reforming (or aromatisation), cyclic and acyclic alkancs containing six to eight| atoms are heated at about 670 K in the presence of palladium, platinum or nickel as i Platinum seems to *be* the best catalyst and so the process is sometimes called **platforming**.

Following are the two examples of reforming.

Hexane when passed over Cr_2O_3 supported over alumina at 670 K undergoes cyclization and finally to aromatisation, forming cyclohexane and benzene respectively.

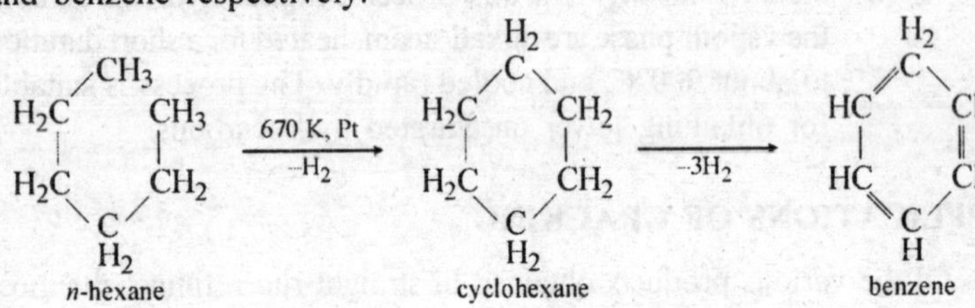

$$\text{n-hexane} \xrightarrow[-H_2]{670\text{ K, Pt}} \text{cyclohexane} \xrightarrow{-3H_2} \text{benzene}$$

Under similar conditions, *n*-heptane yields methylcyclohexane and finally toluene.

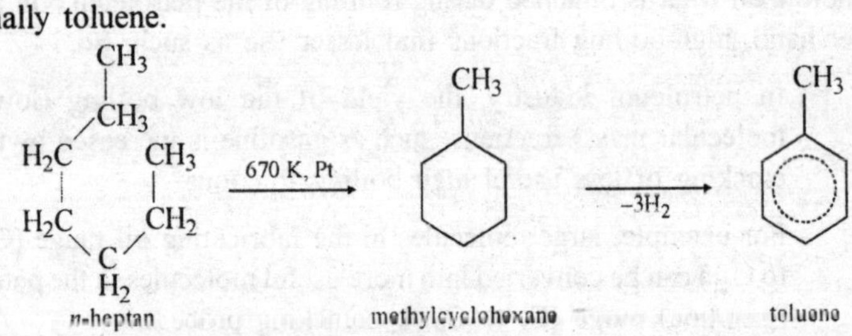

$$\text{n-heptan} \xrightarrow{670\text{ K, Pt}} \text{methylcyclohexane} \xrightarrow{-3H_2} \text{toluene}$$

Fig. 4.26

Applications of Reforming

The process of reforming has been employed for,

(i) Producing aromatic hydrocarbons (arenes) like benzene, toluene and xylenes. Prior to the development of petroleum sources, the bulk of these arenes were obtained from coal tar. But now they are derived from petroleum by the process of reforming.

(ii) Increasing the proportion of aromatic hydrocarbons in gasoline to increase the octane number of straight-run gasoline. Aromatic hydrocarbons have octane numbers higher than 100, while the straight-run gasoline has the octane number of 55.

The straight-run gasoline (octane number 55) is first distilled to obtain a fraction containing n-hexane, cyclohexane and methyl-cyclopentane. This fraction is then heated with a platinum catalyst to convert the cyclohexane into benzene and toluene (octane number > 100).

FLASH POINT OF A HYDROCARBON

The explosive nature of a hydrocarbon is determined by its volatility. The volatility of any liquid depends upon the temperature.

"The lowest temperature at which a liquid hydrocarbon gives off enough vapour to form an explosive mixture with air is termed as its flash point."

The flash point of any liquid hydrocarbon is so adjusted that it remains safe under the conditions of its use. For example, the minimum flash point permitted in India is 44°C, while in France it is 33 °C and in Britain it is 22°C.

Determination of Flash Point

The flash point of a combustible liquid is determined by using Abel's apparatus shown in Fig. 4.27. The given sample of the combustible liquid, *e.g.*, kerosene, petrol, diesel etc., is taken in a small metal cup and heated slowly through a surrounding water bath. From time to time, the slit in the cover of the metal cup is opened, and a tiny flame introduced momentarily by means of an automatic device. The temperature at which the vapours get ignited with a minute explosion is taken as the flash point of the liquid.

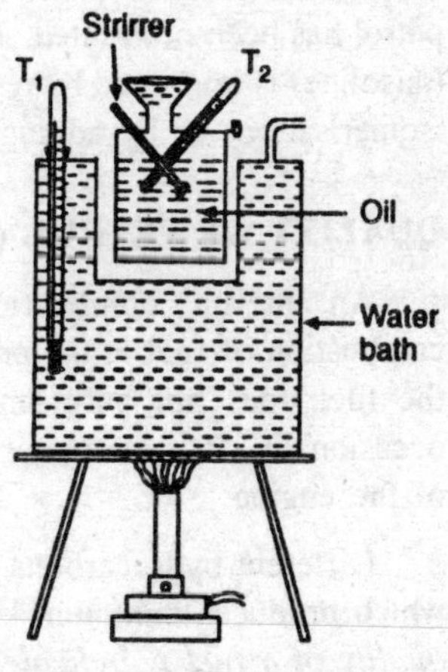

Fig. 4.27

KNOCKING OF FUELS

An internal combustion engine works with a system of pistons^ In the down-stroke of the piston, a mixture of air and gasoline vapour is drawn from the carburetor into the cylinder, and in the upstroke, the mixture is compressed. The ratio of the initial volume to the final volume is called the compression ratio. At the end of the upstroke of the piston, a spark ignites the compressed air-gasoline mixture. As the gases bum, they expand and the flame front moves in a smooth manner and supplics power to the engine.

For maximum efficiency of the engine, a high compression ratio (generally seven to eight) is required. But, as the compression ratio increases, the gasoline-air mixture burns in an explosive manner and produces a metallic sound. This sound is called knocking, and indicates inefficient performance of the gasoline. High compression ratio engines also require less fuel, so gasoline having less knocking tendencies are very valuable.

It has been found that the knocking tendency of the fuels falls off with the nature of the fuel as follows.

Straight chain alkanes > Branched chain alkanes > Alkenes

Knocking may also be prevented or minimised by adding compounds such as tetraethyl lead (TEL) to gasoline. Such compounds are called anti-knocking agents. Due to toxic effects of lead, the use of TEL in petrol has been prohibited. These days, the knocking behavior of petrol (gasoline) is improved by reforming process (cyclisation, aromatisation, isomerisation) or by adding iso compounds.

QUALITY OF PETROL (GASOLINE)–OCTANE NUMBER

An internal combustion engine gets its power by controlled combustion of fuel in the presence of air. If a low quality fuel is used, the fuel does not burn smoothly. An explosive sound is produced occasionally. It is known as knocking. This greatly reduces the power of the engine.

Different hydrocarbons have different knocking tendencies. A fuel which produces minimum knocking is considered as a good fuel. *The quality of a fuel is indicated in terms of its octane number*

. An arbitrary scale of octane number has been set up with n-heptane and 2,2,4-trimethylpentane (iso-octane) as the reference compounds.

2,2,4-trimethylpentane (generally called iso-octane) has excellent anti-knocking properties. So, iso-octane was arbitrarily assigned an octane number of 100, whereas n-heptane which is very prone to knocking was assigned an octane number of zero (0).

$$CH_3-CH_2-CH_2-CH_2-CH_2-CH_2-CH_3$$

n-hep tan e
Oc tan enumbe. zero

$$CH_3-\underset{\underset{CH_3}{|}}{\overset{\overset{CH_3}{|}}{C}}-CH_2-\overset{\overset{CH_3}{|}}{CH}-CH_3$$

iso-octane (2,2,4-trimethylpentane)
Octane number: 100

All fuels are graded in between these two limiting values by comparing with a suitable mixture of the above two compounds. Thus, *the octane number of any fuel is defined as the percentage of iso-octane in a mixture of iso-octane and n-heptane that has the same knocking as the fuel under examination.* Thus, a fuel having an octane number of 80 behaves in a i mixture having 80% of iso-butane, and 20% of n-heptane. Straight run gasolines may have octane values ranging from 20 to 73. Aviation fuel is rated as 100 octane.

Therefore, the antiknock property of a fuel increases with the increase in its i Branched-chain alkanes, alkenes and aromatic hydrocarbons have higher compounds having higher octane numbers are added to straight-run gasoline to raise Its octane number. These high octane compounds, viz., branched-chain alkanes, alkenes, and hydrocarbons are prepared by catalytic cracking and catalytic reforming. *The octane number* decreases *with an increase in chain length.*

Octane ratings of some compounds are given below.

n-Heptane	0	Methanol	116
n-Pentane	62	Toluene	118
tert-butyl alcohol	98	Methane	122
iso-octane	100	Ethane	101
Benzene	106	Propane	96
Ethanol	112	Butane	89

Recently, the hydrocarbon, 2,2,3-trimethylpentane, called **triptane** has been assigned an octane number 125.

QUALITY OF DIESEL–CETANE NUMBER

The quality of diesel is expressed in terms of its **cetane number**. Cetane number is based on a scale in which

Hexadecane ($C_{16}H_{34}$) called cetane, is given a cetane number of 100

1-methylnaphthalene (C_2H_{10}) is given a cetane number of 0.

The cetane number is defined as the percentage of hexadecane (cetane) by volume in a mixture of hexadecane and 1-methylnaphthalene, which has the same ignition quality as the fuel under testing.

SYNTHETIC PETROL (OR GASOLINE)

Mixture of alkanes resembling petroleum (crude oil) can be obtained from coal by two methods, namely the Fischer-Tropsch process and the Bergius process.

These processes were developed in Germany during World War II, when is petroleum supplies were cut off. Germany produced considerable amounts of fuel from coal by the above processes during that period.

These process have once again drawn much attention due to the existing uncertainties in the world oil markets.

Bergius process : In this process, powdered coal is mixed with heavy oil and heated with hydrogen under high pressure (200-250 atm) and at about 748 K in presence of iron oxide as catalyst.

$$\underset{\text{from coal}}{nC} + (n+1)\,H_2\,(g) \xrightarrow[\text{200 – 250 atm, 748 K}]{\text{iron oxide}} \underset{\text{(mixture of hydrocarbons)}}{C_nH_{2+2}}$$

The vapours on condensation give a liquid resembling crude oil. This is called synthetic petroleum, which on fractional distillation gives petrol (gasoline).

Fisher-Tropsch process : In this process, a mixture of water gas and hydrogen under pressure (5 = 10 atm) is passed over a cobalt catalyst at 450-475 K. The water gas required is obtained by passing steam over red hot coke.

$$C\text{ (red hot)} + H_2(g) \rightarrow \underset{\text{water gas}}{CO + H_2}$$

$$nCO + (2n+1)\,H_2 \xrightarrow[\text{5–10 atm, 450–475 K}]{\text{Co catalyst}} C_nH_{2n+2} + nH_2O$$

The petroleum-like product obtained in the process in fractionally distilled to obtain petrol, middle oil and heavy oil. The middle oil fraction is hydrogenated again to produce more petrol.

PETROCHEMICALS AND THEIR APPLICATIONS

Petroleum is a mixture of a large number of saturated hydrocarbons. Many lower saturated and unsaturated hydrocarbons can be produced by cracking process. These lower saturated and unsaturated hydrocarbons are very important raw materials for many useful organic compounds.

Various low molecular mass organic compounds prepared from the lower hydrocarbons obtained from natural gas of/and petroleum, are termed petrochemicals.

Examples:

Some important petrochemicals are,

(i) Methane (CH_4)

(ii) Methanol (methyl alcohol, CH_3OH)

(iii) Ethanol (ethyl alcohol, C_2H_5OH)

(vi) Ethene (ethylene, $CH_2 = CH_2$)

(v) Propene (propylene, $CH_3 - CH = CH_2$)

(vi) Butene (butylene, C_4H_8

(vii) Butadiene (C_4H_6)

(viii) Benzene (C_6H_6)

(ix) Toluene ($C_6H_5CH_3$)

(x) Xylene (C_6H_4 $(CH_3)_2$)

Industrial Uses of Petrochemicals

All petrochemicals are very good raw materials for the preparation of a large number of useful organic compounds such as,

(i) Fertilizers (*e.g.*, urea)

(ii) Insecticides (*e.g.*, DDT, BHC etc).

(iii) Synthetic detergents (*e.g.*, linear alkyl benzenesulphonate)

(iv) Synthetic polymers (*e.g.*, PVC, polythene, polypropene etc).

(v) Synthetic fibres (*e.g.*, polyester)

(vi) Synthetic rubbers (*e.g.*, neorpene)

(viii) Dyes, varnishes and drugs etc.

PRODUCTION OF PETROCHEMICALS

Petrochemicals are obtained from various fractions of petroleum by using process such as, cracking, cyclisation, oxidation, polymerisation, halogenation etc.

Major hydrocarbon petrochemical obtained from petroleum refineries are shown, below.

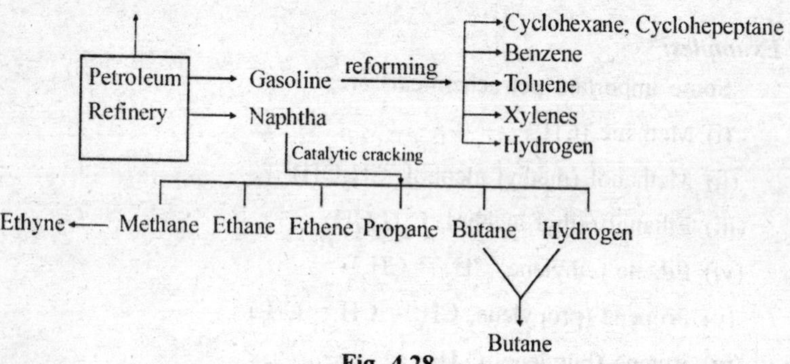

Fig. 4.28

Some Industrially Important Chemicals from Methane and Ethane

Some important chemicals that act as the raw material from manufacturing many industrially important products are obtained from methane and ethane as shown in the following flow sheet charts.

$CH_4 \xrightarrow{H_2O \text{ (steam)}} CO + H_2$ (synthesis gass)

$CH_4 \xrightarrow{Al_2O_3/NH_3} HCN + 3H_2$

$CH_4 \xrightarrow{\text{catalyst, } (O_2)} CH_3OH \xrightarrow{(O)} HCHO \xrightarrow{(O)} HCOOH$

$CH_4 \xrightarrow{Cl_2\ h\nu} CH_3Cl \xrightarrow[-HCl]{Cl_2} CH_2Cl_2 \xrightarrow[-HCl]{Cl_2} CHCl_3 \xrightarrow[-HCl]{Cl_2} CCl_3 \xrightarrow{HF} CCl_2F_3$ (freon)

$CH_4 \xrightarrow{Ni(1273\ K),\ H_2O} CO + 3H_2$

$CH_4 \xrightarrow{1273\ K \text{ (limited air)}} C + H_2$ (carbon black)

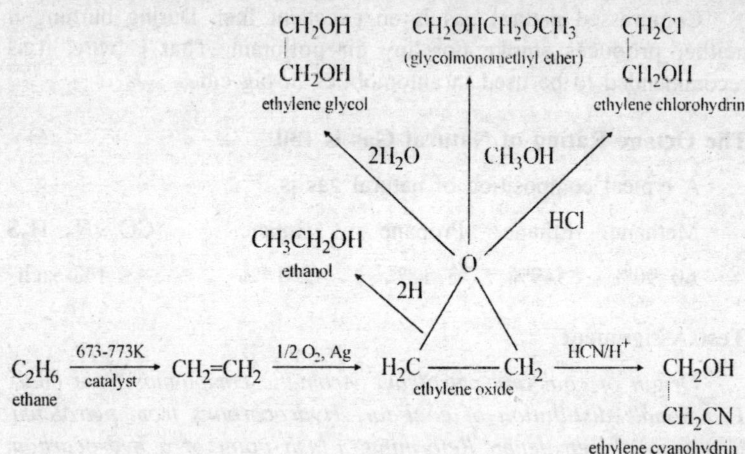

LIQUEFIED PETROLEUM GAS (LPG)

The low boiling fraction of petroleum which leaves the fractionating tower as a gas is known as *petroleum gas*. Petroleum gas is a mixture of ethane, propane and butane. *But the main constituent of petroleum gas is butane (C_4H_{10}).*

Ethane, propane and butane are gases under normal conditions. Butane can be liquefied by applying pressure. *The petroleum gas which is liquefied under pressure is called Liquefied Petroleum Gas (LPG).*

The liquefied petroleum gas (LPG) consists mainly of *butane,* (with a very small amount of ethane and propane). Since the liquefied petroleum gas can be kept in the liquid from only under pressure hence LPG is stored and supplied in gas cylinders. *The gas cylinder we get from the gas agency (Indane, Burshane etc), contains liquefied petroleum gas (LPG).*

LPG is used

(i) As a domestic and industrial fuel

(ii) As a source of low molecular mass petrochemical.

COMPRESSED NATURAL GAS (CNG)

CNG, (Compressed Natural Gas) is highly compressed natural gas. Natural gas occurs deep inside the earth either alone or covering the crude oil in an oil field. Natural gas is a colourless, odourless mixture of methane (the main constituent), ethane, propane and butane.

Compressed natural gas is an excellent fuel. During burning it neither produces smoke nor any air pollutant. That is why, it is recommended to be used in automobiles in big cities.

The Octane Rating of Natural Gas is 130

A typical composition of natural gas is

Methane	Ethane	Propane	Butane	CO_2, N_2, H_2S
60–90%	5–9%	3–18%	2–14%	< 1% each

Test Assignment

Origin of coal and petroleum. Aromatic compounds from coal,. Fractional; distillation of coal tar. Hydrocarbons from petroleum. Cracking of petroleum, Reforming, Flash point of a hydrocarbon, Knocking of fuels, Octane number, Cetane number, Gasoline additives, Petrochemical, LPG, CNG.

General Chemical Properties of the Alkanes

1. *Halogenation :* Chlorination may be brought about by light, hear or catalysts, and the extend of chlorination depends largely on the amount of chlorine used. A mixture of all possible isomeric monochlorides is obtained, but the isomers are formed in unequal amounts, due to the difference of the reactivity of primary, secondary and tertiary hydrogen atoms. Markownikoff (1875) found experimentally that the order of ease of substitution is tertiary hydrogen > secondary > primary, *e.g.*, chlorination of isobutane at 300°C gives a mixture of two isomeric monochlorides.

$$CnH_{2n+2} \xrightarrow[-HX]{+X_2} C_nH_{2n-1}X \xrightarrow[-HX]{X_2} C_2H_{2n}X_2 \xrightarrow[IIX]{X_2} \text{... and so on.}$$

$$\text{(I)}\ \ CH_3CH(CH_3)CH_2Cl \quad \text{and} \quad CH_3CCl(CH_3)CH_3\ \ \text{(II)}$$

Isobutane has on tertiary and nine primary hydrogen atoms. One the basis of equal rates of substitution, the statistical results would be 90 per cent of (1) and 10 per cent of (II). Actually,, the results obtained were 67 percent of (1) and 33 percent of (II). Thus tertiary hydrogen is replaced about 4.5 times as fast as primary hydrogen.

Bromination is similar to chlorination, but not so vigorous. Iodination] is reversible, but it may be carried out in the presence of an oxidising agent, such HIO_3. HNO_3. HgO, etc., which destroys the hydrogen iodide as it is formed and so derives the reaction to the right, *e.g.*,

$$CH_4 + I_2 \rightleftharpoons CH_3I + HI$$

$$5HI + HIO_3 \rightarrow 3I_2 + 3H_2O.$$

Iodides are more conveniently prepared by treating the chloro- or bromo-derivative with sodium, iodide in methanol or acetone solutions. *e.g.*,

$$RCl + NaI \xrightarrow{\text{acetone}} RI + NaCl$$

This reaction is possible because sodium iodide is soluble in methanol or acetone, whereas sodium chloride and sodium bromide are not. This reaction is known as the Finkelstein or Conant-*Finkelstein reaction (1910. 1924).*

Direct fluorination is usually explosive, special conditions are necessary for the preparation of the fluorine derivatives of the alkanes.

2. *Nitration :* Under certain conditions, alkanes react with nitric acid, a hydrogen atom being replaced by a nitro-group, NO_2. This process is known as nitration. Nitration of the alkanes may be carried out in the vapour phase between 150° and 450°C, whereupon a complex mixture of mononitroalkanes is obtained. The mixture consists of all the possible mononitroderivatives and the nitro-compounds formed by every possibility of chain fission of the alkane, *e.g.*, propane gives a mixture of 1-nitropropane 2-nitropropane, nitroethane and nitromethane.

$$CH_3CH_2CH_3 \xrightarrow[400°C]{HNO_3} CH_3CH_2CH_2NO_2 + CH_3\underset{}{\overset{NO_2}{\overset{|}{C}}}HCH_3 + C_2H_5NO_2 + CH_3NO_2$$

As in the case of halogenation, the various hydrogen atoms in propane are not replaced with equal ease.

3. *Sulphonation :* Is the process of replacing a hydrogen atom by a sulphonic acid group. SO_3H. Sulphonation of a normal alkane from hexane onwards may be carried out by treating the alkane with oleum (fuming sulphuric acid) The ease of replacement of hydrogen atoms is. Tertiary very mush greater than secondary,

and secondary greater than primary replacement of a primary hydrogen atom in sulphonation is very slow indeed. Isobutane, which contains a tertiary hydrogen atom, is readily sulphonated to give t-butylsulphonic acid.

$$(CH_3)_3CH + H_2SO_4 \longrightarrow (CH_3)_3CSO_3H + H_2SO_4$$

Sulphuryl chloride, in the presence of light and a catalyst, converts hydrocarbons into sulphonyl chlorides.

I has been shown that in concentrated suphuric acid, hydrocarbons containing a tertiary hydrogen atom undergo hydrogen exchange (Ingold *et al*., 1936). The mechanism is believed to occur via a carbonium ion.

$$R_3CH + 2H_2SO_4 \longrightarrow R_3C^+ + HSO4^- + SO_2 + 2H_2O$$

$$R_3C+ + R_3CH \longrightarrow R_3CH + R_3C+.\ \text{etc.}$$

This reaction is of particular interest since optically active hydrocarbons have been racemised in sulphuric acid *e.g.*,. Burwell *et al*. (194*8) have shown that optically active 3-methyllhepatane is racemised in sulphuric acid.

4. *Oxidation :* All alkanes readily burn in excess of air or oxygen to from carbon dioxide and water. Controlled oxidation, on the other hand, under various conditions, leads to different products. Intensive oxidation gives a mixture of acids consisting of the complete range of C1 to Cn carbon atoms. Less extensive oxidation gives a mixture of products in which no chain fission has occurred. Under moderate conditions, mixed ketones are the major products, and oxidation in the presence of boric acid produces a mixture of secondary alcohols. The oxidation of alkanes in the vapour state occurs via free radicals, *e.g*., alkyl (R), alkylperoxy (ROO) and alkoxy (RO).

Oxidising reagents such as potassium permanganate readily oxidise a tertiary hydrogen atom to a hydroxyl group, *e.g*., isobutane is oxidised to t-butanol.

$$(CH_3)_3CH + [O] \xrightarrow{KMnO_4} (CH_3)_3COH$$

5. *Isomerisation :* Of n-alkanes into branched-chain alkanes in which the side-chain is a methyl group, may be brought about by heating the n-alkane with aluminum chloride at 300°C, *e.g*., n-butane isomerises into isobutane.

According to Pines *et al.* (1946), this isomerisation does not occur unless a trace of water is pressed (to from HCl) from the $AlCl_3$) together with a trace of alkyl halide or an alkane. The isomerisation is believed to be an ionic chain reaction. The alkene 'impurity' is converted into a carbonium ion (by the $AlCl_3$ and HCl), and this initiated the chain reaction, isomerisation then occurring by a 1.2 shift involving a hydride ion in the second step.

$$RCH{=}CH_2 + H^+ \rightleftharpoons RCHCH_3$$

$$RCHCH_3 + CH_3CH_2CH_2CH_3 \longrightarrow RCH_2CH_3 + \overset{CH_3}{\underset{CH_2CHCH_3}{|}} \rightleftharpoons \overset{H}{\underset{CH_2C(CH_3)_2}{|}} \rightleftharpoons (CH_3)_3C$$

$$(CH_3)_3C^+ + CH_3CH_2CH_2CH_3 \rightleftharpoons (CH_3)_3CH + CH_3CH_2CHCH_3$$

The formation of an equilibrium mixture of n- and isobutane from n-butane with the iso-compound predominating may be explained by the fact that the t -carbonium ion is more stable than the s-carbonium ion.

SOME INDIVIDUAL ALKANES

Methane : This occurs in 'natural gas' (q.v) and the gases from oil wells. Methane is the principle product of organic decay in swamps and marshes, the gas being set free by the action of bacteria, this method of formation in nature has given rise to the name 'marsh gas' for methane. Sewage sludge which has been fermented by bacteria yields a gas containing about 70 per cent methane, and this is used as a liquid fuel. Methane also forms about 40 per cent by volume of coal gas..

Methane may be synthesised by striking an electric are between carbon electrodes in an atmosphere of hydrogen, or by heating a mixture of carbon and reduced nickel; at 475°C in the presence of hydrogen.

Sabatier and Senderns (1897) synthesised methane by passing a mixture of hydrogen and carbon monoxide or dioxide over finely divided nickel heated at about 300°C. *e.g.*,

$$CO + 3H_2 \longrightarrow CH_4 + H_2O.$$

Methane may conveniently be prepared by heating a mixture of anhydrous sodium acetate and soda-lime, or by reduction of methyl iodide with dissolving metals or with lithium aluminum hydride or lithium hydride. Other methods of preparation are by the action of water on aluminum carbide or methylmagnesium iodide (a Grignard reagent).

$$Al_4C_3 + 12H_2O \longrightarrow 3CH_4 + 4\,Al(OH)_3$$

$$CH_3MgI + H_2O \longrightarrow CH_4 + MgI(OH)$$

Methane is obtained in vast quantities from natural gas, gas from the oil-wells, and from cracked petroleum (q. r.).

Properties of methane. Methane is a colourless, odourless, non-poisonous gas., its b.p. is – 164* C760 mm. and mp. – 184°C. It is somewhat soluble in water, 100 ml of water dissolving about 5 ml of methane at 20°C, but is quite soluble in ethanol and ether. It burns with a non-luminous flame in air or oxygen, forming carbon dioxide and water.

$$CH_4 + 2O_2 \longrightarrow CO_2 + 2H_2O$$

It explodes violently when mixed with air (or oxygen) and ignited, and this is believed to be the cause of explosions in coal-mines, where methane is known as free-damp. Methane may be catalytically oxidised to methanol and formaldehyde.

Uses of methane. Methane, by heating to 1000°C or by incomplete combustion, produces carbon in a very finely divided state. Carbon prepared this way is known as carbon black, and

$$C + 2H_2O \xleftarrow{O_2} CH_4 \xrightarrow{1000°C} C + 2H_2$$

is used to make paints and printers' ink, it is also used in the rubber industry for motor tyres, etc.

Methane is used as a source of synthesis gas. A mixture of methane and steam is passed over heated nickel (800°C) supported on alumina.

$$CH_4 + H_2O \longrightarrow CO + 3H_2,\ CH_4 + 2H_2O \longrightarrow CO_2 + 4H_2.$$

The reaction to give carbon dioxide occurs to a much less extent than that to give carbon molecule, and the carbon dioxide is removed from the mixture by washing with water under pressure.

Methane is also used as a fuel and as the starting point of many chemicals 2. *Ethane.* C_2H_6, occurs with methane in natural gas and the gases from the oil-wells. It may be prepared by the action of dissolving metals on ethyl iodide, by a Wurtz reaction with Mel, by the electrolysis of sodium acetater, or by the action of water on Et Mgl. The best way of preparing pure ethane in large quantity is by the catalytic reduction of ethylene.

Properties of ethane : Ethane is a colourless gas, b.p. – 89) C, sparingly soluble in water but readily soluble in ethanol. It burns in air or oxygen to form carbons dioxide and water. It reacts with halogens in a similar manner to methane to form substitution products, but a much greater number of products are possible due, firstly to the presence of six hydrogen atoms in ethane compared with four in methane, and secondly, to the fact that isomerism is passible in the substitution products of ethane, and not in those of methane. Thus for example, two dichloropethanes are possible. CH_3CHCl_2 and $ClCH_2CH_2Cl$. Ethane is mainly used as a fuel.

Propane : C_3H_5. is a constituent of natural gas and gas from the oil-wells. It may be prepared by methods used for ethane, using the appropriate starting materials. It is a colourless gas. b.p. – 44.5°C, and is used as a fuel.

Butenes : C_4H_{10} Both n- and isobutane occur in natural gas and petroleum gas, and they may be separated by tractional distillation under pressure. The butenes are used as a liquid fuel and for the preparation of various compounds.

n-Butane. b.p. – 0.5°C, may be prepared by the Wurtz reaction.

$$2CH_3CH_2I + 2Na \longrightarrow CH_3CH_2CH_2CH_3 + 2NaI$$

Solbutane, b.p., – 10.2°C, may be prepared by treating t-butyl iodide with a zinc-copper couple in ethanol.

$$(CH_3)_3 CI \xrightarrow{\text{a H*}} (CH_3)_3 CH + HI$$

Pentanes, C_5H_{12} : All three pentanes occur in natural gas and petroleum gas and are used for preparing chloropentanes and the pentanols.

n-Pentane. $CH_3CH_2CH_2CH_2CH_3$ b.p., 36° C. isopentane, $(CH_3)_2CHCH_2CH_3$. b.p... 28° C. neopentane, $(CH_3)_4C$bp. 94°C.

Petroleum and Naturals Gas

Crude petroleum (mineral oil) is the term usually applied to the gases accruing naturally in the oil fields, the liquid from the wells, and the solids which are dissolved in, or have been separated from, the liquid. The composition of crude petroleum varies with the locality of occurrence, but all contain alkanes (straight and branched-chain from about C_1 to C_{40}), cycloalkanes or napthens, and aromatic hydrocarbons. The low-boiling fractions of almost all petroleums are composed of alkanes, it is the composition of the higher-boiling fractions which differs according to the sources of the petroleum. In addition to hydrocarbons, there are also present compounds containing oxygen, nitrogen, sulphur and metallic constituents.

Distribution and general composition of crude petroleum. If the residue of petroleum, after removal of volatile compounds, contains a large amount of alkanes or wax, the petroleum is classified as paraffinic or paraffin base oil. If naphthenes predominate, the petroleum is classified as asphaltic or asphalt base oil. The crudes from the wells in Pennsylvania. Iran. Irak and Rumaina are paraffinic, those from Baku and Venezuela are asphaltic, and those from Oklahoma. Texas and Mexico are intermediate in composition, and may be classified as paraffinic and asphaltic.

Distillation of petroleum : The crude oil is nearly always associated with water and sand. Hence the crude petroleum discharged from the top of the well contains water and sand in suspension. The mixture is passed, under pressure, into cylindrical tanks, and the gas, oil and solids are drawn off separately.

Table 4.2

Name	*Bp°C*	*Approximate composition*	*Uses*
Light petrol	20– 100	C_5H_{12}-C_7H_{16}	Solvent
Benzine	70 – 90	C_6–C_7	Dry cleaning
Ligroin	80 – 120	C_6–C_8	Solvent
Petroleum (gasoline)	70 – 200	C_6–C_{11}	Motor fuel
Kerosene (paraffin oil)	200 – 300	C_{12}–C_{16}	Lighting
Gas oil (heavy oil)	above 300	C_{13}-C_{18}	Fuel oil
Lubricating oil (mineral oil)	"	C_{16}-C_{20}	Lubricants
Greases, vaseline, petrolatum	"	C_{18}-C_{22}	Pharmaceutical preparations
Paraffin wax (hard wax)	"	C_{20}-C_{30}	Candles, waxed paper. etc.
Residue (asphaltic bitumen)	"	C_{30}-C_{40}	Asphalt tar, petroleum coke

Except for the boiling hydrocarbons, no attempt is made to separate the individual hydrocarbons. The crude oil is fractionated by continuous distillation into four main fractions. Petrol (gasoline), kerosene (kerosine, paraffin oil), gas oil (heavy oil) and lubricating oil. The residue may be fractionated by means of vacuum-distillation to give light, medium and heavy lubricating oils, paraffin wax, and asphaltic bitumen. Each of the four main fractions may be further split up by batch distillation into fractions of narrow boiling range. Recently, it has been possible to isolate individuals by 'super fractionation'. The final number of fractions taken depends on the purpose in view.

Refining of the various fractions : It appears that refining was originally introduced to remove the bad colour and objectionable odour of petrol. Today it is realised that it is more important to remove sulphur compounds which lower the response of petrol to added tetraethyl-lead.

An internal-combustion engine, *i.e.*, one which burns fuel within the working cylinder, is ore efficient the higher the compression ratio. Petrol engines use 'spark ignition', and as the compression ratio increases, a point is reached when 'knocking' is observed, *i.e.*, after passage of the firing spark, instead of all the fuel gas burning smoothly, the end portion burns with explosive violence, given rise to a metallic rattle. The phenomenon of knocking is still not fully understood, but it has been found that, among other factors, the tendency to knock depends on the nature of the petrol. n-Alkanes tend to produce knocking far more than branched-chain alkanes.

Edgar (1927) introduced 2.24 trimethylpentane (incorrectly known as iso-octane), which has higher anti-knock properties, and n-heptane, which has lower anti-knock properties than any commercial petrol, as standards for rating fuels Iso-octane is arbitrarily given the value of 100 and -heptane, 0 and the octane number of any fuel is the per cent of iso-octane' in a mixture of this compound and n-heptane which will knock under the same conditions as the fuel being tested.

Alkanes and aromatic compounds have high octane numbers. Tetraethyl-lead also raises the octane number of a given petrol, but if sulphur compounds are present, the response to this 'dope' is lowered.

Hence, it is very important to remove sulphur compounds from petrol. The method of refining depends on the particular fraction concerned, and it is not practicable to refine before distillation (of the petroleum).

Cracking

The thermal decomposition of organic compounds is known as pyrolysis, pyrolysis, when applied to alkanes, is known as creaking.

When heated to about 500-600°C, alkanes are decomposed into smaller molecules and the products obtained from a given alkane depend on.

(i) the structure of the alkane.

(ii) the pressure under which cracking is carried out, and

(iii) the presence or absence of catalysts such as silica-alumina, silica-alumina-thoria, silica-alumina-zirconia.

The mechanism of cracking is still observe. Many theories have been suggested, and one that is highly favoured is a free-radical mechanism, evidence for which has been obtained from the observation that at cracking temperatures many hydrocarbon produces alkyl free radicals.

When petroleum is cracked, of all the compounds produced., the most important are those containing up to four carbon atoms, methane, ethane ethylene, propane, propene, butane, butene and isobutene. All of these have found wide application as the materials for the preparation of a large number of chemicals.

By using suitable catalysts, alkanes containing six or more carbons atoms may be catalytically cyclised, *e.g.*, n-hexane. Under pressure, passed over chromic oxide carried on an alumina support and heated at 480-550°C, gives benzene.

$$C_6H_{14} \longrightarrow C_6H_6 + 4H_2$$

Therefore are two main types of cracking. (i) liquid phase, and (ii) vapour phase.

(i) *Liquid phase cracking* : Heavy oil (from the petroleum distillation) is cracked by heating at a suitable temperature (475-530°C) and under pressure (100-1000 1b/sq. in), by means of which the cracked metrical is maintained in the liquid condition. The heavy oil is converted into gasoline to the extent of 60-65 per cent of the oil (by volume), and has an octane number of 65-70. If attempts are made to increase the yield of gasoline, the octane number decreases.

(ii) *In vapour phase cracking :* The cracking temperature is 600°C and the pressure is 3-5-10.5kg/cm^2. The cracking stock any be gasoline, kerosene, gas oils,, but not the heavy oils, since these cannot be completely vaporised under the above conditions.

Reforming is the process whereby straight-run gasoline is cracked in order to raise the octane number. The gasoline is heated to about 600°C and under pressure of 28-52.5kg/cm^2 and the yield varies from 60 to 90 percent, the greater the yield , the lower is the octane number. Catalysts – the oxides of silicon and aluminium, plus small amounts of other oxides such as magnesia, zirconia, etc. – are usually employed in the reforming process. The chief chemical changes that occur are isomerisation, dehydration and cyclisation.

Catalytic cracking is also increasing in use, since it has been found that catalytically cracked gasoline contains few alkanes that readily polymerise. Gum formation in cracked gasolines prevented by the addition of inhibitors, which are mainly phenols or aromatic amines, *e.g.*, catechol, p-benzyl-aminophenol, naphthylamine.

Treatment of natural gas : When natural gas does not contain hydrocarbons above ethane, it is said to be 'lean' or 'dry 0146, when it contains the higher hydrocarbons (up to hexane) it is said to be 'rich' or 'wet'. The alkanes may be separated by fractional distillation under increased pressure, thereby giving methane, ethane, propane, n-and isobutanes, n, iso- and neopentanes, and hexane. These gases are used for various purpose.

On the other hand, natural gas itself may be used in the manufacture of various compounds. Oxidation of natural gas under carefully controlled conditions produces a complex mixture of compounds, among which are formaldehyde, acetaldehyde, acetic acid acetone, methanol, ethanol, propanols and butanols. These are separated by distillation, solvent extraction, etc.

We has is also used as a source of gasoline. The vapours of the liquid hydrocarbons (penances and hexane) in the wet gas are removed by various methods, *e.g.*, compression and cooling. The liquid product obtained from wet gas is known as 'natural' of 'casinghead' gas, and is 'wild' because of the dissolved gases, in it. These gasses may be removed by distillation under pressure, and the resulting liquid is known as 'stabilised natural gasoline', this has very high anti-knock properties.

$$xCO + yH_2 \longrightarrow \text{mixture of hydrocarbons} + \text{water}$$

Various metals and oxides have been used as catalysts. One of the best is, cobalt (100 parts), thoria (5 parts) magnesia (8 parts) and kieselguhr (200 parts). When synthesis gas is passed over the catalyst at moderate pressure (9-11 atm), more of the hugh-boiling fraction is obtained than when the process is carried out at atmospheric pressure. The liquid products are fractionally distilled, and refined in the same way as are the petroleum fractions, furthermore the higher-boiling fractions are cracked.

Gasoline from the synthine process costs more than that from petroleum. The Fisher-Tropsch oils appear to be more valuable as chemical raw materials than as fuels, and have been used for the production of higher alkanes, monocarboxylic acids, detergents and in the oxo-process.

GASOLINE ADDITIVES

Certain compounds added to gasoline for improving its combustion in an internal combustion engine are called *gasoline additives*. Tetraethyl lead ($(C_2H_5)_4Pb$) is the main additive. A small quantity of tetraethyl lead (TEL) improves the anti-knock quality of the gasoline. Compounds which improve the octane number of a fuel are called *antiknock compounds*.

Tetraethyl lead (TEL, $(C_2H_5)_4Pb$) is a common antiknock compound. Gasoline containing tetraethyl lead is called *leaded* gasoline or ethyl gasoline, or simply as *leaded petrol*.

A gasoline-engine (internal combustion engine) discharges various substances into the atmosphere. These are carbon dioxide, water, carbon monoxide, nitrogen oxides, unburnt hydrocarbons and from leaded gasoline certain compounds of lead. About 0.1 g of lead per litre of the gasoline burnt escapes into the atmosphere. Most of it finds its way into the soil.

Although, many compounds present in the exhaust of any automobile are harmful, lead is extremely toxic and poisonous. The increasing emission of such harmful substances into the atmosphere due to the increasing use of automobiles, is a serious pollution problem. Certain devices such as, catalytic converters are now being used to oxidise the unburnt hydrocarbons and carbon monoxide to carbon dioxide and to decompose nitrogen oxides into nitrogen and oxygen.

An important step towards pollution control involves removal of lead from gasoline. This is done by replacing tetraethyl lead in gasoline by certain aromatics and branched-chain hydrocarbons. The gasoline free from tetraethyl lead is called **lead-free gasoline** (or lead-free petrol).

UNSATURATED ALDEHYDES

The carbonyl (C = 0 stretch) absorption band of C, B-unsaturated aldehydes occurs in the region 1705-1680 cm^{-1} (s). It is therefore readily distinguished from the carbonyl group in saturated (acyclic) aldehydes. The molecular formula will also help here.

The t-values of both α- and β-protons in α, β-unsaturated carbonyl compounds occur at lower field than those in alkenes (4.3-5.3), and the value for the β-proton is also dependent on whether it is cis or irons with respect to the carbonyl group. The usual range for α-protons is τ ~ 3.5-4.2 and for β-protons, τ ~ 2.4-3.5. This downfield shift for both α- and β-protons may be attributed to the –I effect of the CO group, and the greater shift for the β-proton may be explained by conjugation which results in a small positive charge on the β-carbon atom, thereby increasing deshielding of a β-proton:

$$—CH = CH—C = O \longleftrightarrow —\overset{+}{C}H—CH = C—O^-$$

Since shifts for protons, attached to an unsaturated carbon atom further along the chain are fairly close to those of protons in alkenes, this may be used to show whether a double bond is α, β with respect to the carbonyl group.

The simplest unsaturated aldehyde is **acraldchyde** (*acrolein, prop-2-en-l-al*), b.p. 52°C. This is conveniently prepared by heating glycerol with potassium hydrogen sulphate, which is the most satisfactory dehydrating agent for this reaction:

$$HOCH_2CHOHCH_2OH \xrightarrow{-2H_2O} CH_2 = CHCHO \ (33–48\%)$$

Acraldehyde may also be readily prepared by the oxidation of allyl alcohol with manganese dioxide or nickel peroxide.

Acraldehyde is prepared industrially in many ways, *e.g.*,

(i) by passing a mixture of acetaldehyde and formaldehyde vapour over lithium phosphate as catalyst:

$$CH_3CHO + CH_2O \longrightarrow CH_2 = CHCHO + H_2O$$

(ii) By the direct oxidation of propene over cuprous oxide on alumina:

$$CH_2 = CHCH_3 + O_2 \longrightarrow CH_2 = CHCHO + H_2O$$

Acraldehyde is unstable, readily polymerising to a white solid. It is an α, β-unsatuiated aldehyde, and undergoes many of the usual reactions of an alkene and an aldehyde, *e.g.*, it adds on two halogen atoms or a molecule of halogen acid (contrary to Markownikoff's rule; *cf.* allyl alcohol):

$$CH_2 = CHCHO + Br_2 \longrightarrow BrCH_2CHBrCHO$$

$$CH_2 = CHCHO + HCl \longrightarrow ClCH_2CH_2CHO$$

The addition of hydrogen bromide is unaffected by the presence of peroxides. Acraldehyde reduces ammoniacal silver nitrate, forms a cyanohydrin with hydrogen cyanide, and a phenyl-hydrazone with phenylhydrazine:

$$CH_2 = CHCHO + [O] \xrightarrow{AgNO_3} CH_2 = CHCO_2H$$

$$CH_2 = CHCHO + HCN \longrightarrow CH_2 = CHCH(OH)CN$$

$$CH_2 = CHCHO + C_6H_5NHNH_2 \longrightarrow CH_2 = CHCH = NNHC_6H_5 + H_2O$$

Many of the reactions of α, β-unsaturated aldehydes are 1,2-additions, but (α, β-unsaturated ketones tend to undergo 1,4-additions. Aldehydes are more reactive than ketones towards nucleophiles, and the kinetically controlled product with aldehydes is 1,2-addition and with ketones 1,4-addition. With the latter, rearrangement occurs if part of the addendum is a proton to give the thermodynamically controlled product.

Acraldehyde undergoes the Tischenko reaction to form allyl acrylate:

$$2CH_2 = CHCHO \xrightarrow{(C_2H_5O)_3Al} CH_2 = CHCO_2CH_2CH = CH_2$$

It does not, however, undergo the normal aldol condensation; instead, in the presence of alkali, cleavage of the molecule takes place:

$$CH_2{=}CHCHO + OH^- \rightleftharpoons HO{-}CH_2{-}\bar{C}HCHO$$

$$\xrightleftharpoons{H_2O} H{-}OCH_2CH_2CHO + OH^2$$

$$\rightleftharpoons \bar{O}{-}CH_2{-}CH_2CHO + H_2O$$

$$\rightleftharpoons CH_2O + \bar{C}H_2CHO$$

$$\xrightleftharpoons{H_2O} CH_3CHO + OH^-$$

The first part of the reaction is the reversal of dehydration, and the second part is the reversal of the aldol condensation. This type of reaction is characteristic of α, β-unsaturated aldehydes.

When acraldehyde is reduced with metal and acid, there are obtained an unsaturated alcohol, a saturated aldehyde, a saturated alcohol and a compound formed by bimolecular reduction:

$$CH_2 = CHCHO \xrightarrow{e;H^+} CH_2 = CHCH_2OH + CH_3CH_2CHO$$
$$+ \; CH_3CH_2CH_2OH + \; CH_2 = CHCHOHCHOHCH = CH_2$$

A good yield of the bimolecular product is obtained by reducing acraldehyde with magnesium amalgam (*cf.* pinacol).

Acraldehyde is reduced by sodium amalgam to n-propanol and by aluminium isopropoxide (Meerwein-Ponndorf-Verley reduction) to allyl alcohol. It should be noted that metal and acid do not usually reduce a double bond; it is only when the double bond is in the α, β-position with respect to a carbonyl group that it is reduced in this way.

The product obtained by the catalytic hydrogenation of α, β-unsaturated carbonyl compounds depends largely on the nature of the catalyst.

Saturated carbonyl compounds are obtained with Pd—C at room temperature, unsaturated alcohols with PtO_2—FeCl—(CH_3CO_2), Zn and saturated alcohols with Ni, Pd—C, or PtO_2 at elevated temperatures and hydrogen under pressure.

α, β-Unsaturated carbonyl compounds may be reduced to unsaturated alcohols by lithium aluminium hydride, sodium borohydride, or best of all, by aluminium hydride.

Acraldehyde adds on to butadiene to form a cyclic compound (see the *Diels-Alder* reaction.

An interesting property of α, β-unsaturated aldehydes, ketones and acids is that they often undergo abnormal ozonolysis:

$$R_1^2C = CR^2CR^3 = O \longrightarrow R_1^2CO + R^2CO_2H + R^3CO_2H$$
$$R_1^2C = CR^2CO_2H \longrightarrow R_1^2CO + R^2CO_2H + CO_2$$

Crotonaldehyde (*but-2-en-l-al*), b.p. 104°C, may be prepared by heating aldol alone, better, with a dehydrating agent, *e.g.*, zinc chloride. The best yield is obtained by distilling aid with acetic acid as catalyst:

$$\underset{cis}{\begin{matrix} H & & CH_3 \\ & C & \\ & \| & \\ & C & \\ H & & CHO \end{matrix}} \qquad \underset{trans}{\begin{matrix} CH_3 & & H \\ & C & \\ & \| & \\ & C & \\ H & & CHO \end{matrix}}$$

Crotonaldehyde closely resembles acraldehyde in its chemical properties. It exists, however in two geometrical isomeric forms, cis and trans: A number of unsaturated aldehydes occur naturally in the essential oils. many accompanying corresponding unsaturated alcohol.

The simplest acetylenic aldehyde is *propargylaldehyde* (propiolaldehyde. propynal). b.p. 60°C. This may be prepared from acraldehyde, the aldehyde group of which is 'protected' by acetal formation.

$$CH_2=CHCHO+2C_2H_5OH \xrightleftharpoons{HCl} H_2O+CH_2=CHCH\,(OC_2H_5)_2$$

$$\xrightarrow{Br_2} BrCH_2CHBrCH(OC_2H_5)_2$$

$$\xrightarrow[KOH]{ethanolic} CH \equiv CCH\,(OC_2H_5)_2 \xrightarrow[HCl]{dilute} CH \equiv CCHO$$

It may also be obtained by the controlled oxidation of propargyl alcohol.

Propargylaldehyde undergoes cleavage when treated with sodium hydroxide (cf. acraldehyde above) :

$$CH \equiv CCHO + NaOH \longrightarrow CH \equiv CH + HCO_2Na$$

UNSATURATED KETONES

The carbonyl infra-red absorption band (C = O stretch) in α, β-saturated ketones occurs in the region 1685-1665 cm^{-1} (s). This distinguishes these compounds from or, α, β-saturated aldfehydes (NMR spectra).

The simplest unsaturated ketone is **methyl vinyiketone** (but-3-en-2-one), b.p. 79°C. This may be prepared by means of an aldol condensation between formaldehyde and acetone, the product being dehydrated by heat.

$$HCHO + CH_3COCH_3 \xrightarrow{NaOH} HOCH_2CH_2COCH_3$$

$$\xrightarrow{-H_2O} CH_2 = CHCOCH_3$$

It may also be prepared by the action of acetic, anhydride on ethylene in the presence of zinc chloride as catalyst:

$$CH_2=CH_2+(CH_3CO)_2O \xrightarrow{ZnCl_2} CH_2=CHCOCH_3+CH_3CO_2H$$

Methyl vinyl ketone can be prepared by the action of acetic anhydride on vinylmagnesium bromide in tetrahydrofuran at –60° to –70°C (Martin, 1957); this is a general method for preparing vinyl ketones:

$$2CH_2 = CHMgBr + 2(CH_3CO)_2O \longrightarrow$$

$$2CH_2 = CHCOOCH_3 + MgBr_2 + (CH_3CO_2)_2Mg \quad (60–80\%)$$

It is manufactured by hydrating vinylacetylene in the presence of dilute sulphuric acid and mercuric sulphate (cf. acetaldehyde):

$$CH_2 = CHC \equiv CH + H_2O \longrightarrow CH_2= CHCOCH_3$$

Methyl vinyl ketone polymerises on standing and is"used commercially as the starting material for plastics.

Mesityl oxide (4-methylpent-3-en-2-one), b.p. 130°C, may be prepared by distilling diacetone cohol (*q.v.*) with a trace of iodine:

$$(CH_3)_2C(OH)C_2COCH_3 \xrightarrow{-H_2O} (CH_3)_2 = CHCOCH_3 \quad (90\%)$$

Stross *et al.* (1947) have shown that mesityl oxide prepared this way is a mixture of mesityl oxide and isomesityl oxide, $CH_2 = C(CH_3)CH_2$ $COCH_3$.

Mesityl oxide may also be prepared as follows:

$$(CH_3)_2C = CH_2 + CH_3COCl \xrightarrow{ZnCl_2} (CH_3)_2CClCH_2COCH_3$$

$$\xrightarrow{-HCl} (CH_3)_2C = CHCOCH_3$$

It is used as a solvent for oils, gums, etc.

Phorone (2,6-dimethylhepta-2,5-dien-4-one), m.p. 28°C, may be prepared by the action of hydrochloric acid on acetone:

$$3CH_3COCH_3 \xrightarrow{HCl} (CH_3)_2C = CHCOCH = C(CH_3)_2 + 2H_2O$$

Catalytic reduction and reduction wrth metal **hydrides** of α, β-unsaturatedketones have been described on. Reduction with metal and acid, or better, with magnesium amalgam, gives bimolecular products (cf. acraldehyde, above):

$$2(CH_3)_2C{=}CHCOCH_3 \xrightarrow{Mg/Hg} \begin{array}{c}(CH_3)_2CCH_2COCH_3 \\ | \\ (CH_3)_2CCH_2COCH_3\end{array}$$

The addition of halogen to α, β-unsaturated ketones occurs by the normal electrophilic mechanism across the 3,4-positions. The addition of halogen acids, however, occurs by nucleophilic attack by the 1,4-mechanism.

The α,β-unsaturated ketones add on ammonia, primary amines and secondary amines to form β-amino-compounds with ammonia, mesityl oxide forms diacetonamine and phorone forms triacetonamine:

$$(CH_3)_2C{=}CHCOCH_3 + NH_3 \longrightarrow (CH_3)_2C(NH_2)CH_2COCH_3$$

$$(CH_3)_2C{=}CHCOCH{=}C(CH_3)_2 + 2NH_3 \longrightarrow$$

$$\left[(CH_3)_2C(NH_2)CH_2COCH_2C(NH_2)(CH_3)_2\right] \xrightarrow{-NH_3} OC\begin{matrix} CH_2{-}C(CH_3)_2 \\ \quad NH \\ CH_2{-}C(CH_3)_2 \end{matrix}$$

All the evidence is in favour of a 1,4-addition mechanism when the addendum is of the type H—Z, the product being an enol which then tautomerises to the stable keto form, the overall result being 3,4-addition, *e.g.*, with halogen acids:

$$\overset{4}{R^1{-}CH}{=}\overset{3}{CH}{-}\overset{2}{CR^2}{=}\overset{1}{O} \xrightarrow{HX} X^- + R^1{-}CH{=}CH{-}CR^2{=}\overset{+}{O}H$$

$$\longleftrightarrow R^1\overset{+}{C}H{-}CH{=}CR^2{-}OH$$

$$\xrightarrow{X^-} R^1{-}CHX{-}CH{=}CR^2{-}O{-}H$$

$$\longrightarrow R^1{-}CHX{-}CH_2{-}CR^2{=}O$$

The 1,4-addition intermediate is the kinetically controlled product and the 3,4-product is the thermodynamically controlled product.

Similarly, α, β-unsaturated ketones add on hydrogen cyanide to form β-cyano-compounds $R^1CH(CN)CH_2COR^2$; and sodium hydrogen sulphite to form β-sulphonic acids. $R^1CH(SO_3Na)CH_2COR^2$.

The reaction with hydroxylamine usually results in a mixture of the oxime. $R^1CH = CHC(=NOH)R^2$ and the β-hydroxylaminoketone, $R^1CH(NHOH)\text{-}CH_2COR^2$ Trialkylboranes also undergo 1,4-addition, *e.g.*,

$$R_1^3B + CH_2 = CHCOR^2 \longrightarrow R_1CH_2CH = CR_2OBR_2^1$$
$$\xrightarrow{H_2O} R^1CH_2CH_2COR^2$$

A number of factors affect the direction of these addition reactions, an important one being steric hindrance. Steric hindrance at the carbonyl group favours 1,4-addition, whereas steric hindrance at the 3- and/or 4-positions favours 1,2-addition.

α,β-Unsaturated ketones (and aldehydes) are more difficult to epoxidise than alkenes, but are epoxidised by alkaline hydrogen peroxide.

One of the most important reactions that α,β-unsaturated ketones undergo is the Michae **condensation** (1887). This is the addition reaction bctwccn an α, β unsaturated keto-compound and a compound with an active methylene group, *e.g.*, malonic ester, acetoacetic ester, etc. The condensation is carried out in the presence of a base, *e.g.*, sodium ethoxide, or secondary amines. The mechanism of the reaction is believed to be (using malonic ester and mesityl oxide as an example):

$$CH_2(CO_2Et)_2 + EtO^- \rightleftharpoons \bar{C}H(CO_2Et)_2 + EtOH$$

$$Me_2C{=}CH{-}\overset{\overset{\displaystyle O}{\|}}{C}Me + \bar{C}H(CO_2Et)_2 \rightleftharpoons \underset{\underset{\displaystyle CH(CO_2Et)_2}{|}}{Me_2C}{-}CH{=}\overset{\overset{\displaystyle O^-}{|}}{C}Me \xrightleftharpoons{EtOH}$$

$$\underset{\underset{\displaystyle CH(CO_2Et)_2}{|}}{Me_2C}{-}CH{=}\overset{\overset{\displaystyle O{-}H}{|}}{C}Me + EtO^- \rightleftharpoons \underset{\underset{\displaystyle CH(CO_2Et)_2}{|}}{Me_2C}{-}CH_2{-}CH_2COMe + EtO^-$$

With this example, after the Michael condensation has occurred, it is followed by an internal Claisen condensation and leads to the formation of dimedone.

The description of the Michael condensation given above is a limited one. The term now includes a large variety of addenda and acceptors, the a, unsaturated compounds may be ketones, esters, cyanides, etc., and the addenda may be malonic ester, E.A.A., cyanoacetic ester, nitroalkanes, etc.

The Michael condensation, as shown above, is reversible. Thus a Michael *retrogression*, *i.e.*, reversal, may occur to produce compounds different from the starting compounds, *e.g.* starting with benzylidenemalonic ester and E.A.A. in the presence of a large amount of sodiun ethoxide results in a mixture containing the reactants, the Michael condensation product, *and* benzylideneacetoacetate and malonic ester (produced by the retrogression).

$$\begin{array}{l} Me_2C\text{—}CH_2COMe \\ \quad | \\ CH(CO_2Et)_2 \end{array} \xrightarrow{EtO^-} \begin{array}{c} Me_2C(CH_2)CO \\ CH \quad CH_2 \\ EtO_2C \quad C(=O)OEt \end{array} \xrightarrow{-EtO^-}$$

$$\begin{array}{c} CH_2\text{—}CO \\ Me_2C \qquad CH_2 \\ CH\text{—}CO \\ | \\ CO_2\text{—}Et \end{array} \xrightarrow[\text{(ii) } H^+]{\text{(i) KOH}} \begin{array}{c} CH_2\text{—}CO \\ Me_2C \qquad CH_2 \\ CH_2\text{—}CO \end{array}$$

If the reaction is carried out with the last two compounds as starting materials, the same mixture is obtained. Thus:

$$PhCH{=}C(CO_2Et)_2 + MeCOCH_2CO_2Et$$

$$\xrightleftharpoons{EtO^-Na^+} \begin{array}{c} PhCHCH(CO_2Et)_2 \\ | \\ MeCOCHCO_2Et \end{array} \xrightleftharpoons{EtO^-Na^+}$$

$$\begin{array}{c} PhCH\text{—}CH(CO_2Et)_2 \\ | \\ MeCO\,CCO_2Et \end{array} \rightleftharpoons \begin{array}{c} PhCH \\ \| \\ MeCOCCO^2Et \end{array} + \begin{array}{c} \bar{C}H(CO_2Et)_2 \\ \updownarrow \text{EtOH} \\ CH_2(CO_2Et)_2 + EtO^- \end{array}$$

COMMON REACTIONS OF ALDEHYDES AND KETONES

1. Catalytic hydrogenation readily converts aldehydes and ketones into primary and secondary alcohols, respectively. Reduction with dissolving metals also produces alcohols:

$$>C{=}O + 2e \longrightarrow >\bar{C}\text{—}O^- \longrightarrow >CHOH$$

This reaction, however, may also lead to the formation of a 1,2-glycol:

$$2\rangle C{=}O \xrightarrow{e;\,H^+} \rangle C(OH)C(OH)\langle$$

Reduction of both aldehydes and ketones to alcohols may be effected by means of the *Meerwein-Ponndorf-Verley reduction* (1925, 1926). The carbonyl compound is heated with is removed from the equilibrium mixture by slow distillation:

$$R_2CO + (CH_3)_2CHOAl/_3 \rightleftharpoons R_2CHOAl/_3 +$$

$$CH3COCH_3 \xrightarrow[H_2SO_4]{\text{dilute}} R_2CHOH\ (g.)$$

This reducing agent is specific for the carbonyl group, and so may be used for reducible, *e.g.*, a double bond or a nitro-group. The equilibrium is displaced to the right by distillation of the acetone (which is the component with the lowest b.p. in the system). Of all the metal alkoxides that may be used for this to form coordination compounds.

Aldehydes and ketones are readily reduced to alcohols by various metallic hydrides. *Clemmensen reduction* (1913): the carbonyl compound is reduced with *amalgamated zinc* and concentrated hydrochloric acid:

$$R^1COR^2 \xrightarrow{e;H^+} R^1CH_2R^2$$

The Clemmensen reduction does not appear to work well for aldehydes, but is reasonably good for many ketones.

Several mechanisms have been proposed for the Clemmensen reduction, *e.g.*,

$$R{-}\overset{\overset{O}{\|}}{C}{-}R + Zn + Cl^- \rightleftharpoons R{-}\underset{\underset{ZnCl}{|}}{\overset{\overset{O^-}{|}}{C}}{-}R \overset{H^+}{\rightleftharpoons} R{-}\underset{\underset{ZnCl}{|}}{\overset{\overset{OH}{|}}{C}}{-}R \xrightarrow[-H_2O]{H^+}$$

$$R{-}\underset{\underset{ZnCl}{|}}{\overset{+}{C}}{-}R \xrightarrow{Zn} {}^+ZnCl + R{-}\overset{\overset{Zn^-}{|}}{C^+}{-}R \xrightarrow[Cl^-]{Zn} R{-}\overset{\overset{ZnCl}{|}}{\underline{C}^+}{-}R$$

$$\xrightarrow{H^+} R{-}\overset{\overset{ZnCl}{|}}{CH}{-}R \xrightarrow{H^+} R{-}CH_2{-}R$$

2. Aldehydes and ketones add on sodium hydrogen sulphite to form *bisulphite* compounds:

$$\rangle C{=}O + NaHSO_3 \rightleftharpoons \rangle C\langle^{OH}_{SO_3Na} \quad \text{or} \quad \left[\rangle C\langle^{OH}_{SO_3}\right]^- Na^+$$

These bisulphite compounds are *hydroxysulphoric acid salts*, since the sulphur atom is *directly* attached to the carbon atom. This structure is supported by work with isotope ^{34}S (Sheppard *et al.*, 1954).

The position of equilibrium lies largely to the right for most aldehydes and to the left for most ketones. *e.g.*, Hooper (1967) showed from NMR studies of solutions of ketones in aqueous sodium hydrogen sulphite that the amounts of *unreacted* ketone were: Me_2CO, 24·6; MeCOEt, 71·8 MeCOisoPr, 100 per cent. These results may be partly explained in terms of steric effects.

Bisulphite compounds are usually crystalline solids, insoluble in sodium hydrogen sulphite solution. Since they regenerate the carbonyl compound when heated with dilute acid or sodium carbonate solution, their formation affords a convenient means of separating carbonyl compounds from non-carbonyl compounds.

The mechanism of bisulphite formation has been the subjected of much discussion. Hinshelwood *et al.* (1958), from a study of the *reverse* reaction, have proposed the following mechanism.

$$HSO_3^- \rightleftharpoons H^+ + SO_3^{2-}$$

$$>C=O + SO_3^{2-} \rightleftharpoons >C(OH)(SO_3^{2-}) \underset{}{\overset{HSO_3^-}{\rightleftharpoons}} >C(OH)(SO_3^-) + SO_3^{2-}$$

3. Aldehydes and ketones add on hydrogen cyanide to form *cyanohydrins*. The carbonyl compound is treated with sodium cyanide and dilute sulphuric acid:

$$>C=O + HCN \rightleftharpoons >C(OH)(CN) \qquad (g. - v.g.)$$

Lapworth (1903, 1904) showed that the addition of hydrogen cyanide to carbonyl compounds is accelerated by bases and retarded by acids, and he concluded that the addendum was the cyanide ion, and proposed the following mechanism:

$$R_2C=O + CN^- \underset{}{\overset{slow}{\rightleftharpoons}} R_2C(O^-)(CN) \xrightarrow{H^+;\ fast} R_2C(OH)(CN)$$

Cyanohydrins are important compounds in organic synthesis since they are readily hydrolysed to α-hydroxy-acids:

$$RCH(OH)CN \xrightarrow{H^+} RCH(OH)CO_2H$$

All aldehydes form cyanohydrins; only the ketones acetone, butanone, pentan-3-one, and pinacolone form cyanothydrins.

4. Aldehydes and ketones combine with a variety of compounds of the type Z—NH_2 (Z = OH, NH_2, etc.). The general reaction may be written

$$>C{=}O + NH_2{-}Z \longrightarrow \left[>C\begin{matrix} OH \\ NHZ \end{matrix} \right] \xrightarrow{-H_2O} >C{=}N{-}Z$$

Oximes are formed with hydroxylamine:

$$R_2CO + NH_2OH \longrightarrow \left[R_2C\begin{matrix} OH \\ NHOH \end{matrix} \right] \xrightarrow{-H_2O} R_2C{=}NOH \quad (v.g.)$$

Oximes are usually well-defined crystalline solids, and may be used to identify carbonyl compounds.

The absorption region of oximes (C═N stretch) is 1690–1630 cm^{-1} (v) and that for the O—H stretch is 3650–3500 cm^{-1}(s) (*cf.* alcohols).

Aldoximes form cyanides when boiled with acetic anhydride, whereas ketoximes form the acetyl derivative of the oxime:

$$RCH{=}NOH \xrightarrow[-H_2O]{(CH_3CO)_2O} RCN$$

$$R_2C{=}NON \xrightarrow{(CH_3CO)_2O} R_2C{=}NOCOCH_3$$

Aldehydes and ketones combine with *N*-substituted hydroxylamines to form *nitrones*:

$$>C{=}O + RNHOH \longrightarrow H_2O + >C{=}\overset{+}{N}\begin{matrix} R \\ O^- \end{matrix} \longleftrightarrow >\overset{+}{C}{=}\ddot{N}\begin{matrix} R \\ O \end{matrix}$$

Nitrones readily undergo 1,3- addition reactions, *e.g.*,

$$\begin{matrix} R \\ H \end{matrix}>C{=}\overset{+}{N}\begin{matrix} R \\ O^- \end{matrix} + HCN \longrightarrow \begin{matrix} R \\ H \end{matrix}>\underset{CN}{C}{-}N\begin{matrix} R \\ OH \end{matrix} \xrightarrow{-H_2O} \begin{matrix} R \\ NC \end{matrix}>C{=}N{-}R \quad \text{cyno-imine}$$

Hydrazones and *azines* are formed with hydraxine:

$$>C{=}O + NH_2NH_2 \longrightarrow \left[>C\begin{matrix} NHNH_2 \\ OH \end{matrix} \right] \xrightarrow{-H_2O} >C{=}NNH_2 \xrightarrow{>CO} >C{=}NN{=}C<$$

By using suitable derivatives of hydrazine, more well-defined crystalline products are obtained (and azine formation is avoided), *e.g.*, *phenylhydrazine* forms *phenylhydrazones*:

$$C_6H_5NHNH_2 + O{=}C\langle \longrightarrow H_2O + C_6H_5NHN{=}C\langle$$

Semicarbazide forms *semicarbazones*:

$$NH_2CONHNH_2 + O{=}C\langle \longrightarrow NH_2CONHN{=}C\langle + H_2O$$

The mechanism of the formation of semicarbazones and oximes has been studied in great detail.

These reactions (as are most reactions of this type, *i.e.*, $\rangle CO + NH_2{-}Z$) are catalysed by acids, and it has been found that the rates of formation of the products reach a maximum at some particular pH. According to Jencks *et al.* (1959, 1962), the pH optima observed are due to transitions in the rate-determining steps with changing concentration of acid. Let us consider the following mechanism (Z = —NHCONH$_2$ or —OH).

$$R_2C{=}O + NH_2{-}Z \longrightarrow R_2C\langle \longrightarrow R_2C\langle^{OH}_{NHZ} \quad (I)$$

$$\longrightarrow R_2C{=}NH + H_2O \quad (II)$$

Jencks showed, by means of ultraviolet and infra-red studies, that at pH 7 the formation of (I) is fast, and that as the acidity increases, the overall rate of the reaction increases, since the dehydration step, which is the r/d step, also increases (through protonation of the OH, etc.). However, as the acidity increases, the addition step becomes progressively shower because Z—NH$_2$ is decreasing in concentration due to its conversion into its conjugate acid, Z—NH$_3^+$. The latter no longer has a lone pair of electrons on the N atom and so is not a nucleophile. Hence, at a sufficiently high acidity, the addition step is slowed down to such an extent that this now becomes the r/d step in the reaction.

Oximes and hydrazones regenerate the carbonyl compound when refluxed with dilute hydrochloric acid. Regeneration from phenylhydrazones by this method is usually difficult. One fairly successful method is by exchange of the phenylhydrazino group with another oxo compound, *e.g.*, pyruvic acid. The most satisfactory method, however, appears to be the use of acetylacetone in faintly acid solution (Ried *et al.*, 19862); the product is 3,5-dimethyl-1-phenylpyrazole.

$$>C{=}NNHPh + MeCOCH_2COMe \longrightarrow >C{=}O + \text{(3,5-dimethyl-1-phenylpyrazole: N–N(Ph), Me, Me)}$$

Wolff-Kishner reduction (1912). When hydrazones (or semicarbazones) are heated with sodium ethoxide 180°C, nitrogen is eliminated, and a hydrocarbon is obtained, *i.e.*, by this means the carbonyl group is converted into the methylene group:

$$>C{=}NNH_2 \longrightarrow >CH_2 + N_2 \quad (g.\text{–}v.g.)$$

The yield of hydrocarbon is better than that obtained by the Clemmensen reduction, but both the Wolff-Kishner and the Clemmensen reduction are generally unsuccessful with sterically hindered ketones. Barton *et al.* (1954, 1955), however, have overcome this problem by using a modification of the improved technique of the Wolff-Kishner reduction introduced by Huang-Minlon (1946). All the methods described use elevated temperatures (180–200°C), but Cram *et al.* (1962) have shown that the reaction proceeds at room temperature if dimethyl sulphoxide is used as solvent.

EXERCISES

1. Show, by equations, how acetaldehyde may be converted into: (i) EtOH; (ii) CH_3COCl; (iii) CHI_3; (iv) MeCOEt; (v) CH_3CO_2Et; (vi) $CH_3CHOHCO_2H$; (vii) $EtNH_2$; (viii) $CH3CHBrCHBrCO_2H$.
2. Convert $ClCH_2CH_2CHO$ into HOCH2CHOHCHO.
3. Draw the energy profile for the following, and give your reasons.
   $CH_3COCH_3 + NaOH + Br_2 \rightarrow CH_3COCH_2Br + H_2O + NaBr$
4. Acetone, in dilute aqueous solution is 100 per cent unhydrated. When acetone is dissolved in water enriched with ^{18}O, recovered acetone contains ^{18}O. Explain.
5. Convert :
   (*i*) n-butanol into 2-ethylhexan-1-ol;
   (*ii*) bromocyclohexane into cyclopentanecarboxylic acid.
6. Calculate *DH* for each reaction:

$$CH_3CHO + OH^- \rightleftharpoons \overline{C}H_2CHO + H_2O$$

$$CH_3COCH_3 + \overline{OH} \rightleftharpoons \overline{C}H_2COCH_3 + H_2O$$

Are these reactions equally thermodynamically favourable? Are the experimental results in agreement with your calculations? If not, why not?

7. Compound $C_5H_{10}O$ had a band at 1715 cm^{-1} in its i.r. spectrum, and showed two signals, a triplet and a quartet, in its N.M.R. spectrum. What is the compound?
8. One of the i.r. bands of compound A, C_3H_6O, shifts to a higher frequency when its solution in CCl_4 is progressively diluted. Write out possible acyclic structures [4] for the formula C_3H_6O, and suggest which is most likely for A. Using i.r. ad N.M.R. spectroscocpy, how would you distinguish between the four structures?
9. The bromination of acetone is catalysed by acids, and the rate law is zero order with respect to bromine. Discuss this statement.
10. Suggest what compounds the letters stand for in the following equations:

(*i*) $MeCOMe + CH{\equiv}CH \xrightarrow[(ii)\ A]{(i)\ NaNH_2} B$

(*ii*) $PhCOMe + ClCH_2CO_2Et \xrightarrow{NaNH_2} C$

(*iii*) $Et_2CO + NH_4Cl + NACN \xrightarrow{NH_4OH} F$

11. Complete the following equations and give reasons for your answer.

(*i*) $Me_2C(OH)CH_2OH \xrightarrow{H^+} ?$

(*ii*) $PhMeC(OH)C(OH)MePh \xrightarrow{H^+} ?$

12. Describe paths that could lead to the formation of the following fragment ions:

(*i*)

m/e	29	43	44	71	72
n-Butyraldehyde (%)	54	79	100	5	73
Isobutyraldehyde (%)	23	100	—	1	53

(*ii*)

m/e	29	43	57	58	71	86
Pentan-2-one (%)	16	100	26	9	9	25
Pentan-3-one (%)	99	—	100	—	1	17